Metal Cutting

Metal Cutting

Fourth Edition

Edward M. Trent
Department of Metallurgy and Materials
University of Birmingham, England

Paul K. Wright
Department of Mechanical Engineering
University of California at Berkeley, U.S.

Boston Oxford Auckland Johannesburg Melbourne New Delhi

Copyright © 2000 by Butterworth–Heinemann

 A member of the Reed Elsevier group

All rights reserved.

∞ Recognizing the importance of preserving what has been written, Butterworth–Heinemann prints its books on acid-free paper whenever possible.

 Butterworth–Heinemann supports the efforts of American Forests and the Global ReLeaf program in its campaign for the betterment of trees, forests, and our environment.

Library of Congress Cataloging-in-Publication Data

Trent, E. M. (Edward Moor)
    Metal cutting / Edward M. Trent, Paul K. Wright.– 4th ed.
        p. cm.
        Includes bibliographical references and index.
        ISBN 0-7506-7069-X
            1. Metal-cutting. 2. Metal-cutting tools. I. Wright, Paul Kenneth. II. Title.

TJ1185.T73 2000
671.5'3—dc21

99-052104

The publisher offers special discounts on bulk orders of this book.
For information, please contact:
Manager of Special Sales
Butterworth–Heinemann
225 Wildwood Avenue
Woburn, MA 01801–2041
Tel: 781-904-2500
Fax: 781-904-2620

For information on all Butterworth–Heinemann publications available, contact our World Wide Web home page at: http://www.bh.com

10 9 8 7 6 5 4 3 2

Printed in the United States of America

TABLE OF CONTENTS

FOREWORD

Dr. Edward M. Trent who died recently (March, 1999) aged 85, was born in England, but was taken to the U.S.A. as a baby when his parents emigrated to Pittsburgh and then to Philadelphia. Returning to England, he was a bright scholar at Lansdowne High School and was accepted as a student by Sheffield University in England just before his seventeenth birthday, where he studied metallurgy. After his first degree (B.Sc.), he went on to gain his M.Sc. and Ph.D. and was awarded medals in 1933 and 1934 for excellence in Metallurgy. His special research subject was the machining process, and he continued in this work with Wickman/Wimet in Coventry until 1969. Sheffield University recognized the importance of his research, and awarded him the degree of D.Met. in 1965.

Prior to the 1950's, little was known about the factors governing the life of metal cutting tools. In a key paper, Edward Trent proposed that the failure of tungsten carbide tools to cut iron alloys at high speeds was due to the diffusion of tungsten and carbon atoms into the workpiece, producing a crater in the cutting tool, and resulting in a short life of the tool. Sceptics disagreed, but when tools were covered with an insoluble coating, his ideas were confirmed. With the practical knowledge he had gained in industry, and his exceptional skill as a metallographer, Edward Trent joined the Industrial Metallurgy Department at Birmingham University, England in 1969 and was a faculty member there until 1979. Just before his retirement, he was awarded the Hadfield Medal by the Iron and Steel Institute in recognition of his contribution to metallurgy.

Edward Trent was thus a leading figure in the materials science aspects of deformation and metal cutting. As early as 1941, he published interesting photomicrographs of thermoplastic shear bands in high tensile steel ropes that were crushed by hammer blows. One of these is reproduced in *Figure 5.6* of this text. Such studies of adiabatic shear zones naturally led him onward to the metal cutting problem. It is an interesting coincidence that also in the late 1930s and early 1940s, another leader in materials science, Hans Ernst, curious about the mechanism by which a cutting tool removes metal from a workpiece, carried out some of the first detailed microscopy of the process of chip formation. He employed such methods as studying the action of chip formation through the microscope during cutting, taking high-speed motion pictures of such, and making photomicrographs of sections through chips still attached to workpieces. As a result of such studies, he arrived at the concept of the "shear plane" in chip formation, i.e. the very narrow shear zone between the body of the workpiece and the body of the chip that is being removed by the cutting tool, which could be geometrically approximated as a plane. From such studies as those by Ernst, Trent and others, an understanding emerged of the geometrical nature of such shear zones, and of the role played by them in the plastic flows involved in the chip formation process in metal cutting. That understanding laid the groundwork for the development, from the mid-1950s on, of analytical, physics-based models of the chip formation process by researchers such as Merchant, Shaw and others.

These fundamental studies of chip formation and tool wear, and the metal cutting technology resulting from it, are still the base of our understanding of the metal cutting process today. As the manufacturing industry builds on the astounding potential of digital computer technology, born in the 1950s, and expands to Internet based collaboration, the resulting global enterprises still depend on the "local" detailed fundamental metal cutting technology if they are to obtain increasingly precise products at a high quality level and with rapid throughput. However, one of the most important strengths that the computer technology has brought to bear on this situation is the fact that it provides powerful capability to integrate the machining performance with the performance of all of the other components of the overall system of manufacturing. Accomplishment of such integration in industry has enabled the process of performing machining operations to have full online access to all of the total database of each enterprise's full system of manufacturing. Such capability greatly enhances both the accuracy and the speed of computer-based engineering of machining operations. Furthermore, it enables each element of that system (product design, process and operations planning, production planning and control, etc.), anywhere in the global enterprise to interact fully with the process of performing machining and with its technology base. These are enablements that endow machining technology with capability not only to play a major role in the functioning and productivity of an enterprise but also to enable the enterprise as a whole to utilize it fully as the powerful tool that it is.

M. Eugene Merchant and Paul K. Wright, June 1999

PREFACE

A new edition of *Metal Cutting* has been prepared. Every chapter is updated emphasizing new information on machinability and tool materials. Today's interests in "dry-machining", "high speed machining" and "computer modeling" are also featured in brand new chapters.

Chapters 1 through 5 now contain the recent economic trends in the machine tool and metal cutting industry. There is more information on the essential features of metal cutting, and new experimental results on the stresses and temperatures that occur during cutting.

Chapters 6 through 8 focus on the materials science aspects of cutting tool materials. Photographs of tool wear mechanisms and flow patterns in cutting are presented. New developments in coatings for all types of tool material are described in detail. The new trends in diamond coated tools are shown and Chapter 8 concludes with a cross-comparison of all tooling types.

Chapters 9 and 10 provide new information on "machinability" and the selection of coolants and lubricants. These chapters deal with issues that are highly relevant for the day-to-day practitioner of machining. The information includes recommended tooling selection for different work materials, how to select "free cutting" grades of steel, and how to cut aerospace alloys. Specific recommendations on "dry machining" are given at the end of Chapter 10. These recommendations have a broad scope, not limited to cutting fluids alone. An improvement in the "dry-machining" of steel can be achieved by the commercial introduction of special deoxidation treatments during steel making. Calcium deoxidation modifies the non-metallic inclusions, the action of which at the tool/work interface can greatly increase tool life without damage to material properties.

Chapters 11 through 15 are all new in this edition of *Metal Cutting*. They reflect two essential aspects of 21st century manufacturing. The first trend is towards high speed machining of aerospace and other difficult to machine alloys. The economics of this trend are first discussed in Chapter 11. Next, the material behavior at high strain-rates is reviewed before going into details on the physics of high-speed machining of aluminum, stainless steel and titanium alloys. The second trend is the use of the computer to support analytical and computer modeling of cutting. A review of models, techniques and recent findings is given in Chapter 12.

Chapter 13, the Conclusion, provides more information and ideas on the economic importance of machining. It will be shown that machining is a fundamental base for all 21st century manufacturing methods and that there is a subtle but crucial link between precision machining and the semiconductor and biotech industries.

Chapter 14 contains some review exercises for new students to the field.

Chapter 15 contains an extensive Bibliography to other books. These are complementary to this text. Also, references to other books on forming and rapid prototyping are given, together with some economic references.

A Web Site for this edition of *Metal Cutting* has been created which will continue to be updated with links to other Web Sites and their URLs. Chapter 15 will therefore be an open-ended resource and we invite colleagues, cutting tool manufacturers and others to add data, information, interesting homework and assignments.

It is especially interesting that the technology of metal cutting continues to have such an enormous historical and economic range. While metal cutting has roots going back to the Industrial Revolution, it keeps extending its frontiers in response to the everyday needs of a wide range of contemporary industries. The economic importance of metal cutting using machine tools cannot be underestimated. *Today in industrialized countries, the cost of machining amounts to more than 15% of the value of all manufactured products in those countries.*[1] Many examples of this economic importance are apparent at the time of writing, around the year 2000.

First, for standard consumer-products such as automobiles, aircraft and household appliances, metal cutting remains a core production method. The increased international competition in these basic, consumer-product-oriented industries has demanded greater efficiency and productivity. Responding to this challenge, the advances in new tooling, high speed machining, and modeling are discussed in Chapters 8 through 12. We are grateful to our colleagues at The Boeing Company, Dr. Donald Sandstrom in particular, for the many new insights that are beginning to emerge on high speed machining. The ideas on high speed machining and tool materials have also been formed during a long correspondence and association with Dr. R. Komanduri at Oklahoma State University. In addition, Dr. J.T. Black of Auburn University has contributed many concepts, during a collaboration and friendship that spans several decades, and he has been particularly helpful in providing materials and the overall structure for Chapter 12 on modeling.

For higher cutting speeds and hence production rates, ceramic tools, sialon tools and alumina-based tools containing zirconia or silicon carbide "whiskers" have been developed. Meanwhile, ultra-hard polycrystalline diamond, CVD coated diamond and cubic boron nitride have extended their range of application for cutting very hard and abrasive materials. These trends are considered in detail in Chapter 8. At the same time, high speed steel and cemented carbide continue as the tool materials for the bulk of industrial metal cutting operations. Developments of these tools are reported in Chapters 6 and 7, and relate mostly to PVD and CVD coatings.

Second, metal cutting is more ubiquitous in industrial society than it may appear from the above description of traditional manufacturing industries. All forgings, for example used in cars and trucks, and many sheet metal products, for example used in steel furniture and filing cabinets, are formed in dies that have been machined. In fact, most of today's electronic products are packaged in a plastic casing that has been injection-molded into a die. Cell phones, computers, music-systems, the "Walkman" and all such products, thus depend on the metal cutting of dies.

While the initial prototype of a new cell phone might well be created with one of the newer rapid prototyping processes that emerged after 1987 - stereolithography (SLA), selective laser sintering (SLS), or fused deposition modeling (FDM) - the final plastic products, made in batch sizes in the thousands or millions (*see* Chapter 13) will be injection-molded in a die that has been cut in metal with great precision and surface finish constraints. The machining of such dies, usually from highly alloyed steels, requires some of the most exacting precisions and surface qualities in metal cutting technology. It prompts the need for new cutting tool designs; novel manufacturing software that can predict and correct for tool deflections and deleterious burrs; and new CAD/CAM procedures that incorporate the physics and knowledge-bases of machining into the basic geometrical design of a component. In these regards we are grateful to the recent discussions and contributions from our colleagues: at the University of California, Berkeley - Dr. David Dornfeld, Dr. Paul Sheng, Dr. Carlo Sequin, Dr. Frank Wang, Dr. Sung Ahn, Dr. Kyriakos Komvopoulos, Dr. Bruce Kramer (concurrently with the National Science Foundation) and Dr. Jami Shah (concurrently with Arizona State University); at the University of Illinois, Urbana-Champaign - Dr. Richard DeVor, Dr. James Stori and Dr. Shiv Kapoor; at M.I.T. - Dr. Sanjay Sarma; at Siemens - Dr. Steven Schofield; at the University of Kentucky - Dr. I. S. Jawahir; at Ford/Visteon - Mr. Charles Szuluk, Dr. Shuh-Yuan Liou, Dr. Richard Furness and Dr. Shounak Athavale; at the University of New South Wales - the late Dr. Peter Oxley.

Third, the worldwide semiconductor industry also depends on precision metal cutting as a core production method for the semiconductor manufacturing equipment, which in turn fabricates the semiconductors. The crucial link between the machine tool industry and the semiconductor and electronics industry is a subtle one that is further expanded with some interesting Tables in the Conclusion - Chapter 13. For the purposes of this Preface, the key point is: *the machine tool industry is a key building block for industrial society, since it provides the base upon which other industries perform their production.*

This third point can be amplified by considering some basic data in this paragraph, followed by some historical comparisons in the next two paragraphs. At the time of writing, around the year 2000, the line-widths (i.e. the length of the field-effect transistor gate) in a typical logic-transistor of an Integrated Circuit (IC) is about 0.35 microns: during the months this book is going to press, the gate length will reduce to 0.18 microns in many applications. Computer-aided design tools, automated process technologies, advanced clean room systems, and rigorous testing equipment have helped bring semiconductor fabrication to these sub-micron levels. These demands for ever smaller and more powerful semiconductors produce a corresponding demand for advanced manufacturing equipment. It includes advanced lithography equipment, specialized ion-beam machines, chemical-mechanical polishing equipment to achieve ultra-flat surfaces, lasers and high vacuum systems. The demands from the lithography process, which accounts for about 35% of the cost of semiconductors, are perhaps the most demanding. As specific examples: highly precise, magnetically levitated stages are needed for these new generations of semiconductors; lenses are being replaced by mirrors which are held on ultra-precise adjustment stages; the mirrors themselves must be ground and polished to perfection. All such sub-components must be fabricated in a machine shop by metal cutting. Furthermore, the new levels of ultra-precision demand an even greater understanding of the detailed processes going on at the all-important cutting edge. During precision machining, with small undeformed chip-thicknesses, the flow patterns at the cutting edge govern the quality and integrity of the surface finish.

The design of tool angles and the selection of the correct machining parameters have become even more important than in previous decades.

It is interesting to make some historical comparisons from 200 years ago with the above demands in semiconductors today. It is well known that the technical roots of the first Industrial Revolution began with James Watt's steam engine in 1769. Thereafter, a complex mix of technical, economic and political factors accelerated the Revolution over a fifty year period between approximately 1770 and 1820. The period from 1820 to 1910 then saw the rise and consolidation of many industries including the basic machine tool industry. But in particular, the machine tool industry remained unique - it was always the essential base upon which these other industries critically depended. Increased standardization, improved precision, more reliable cutting tools and more powerful machines, provided a base for all other metal-product type industries. These secondary industries - Samuel Colt's gun-making, and Henry Ford's automobile industry for example - *could only expand because of the availability of reliable machine tools, and this statement remains critically true today for all industries.*

One can thus conjecture a similar historical trend, given that society is now in the middle of a second Industrial Revolution, often called the Information Age Revolution. Rather than beginning with the steam engine in 1769, this new Age begins with the invention of the transistor in 1947. This is followed by a period of approximately fifty years of rapid growth in other transistor technologies, integrated circuits (beginning in 1958), microprocessors (beginning in 1971) all of which is fueled by the economic and political importance of computing and networking. Research shows[2] that the semiconductor industry is now moving into an analogous and a very necessary period of consolidation in equipment refinement and productivity gains. During this period, the 21st century semiconductor equipment industries will be the equivalent of the 19th century machine tool industries - they will be the important base upon which the rest of the industry will depend upon for growth. This "new machine tool industry of the 21st century" is redefined as "a combination of the semiconductor equipment manufacturers supported by the classical base of machine tools, cutting tools and metal cutting theory."

In summary, the metal cutting process is a basic building-block for consumer product manufacturing of all kinds. But it cannot be emphasized enough that metal cutting is fundamentally related to all other manufacturing processes - not just the obviously metal-based production of automobiles, airplanes, and humble products such as lawn mowers and washing machines. All aspects of CAD/CAM, rapid prototyping, die making, and complex equipment making - especially for semiconductors - have metal cutting at their core.

Edward M. Trent and Paul K. Wright, 1999

1) Merchant, M.E., *Machining Science and Technology, 2,*157 (1998)
2) Leachman, R.C. and Hodges, D.A., *IEEE Transactions on Semiconductor Manufacturing*, **9**, (2) 158 (1996)

ACKNOWLEDGMENTS

We have been fortunate in having many able and congenial colleagues during the many years in which we have been actively interested in the subject of metal cutting. Without their collaboration and the contributions which they have made in skill and ideas this book would not have been written. We would like here to acknowledge the part which they have played in developing our understanding of the metal cutting process. We also acknowledge that there are some differences of opinion on precise issues of chip formation, seizure effects, how diffusion occurs at the chip-tool interface, and how segmented chip forms are triggered at certain speeds with different work materials. These different views make the topic of metal cutting lively and engaging for students.

The metallurgical emphasis of this book has its "roots" at Sheffield University, England in the Metallurgy Department where two people in particular - Dr. Edwin Gregory and Mr. G.A. de Belin - were responsible for a very high level of teaching in the techniques of metallography.

These traditions were continued in the *"Machining Research Group"* in the Department of Metallurgy and Materials of the University of Birmingham, England, under the late Professor E.C. Rollason and late Professor D. V. Wilson. In that work we acknowledge in particular the contributions of Mr. E. Lardner, Dr. D.R. Milner, and the late Professor G.W. Rowe.

Many of the specific photographs that are shown, come from the detailed experimental work that was carried out by our colleague Edward Smart. His great devotion to the experimental aspects of machining research, inspired many students in their careers. We miss his cheerful countenance and hope this book continues to keep his memory alive.

The most important element in the evidence presented here comes from such direct observation using optical metallography. The field of useful observation is being greatly extended by electron microscopy and by instruments such as the microprobe analyser. Computer modeling of machining is now key to the modern analysis. *Nevertheless, a high level of optical metallography and experimental work remains at the center of all investigations - in such work the behavior of tool materials and work materials are directly observed.*

The resources of Wickman-Wimet Ltd. under the former Research Director, Mr. A.E. Oliver, are acknowledged.

A long association with Kennametal Inc. is also acknowledged and we thank Dr. Yefim Val for his many years of support and interest.

Similarly, Cincinnati Milacron was a great source of support for the work and we acknowledge Mr. R. Messinger, Dr. R. Kegg and Mr. L. Burnett.

Thanks are due to the Association for Manufacturing Technology for material in Chapter 1 - the associations with Mr. Charles Carter have been most valuable.

Collaboration with Sandia National Laboratories has been invaluable and we thank Mr. L. Tallerico, Dr. R. Stoltz, Mr. A. Hazleton and Mr. A. West for their support.

Funding from the Ford Motor (and particularly the Visteon organization) is gratefully acknowledged and we particularly thank Mr. Charles Szuluk for his long term interest in integrated manufacturing.

We are also indebted to the following: the firms of Fagersta and Speed Steel (Sweden) for advice and data on properties and performance of high speed steel tools; Dr J. Lumby and Lucas Industries for information on, and permission to publish an electron micrograph of, sialon; Dr P. Heath and DeBeers Industrial Diamonds for information on cubic boron nitride and polycrystalline diamond and permission to publish photomicrographs of structures and of tool wear.

Support from the National Science Foundation has been invaluable in developing some of the fundamental ideas on stress analysis and tooling materials. Many colleagues have served at NSF during the course of the research: Dr. C. Astill, Dr. B von Turkovich, Dr. W. Spurgeon, Dr. J. Meyer, Dr. T. Woo, Dr. S. Settles, Dr. J. Lee, Dr. M. DeVries, Dr. W. DeVries, Dr. L. Martin-Vega, Dr. C. Srinivasan, Dr. B. Chern, Dr. B. Kramer, and Dr. G. Hazelrigg have all played a role in this work. Support from DARPA and ONR is also gratefully acknowledged as is the personal interest of Dr. W. Isler, Dr. E. Mettala and Dr. J. Sheridan.

The authors are also grateful to their colleagues Dr. E. Amini, Dr. A. Bagchi, Dr. D.A. Dearnley, Dr. B.W. Dines, Dr. R. M. Freeman, Dr. J. Hau-Bracamonte, Dr. R. Komanduri, Dr. Y. Naerheim, Dr. R. Milovic, Mr. M.E. Mueller, Dr. M. Samandi, Dr. D. Sandstrom, Dr. J.A. Stori, Mr. K.F. Sullivan, Dr. J. Wallbank, and Dr. M. L. H. Wise for photomicrographs and graphs acknowledged in the captions to the illustrations which they have contributed.

As well as the specific photomicrographs and graphs acknowledged in the captions, the ideas in the book have evolved over many decades from close friendships, co-authorships, joint projects and informal discussions at various conferences. In this regard, we acknowledge a long association concerning manufacturing and metal cutting with the following colleagues:

• Professor T.H.C. Childs and Dr. P. Dearnley - part of the original group at Birmingham and now at the University of Leeds;

• Professor D. Tabor, Dr. N. Gane, Dr. J. Williams, Dr. D. Doyle and Dr. J. G. Horne from the "transparent sapphire tool group" at the Cavendish Laboratory, Cambridge University;

• Mr. L.S. Aiken, Mr. W. Beasley, Mr. P.D. Smith, Dr. J.L. Robinson, Dr. P.S. Jackson, Dr. A.W. Wolfenden, Professor G. Arndt, Professor R.F. Meyer, and the late Professor J.H. Percy in New Zealand;

• Professor P.L.B Oxley, Professor E.J.A. Armarego, Professor R.H. Brown and Dr. M.G. Stevenson in Australia;

- Dr. A. Bagchi, Dr. A.J. Holzer, Dr. D.A. Bourne, Dr. C.C. Hayes, Dr. R.S. Rao, Dr. J.G. Chow, Dr. S. C. Y. Lu, Dr. D.W. Yen, Dr. C. King, Mr. H. Kulluk, the late Dr. J.L. Swedlow, Dr. M.R. Cutkosky and Dr. F. B. Prinz (both now at Stanford), Professor S. Finger, Professor M. Nagurka, Professor P. Khosla, Dr. L. Weiss, Professor R. Sturges, Professor W. Sirignano and Professor R. Reddy, from the research carried out at Carnegie Mellon University;
- Professor J.T. Schwartz, Professor K. Perlin, Mr. F.B. Hansen, Mr. L. Pavlakos, Dr. J. Hong, Dr. X. Tan and Mr. I. Greenfeld of the Courant Institute at New York University;
- Professor B.F. von Turkovich at the University of Vermont; Professors R. Komanduri and J. Mize, now at Oklahoma State University; Professor M.C. Shaw, at Arizona State University; Dr. O. Richmond, Dr. M. Devenpeck and Dr. E. Appleby at Alcoa Research Laboratories; Professor J.T. Black at Auburn University; Professor S. Kalpakjian at the Illinois Institute of Technology; Professor H. Voelcker at Cornell University; Professor F. Ling at the University of Texas, Austin; Professors N.P. Suh and D. Hardt at M.I.T.; Professors T. Kurfess, S. Liang, S. Melkote and J. Colton at Georgia Tech.; Professors S.K. Gupta and D. Nau at the University of Maryland; Professor W. Regli at Drexel; Professor S. Ramalingam at Minnesota; Dr. R. Woods at Kaiser Aluminum; Professor D. Williams at Loughborough University, England; Professor Nabil Gindy, Dr. T. Ratchev and colleagues at Nottingham University; Professor M. Elbestawi and his group at McMaster University; Professor Y. Altintas at the University of British Columbia; Professors Y. Koren, G. Ulsoy and J. Stein at the University of Michigan; Professors K. Weinmann and J. Sutherland at Michigan Technological University; Professor K.P. Rajurkar at the University of Nebraska; Professor J. Tlusty at the University of Florida; Professors A. Lavine and D. Wang at UCLA; Professors I. Jawahir, A. T. Malc and O. Dillon at the University of Kentucky; Professor Tony Woo at the University of Washington; Professor S. Settles now at the University of Southern California; Professor W. DeVries at Iowa State University; Professors M. DeVries and R. Gadh at the University of Wisconsin; and the many colleagues at Berkeley, Illinois, M.I.T., Boeing and Ford/Visteon mentioned in the Preface.

The resources of the University of California, Berkeley are acknowledged and thanks are due to Professor David Hodges, Professor Paul Gray, Professor Dan Mote, Professor David Bogy, Professor Robert Cole, Professor Shankar Sastry and Professor Robert Brodersen for their collaborations and support.

Professors Erich Thomsen, Joseph Frisch and the late Shiro Kobayashi were the founders of manufacturing related research at Berkeley and we thank them for their support and interest in this work.

For their detailed assistance in the preparation of this edition we thank Bonita Korpi, William Chui, Eric Mellers and Zachary Katz. Thanks are also due to V. Sundararajan, C. Smith, G. Sun, S. Roundy, J. Kim, J. Brock, J. Plancarte, R. Inouye, D. Chapman, R. Hillaire, K. Urabe, N. An, M. McKenzie, L. Marchetti, J. Smith and S. McMains.

The authors wish to thank the following organizations for permission to reproduce the illustrations listed.
- *The Metals Society* - Figures 3.7, 3.9, 3.11, 3.22, 5.11, 5.12, 5.14, 6.6, 6.11, 6.12, 6.13, 6.14, 6.15, 6.21, 6.22, 6.26, 7.9, 7.15, 7.17, 7.20, 7.24, 7.32, 9.2, 9.9b, 9.18, 9.27, 9.33, 9.36, 9.43, 10.11, 10.12, 10.15, 10.17.
- *International Journal for Production Research* - Figures 5.13 to 5.18 , 9.37, 9.38, 9.39.

- *International Journal for Machine Tool Design and Research* - Figures 9.13, 10.1, 10.2, 10.3, 10.4, 10.5.
- *American Society for Metals* - Figures 6.27, 9.25
- *American Society of Mechanical Engineers* - Figures 5.7, 5.8.
- *Proceedings of the Royal Society* - Figures 4.24 to 4.30
- The following photomicrographs were taken in the laboratories of Wickman Wimet Ltd. who granted permission for their reproduction: Figures 3.7, 3.9, 3.10, 3.11, 7.2, 7.6, 7.8, 7.9, 7.10, 7.14, 7.15, 7.19, 7.20, 7.21, 7.25, 7.26, 7.27, 7.32, 9.2, 9.26, 9.31, 9.32, 9.36, 9.42, 10.11, 10.12, 10.17.

Our wives, Enid and Terry, and our families have contributed in innumerable ways to the writing of the book, but especially by spending many hours discussing the presentation and overall scope. We are very grateful to them for this assistance and for their patience during the months that we have been preoccupied with the work.

CHAPTER 1   # INTRODUCTION: HISTORICAL AND ECONOMIC CONTEXT

1.1   THE METAL CUTTING (or MACHINING) PROCESS

It would serve no useful purpose to attempt a precise definition of *metal cutting* or *machining*. In this book the term is intended to include operations in which a thin layer of metal, the *chip* or *swarf*, is removed by a wedge-shaped tool from a larger body. There is no hard and fast line separating *chip-forming* operations from others such as the shearing of sheet metal, the punching of holes or the cropping of lengths from a bar. These also can be considered as metal cutting, but the action of the tools and the process of separation into two parts are so different from those encountered in chip-forming operations, that the subject requires a different treatment.

There is a great similarity between the operations of cutting and grinding. Our ancestors ground stone tools before metals were discovered and later used the same process for sharpening metal tools and weapons. The grinding wheel does much the same job as the file, which can be classified as a cutting tool, but has a much larger number of cutting edges, randomly shaped and oriented. Each edge removes a much smaller fragment of metal than is normal in cutting, and it is largely because of this difference in size that conclusions drawn from investigations into metal cutting must be applied with reservations to the operation of grinding.[1,2]

In the engineering industry, the term *machining* is used to cover chip-forming operations, and this definition appears in many dictionaries. Most machining today is carried out to shape metals and alloys (many plastic products are also machined), but the lathe was first used to turn wood and bone. The term metal cutting is used here because research has shown certain characteristic features of the behavior of metals during cutting which dominate the process and, without further work, it is not possible to extend the principles described here to the cutting of other materials.

While metal cutting is commonly associated with big industries (automotive, aerospace, home appliance, etc.) that manufacture big products, the machining of metals and alloys plays a crucial role in a range of manufacturing activities, including the ultraprecision machining of extremely delicate components (*Figure 1.1*).

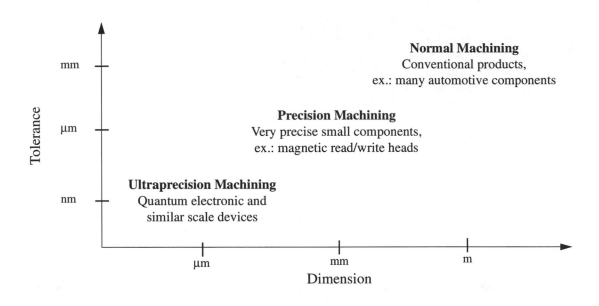

FIGURE 1.1   Three relatively distinct manufacturing paradigms (Adapted from Wirtz, 1991)

1.2   A SHORT HISTORY OF MACHINING

Before the middle of the 18th century, wood was the main material used in engineering structures. To shape wooden parts, craftsmen used machine tools - the lathe among them - which were typically constructed of wood as well. The boring of cannons and the production of metal screws and small instrument parts were the exceptions: these processes required metal tools. It was the steam engine, with its large metal cylinders and other parts of unprecedented dimensional accuracy, which led to the first major developments in metal cutting in the 1760s.

The materials which constituted the first steam engines were not very difficult to machine. Gray cast iron, wrought iron, brass and bronze were readily cut using hardened carbon steel tools. The methods of heat treatment of tool steel had been evolved by centuries of craftsmen, and reasonably reliable tools were available, although rapid failure of the tools could be avoided only by cutting very slowly. It required 27.5 working days to bore and face one of Watt's large cylinders.[3]

At the inception of the steam engine, no machine tool industry existed. The century from 1760 to 1860 saw the establishment of enterprises devoted to the production of machine tools. Maudslay, Whitworth, and Eli Whitney, among many other great engineers, generated, in metallic components, the cylindrical and flat surfaces, threads, grooves, slots and holes of the many shapes required by developing industries.[4] The lathe, planer, shaper, milling machine, drilling machine and power saws all developed into rigid machines capable, in the hands of good craftspeople, of turning out large numbers of very accurate parts that had never before been possible.

By 1860 the basic problem of how to produce the necessary shapes in the existing materials had been solved. There had been little change in the materials which had to be machined - cast iron, wrought iron and a few copper based alloys. High carbon tool steel, hardened and tempered by the blacksmith, still had to answer all the tooling requirements. The quality and the consistency of tool steels had been greatly improved by a century of experience with the crucible steel process. Yet even the best carbon steel tools, pushed to their functional limits, were increasingly insufficient for manufacturers' needs, constraining production speed and hampering efficiency.

From the mid-1880s on, innovative energies in manufacturing shifted from developing basic machine tools and producing highly-accurate parts to reducing machining costs and cutting new types of metals and alloys. With the Bessemer and Open Hearth steel making processes, steel rapidly replaced wrought iron as the workhorse of construction materials. Industry required ever greater tonnages of steel (steel production soon vastly exceeded the earlier output of wrought iron), and required it machined to particular specifications.

Alloy steels proved much more difficult than wrought iron to machine, and cutting speeds had to be lowered even further to maintain reasonable tool life. Towards the end of the 19th century, both the labor and capital costs of machining were becoming very great. The incentive to reduce costs by accelerating and automating the cutting process became more intense, and, up to the present time, still acts as the major driving force behind technological developments in the metal cutting field.

The discovery and manipulation of new cutting tool materials has been perhaps the most important theme in the last century of metal cutting. Productivity could not have significantly increased without the higher cutting speeds achievable using high-speed steel and cemented carbide tools, both important advances over traditional carbon steel technology. The next major step occurred with the development of ceramic and ultra-hard tool materials. Recently, a group of new techniques, including electrical discharge and water-jet machining, have joined ceramic and ultra-hard materials at the forefront of metal cutting technology.

Machine tool manufacturers have created machines capable of maximizing the utility of each generation of cutting tool materials. Designers and machinists have optimized the shapes of tools to lengthen tool life at high cutting speeds, while lubricant manufacturers have developed new coolants and lubricants to improve surface finish and permit increased rates of metal removal. Tool control has also advanced considerably since the days of manually operated machines. Automatic machines, computer numerically controlled (CNC) machines and transfer machines produce better tool efficiency, greatly increasing output per employee.

Increasingly, the process of metal cutting is integrated with computer software and hardware that control machine tools. The age of "mechatronics" accompanies a trend toward integrated manufacturing systems composed of cells and modules of machines rather than individual, stand-alone units. Machining today requires a wider range of skills than it did a century ago: computer programming and knowledge of electronic equipment, among others. Nevertheless, knowledge of the physical realities of the tool-work interface is as important as ever.

One last note should be added to our understanding of the evolution of machine tool technology concerning the double role of basic metal producers. Many new alloys have been developed to meet the increasingly severe conditions of stress, temperature and corrosion imposed by the needs of our industrial civilization. Some of these materials, like aluminum and magnesium, are easy to machine, but others, such as high-alloy steels and nickel-based alloys, become more difficult to cut as their useful properties (i.e. strength, durability, etc.) improve. The machine tool and cutting tool industries have had to develop new strategies to cope with these new metals. At the same time, basic metal producers have responded to the demands of production engineers for metals which can be cut faster. New heat treatments have been devised, and the introduction of alloys like free-machining steel and brass has made great savings in production costs.

1.3   MACHINING AND THE GLOBAL ECONOMY

Today, metal cutting is a significant industry in most economically developed countries, though small in comparison to the customer industries it serves. The automobile, railway, ship-building, aircraft manufacture, home appliance, consumer electronics and construction indus-tries - all these have large machine shops with many thousands of employees engaged in machining. Worldwide annual consumption of machine tools (metal cutting + metal forming units) over the last several years has been on the order of $35 - $40 billion per year.[5] As shown on the left of *Figure 1.2*, the U.S. is currently the world's largest consumer, purchasing over $7 billion in new machine tools in 1996.[6]

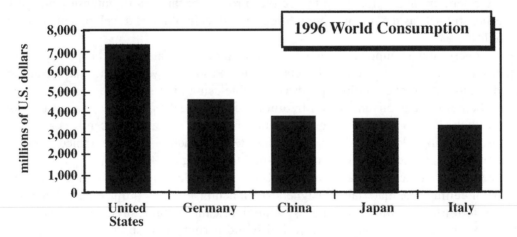

FIGURE 1.2   Top five machine tool consumers (Gardner Publications)

The U.S., once the world leader in machine tool production, suffered from foreign competi-tion in the 1970s and 1980s and has struggled to regain market position ever since. While domestic demand fell during the recession of 1982 - 1983, Asian and European demand, espe-cially for cheap, reliable CNC technology, exploded. Although this type of machine control was originally developed in the United States, Japanese firms successfully commercialized and exported the technology, taking a pronounced lead in the machine tool industry by the late 1980s.

As U.S. consumption grew after the 1982 - 1983 recession, so did imports from Japan and Germany *(Figure 1.3)*. Despite this competition from Japan and Germany, the U.S. remains a major machine tool producer *(Figure 1.4)*, with shipments of more than $4.5 billion and exports of $1.2 billion in 1996.[6]

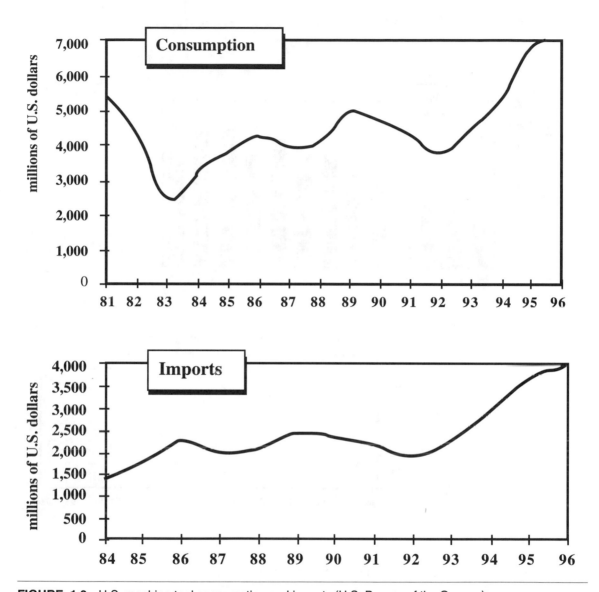

FIGURE 1.3 U.S. machine tool consumption and imports (U.S. Bureau of the Census)

The U.S. machine tool industry is also an important job provider for skilled, technical workers, employing between 50,000 and 100,000 workers during the last decade in firms of all sizes.[6] The recent resurgence of the U.S. machine tool industry can be seen in *Figure 1.5* by the growth in average revenue and number of employees in machine tool companies. This resurgence during the mid-1990s can be attributed to several factors, including:

- Improved business practices in the larger, better-established machine tool companies
- The entry into the market of smaller, newer U.S.-based machine tool companies supplying high quality, "easy-to-buy-and-maintain" machine tools in the $50,000-$100,000 price range.
- More favorable exchange rates in the period of the mid-1990s.

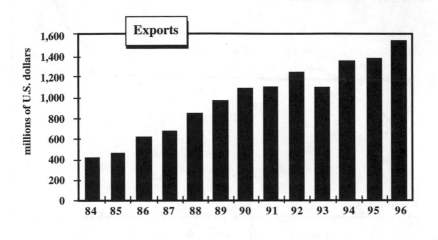

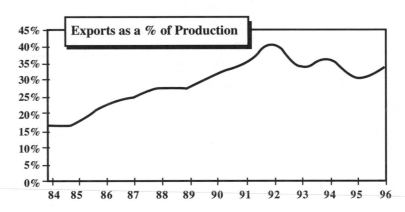

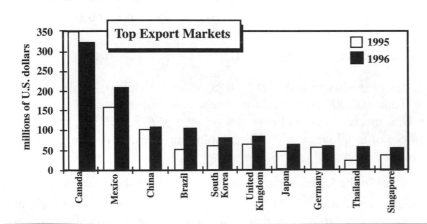

FIGURE 1.4   U.S. machine tool exports (U.S. Bureau of the Census)

Average Machine Tool Company		
	Revenue (millions of U.S. Dollars)	**Employment** total employees
1995	$110	96,000
1996	$127	122,000
Increase	$17	26,000
% Increase	15 %	27%

FIGURE 1.5   U.S. machine tool companies, 1995-96 statistics (U.S. Bureau of the Census)

1.4   SUMMARY AND CONCLUSION

1.4.1 Economics

To summarize the economic importance, the cost of machining amounts to more than 15% of the value of all manufactured products in all industrialized countries.[7] *Metcut Research Associates* in Cincinnati, Ohio, estimates that, in the U.S., the annual labor and overhead costs of machining are about $300 billion dollars per year (this excludes work materials and tools). U.S. consumption of new machine tools (CNC lathes, milling machines, etc.) is about $7.5 billion dollars per year. Consumable cutting tool materials have U.S. sales of about $2 to 2.5 billion dollars per year.[8,9] For comparison purposes, it is of interest to note a ratio of $\{300 \rightarrow 7.5 \rightarrow 2.5\}$ billion dollars for {labor costs $\rightarrow$ fixed machinery investments $\rightarrow$ disposable cutting tools}.

1.4.2 Sociology

Progress in machining is achieved by the ingenuity, logical thought and dogged worrying of many thousands of practitioners engaged in the many-sided arts of metal cutting. The machinist operating the machine, the tool designer, the lubrication engineer, and the metallurgist are all constantly probing for solutions to the challenges presented by novel materials, high costs, and the needs for faster metal removal, greater precision and smoother surface finish. However competent they may be, there can be few craftspeople, engineers or scientists engaged in this field who do not feel that they would be better able to solve their problems if they had a deeper knowledge of what was happening at the cutting edge of the tool.

1.4.3 Technology

It is what happens in a very small volume of metal around the cutting edge that determines the performance of tools, the machinability of metals and alloys, and the final qualities of the machined surface. During cutting, the interface between tool and work material is largely inaccessible to observation, but indirect evidence concerning stresses, temperatures, metal flow and other interactions has been contributed by many researchers. This book will endeavor to summarize the available knowledge contained in published work, the authors' own research, and that of many of their colleagues to provide a thorough understanding of metal cutting.

1.5 REFERENCES

1. Doyle, E.D. and Aghan, R.L., *Metall. Trans.* B, **6B**, 143 (1975)
2. Rabinowicz, E., *Wear*, **18**,169 (1971)
3. Rolt, L.T.C., *Tools for the Job*, Batsford (1965)
4. Cookson, J.O. and Sweeney, G., *ISI Publication No 138,* p. 83 (1971).
5. *A Technology Roadmap for the Machine Tool Industry*, The Association for Manufacturing Technology, McLean, VA (1996)
6. *1997-98 Economic Handbook of the Machine Tool Industry*, The Association for Manufacturing Technology, McLean, VA (1997).
7. Merchant, M.E., *Machining Science and Technology*, **2**, 157 (1998).
8. Komanduri, R., *Tool Materials*, Encyclopedia of Chemical Technology, Fourth Edition, John Wiley and Sons Inc., Volume 24, 390 (1997)
9. Huston, M.F., and Knobeloch, G.W., *Cutting Materials, Tools and Market Trends,* in the Conference on High-Performance Tools, Dusseldorf, Germany, 21 (1998)

CHAPTER 2

METAL CUTTING OPERATIONS AND TERMINOLOGY

2.1 INTRODUCTION

Of all the processes used to shape metals, it is in machining that the conditions of operation are most varied. Almost all metals and alloys are machined - hard or soft, cast or wrought, ductile or brittle, with high or low melting point. Most shapes used in the engineering world are produced by machining. As regards size, components from watch parts to aircraft wing spars, over 30 meters long, are machined. Many different machining operations are used; cutting speeds may be as high as 3,500 m min^{-1} (11,500 ft/min) for aluminum wing panels, or as low as a few centimeters per minute; cutting time may be continuous for several hours or interrupted in fractions of a second.

It is important that this great variability be appreciated. Some of the major variables are, therefore, discussed and the more significant terms defined by describing briefly some of the more important metal cutting operations with their distinctive features.

2.2 TURNING

The basic operation of turning, (also called semi-orthogonal cutting in the research laboratory), is also the one most commonly employed in experimental work on metal cutting. The work material is held in the chuck of a lathe and rotated. The tool is held rigidly in a tool post and moved at a constant rate along the axis of the bar, cutting away a layer of metal to form a cylinder or a surface of more complex profile. This is shown diagrammatically in *Figure 2.1*.

The *cutting speed* (*V*) is the rate at which the uncut surface of the work passes the cutting edge of the tool, usually expressed in units of ft/min or m min^{-1}. The *feed* (*f*) is the distance moved by the tool in an axial direction at each revolution of the work.

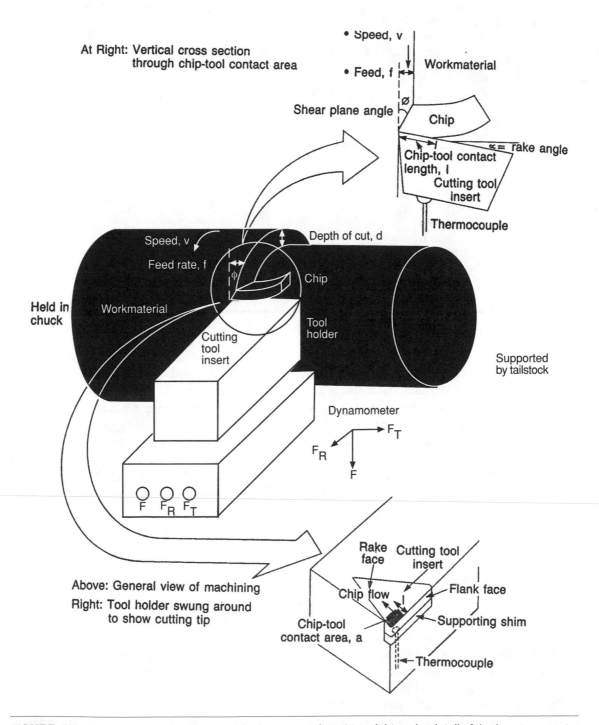

FIGURE 2.1   Lathe turning showing a vertical cross-section at top right and a detail of the insert geometry at bottom right. The dynamometer platform and the remote thermocouple on the bottom of the insert are not used during today's production machining. However they are useful in the research laboratory for routine measurement of cutting forces and overall temperature.

The *depth of cut (w)* is the thickness of metal removed from the bar, measured in a radial direction.[†] The product of these three gives the rate of metal removal, a parameter often used in measuring the efficiency of a cutting operation.

$$Vfw = \text{rate of removal} \tag{2.1}$$

The cutting speed and the feed are the two most important parameters which can be adjusted by the operator to achieve optimum cutting conditions. The depth of cut is often fixed by the initial size of the bar and the required size of the product.

Cutting speed is usually between 3 and 200 m min^{-1} (10 and 600 ft/min). However, in modern high speed machining, speeds may be as high as 3,500 m min^{-1} when machining aluminum alloys. The rotational speed (RPM) of the spindle is usually constant during a single operation, so that when cutting a complex form the cutting speed varies with the diameter being cut at any instant. At the nose of the tool the speed is always lower than at the outer surface of the bar, but the difference is usually small and the cutting speed is considered as constant along the tool edge in turning. Recent computer-controlled machine tools have the capacity to maintain a constant cutting speed, V, by varying the rotational speed as the work-piece diameter changes.

Feed may be as low as 0.0125 mm (0.0005 in) per rev. and with very heavy cutting up to 2.5 mm (0.1 in) per rev. Depth of cut may vary from zero over part of the cycle to over 25 mm (1 in). It is possible to remove metal at a rate of more than 1600 cm^3 (100 in^3) per minute, but such a rate would be very uncommon and 80-160 cm^3 (5-10 in^3) per minute would normally be considered rapid. *Figure 2.2* shows some of the main features of a turning tool. The surface of the tool over which the chip flows is known as the *rake face*. *The cutting edge* is formed by the intersection of the rake face with the *clearance face* or *flank* of the tool. The tool is so designed and held in such a position that the clearance face does not rub against the freshly cut metal surface. The *clearance angle* is variable but is often on the order of 6-10°. The rake face is inclined at an angle to the axis of the bar of work material and this angle can be adjusted to achieve optimum cutting performance for particular tool materials, work materials and cutting conditions. The *rake angle* is measured from a line parallel to the axis of rotation of the work-piece (*Figures 2.1 and 2.2*). A *positive rake* angle is one where the rake face dips below the line. Early metal cutting tools had large positive rake angles to give a cutting edge which was keen, but easily damaged. Positive rake angles may be up to 30°, but the greater robustness of tools with smaller rake angle leads in many cases to the use of zero or *negative rake* angle. With a negative rake angle of 5° or 6°, the included angle between the rake and clearance faces may be 90°, and this has advantages. The tool terminates in an *end clearance face*, which is also inclined at such an angle as to avoid rubbing against the freshly cut surface. The *nose* of the tool is at the intersection of all three faces and may be sharp, but more frequently there is a *nose radius* between the two clearance faces.

[†].The feed rate (f) during turning is also called the undeformed chip thickness, (t_1) in *Figures 2.1 and 3.1*. The depth of cut, w, in turning is also referred to as the undeformed chip width. Since machining has developed from a practitioner's viewpoint, the terminology is not really consistent from one operation to another. For example, in *Figure 2.5*, for end milling, the term depth of cut is used in a different way. This is unfortunately confusing for a new student of the field. Perhaps the best way to accommodate these inconsistencies is to always view the "slice" of material being removed as the "undeformed chip thickness" (t_1) and the direction normal to this (into the plane of the paper in *Figure 2.1*) as the "undeformed chip width".

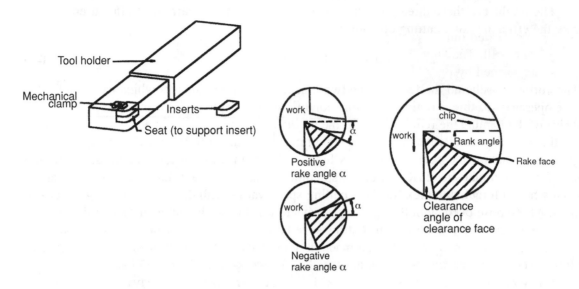

FIGURE 2.2   Cutting tool terminology

This very simplified description of the geometry of one form of turning tool is intended to help the reader without practical experience of cutting, follow the terms used later in the book. The design of tools involves an immense variety of shapes and the full nomenclature and specifications are very complex.[1,2]

It is difficult to appreciate the action of many types of tools without actually observing or, preferably, using them. The performance of cutting tools is very dependent on their precise shape. In most cases there are critical features or dimensions which must be accurately formed for efficient cutting. These may be, for example, the clearance angles, the nose radius and its blending into the faces, or the sharpness of the cutting edge. The importance of precision in tool making, whether in the tool room of the user, or in the factory of the tool maker, cannot be over estimated. This is an area where excellence in hand-skills is still of great value despite the well-controlled automated aspects of sintered tool material manufacturing. A number of other machining operations are discussed briefly, with the objective of demonstrating some characteristic differences and similarities in the cutting parameters.

2.3   BORING OPERATIONS

Essentially the conditions of boring of internal surfaces differ little from those of turning, but this operation illustrates the importance of rigidity in machining. Particularly when a long cylinder with a small internal diameter is bored, the bar holding the tool must be long and slender and cannot be as rigid as the thick, stocky tools and tool post used for most turning. The tool tends to be deflected to a greater extent by the cutting forces and to vibrate. Vibration may affect not only the dimensions of the machined surface, but its roughness and the life of the cutting tools.

2.4   DRILLING

In drilling, carried out on a lathe or a drilling machine, the tool most commonly used is the familiar twist-drill. The "business end" of a twist drill has two cutting edges. The rake faces of the drill are formed by part of each of the *flutes*, *Figure 2.3*, the rake angle being controlled by the *helix angle* of the drill. The chips slide up the flutes, while the end faces must be ground at the correct angle to form the clearance face.

An essential feature of drilling is the variation in cutting speed along the cutting edge. The speed is a maximum at the periphery, which generates the cylindrical surface, and approaches zero near the center-line of the drill, the *web*, *Figure 2.3*, where the cutting edge is blended to a chisel shape. The rake angle also decreases from the periphery, and at the chisel edge the cutting action is that of a tool with a very large negative rake angle.

The variations in speed and rake angle along the edge are responsible for many aspects of drilling which are peculiar to this operation. Drills are slender, highly stressed tools, the flutes of which have to be carefully designed to permit chip flow while maintaining adequate strength. The helix angles and other features are adapted to the drilling of specific classes of material.

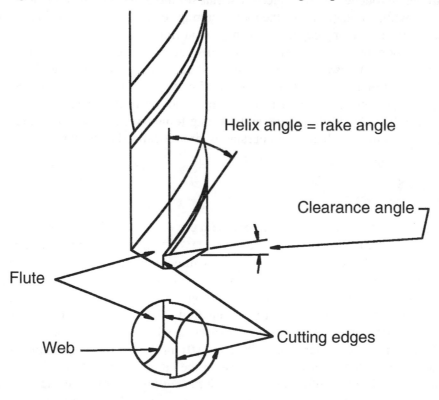

FIGURE 2.3   The twist drill

2.5   FACING

Turning, boring and drilling generate cylindrical or more complex surfaces of rotation. Facing, also carried out on a lathe, generates a flat surface, normal to the axis of rotation, by feeding the tool from the surface towards the center or outward from the center. In facing, the depth of cut is measured in a direction parallel to the axis and the feed in a radial direction. A characteristic of this operation is that the cutting speed varies continuously, approaching zero towards the center of the bar.

2.6   FORMING AND PARTING OFF

Surfaces of rotation of complex form may be generated by turning, but some shapes may be formed more efficiently by use of a tool with a cutting edge of the required profile, which is fed into the peripheral surface of the bar in a radial direction. Such tools often have a long cutting edge. The part which touches the workpiece first cuts for the longest time and makes the deepest part of the form, while another part of the tool is cutting for only a very short part of the cycle. Cutting forces, with such a long edge, may be high requiring the feed to be set low.

Many short components, such as screws and bolts, are made from a length of bar, the final operation being parting off, as a thin tool cuts into the bar from the periphery to the center or to a central hole. The tool must be thin to avoid waste of material, but it may have to penetrate to a depth of several centimeters. These slender tools must not be further weakened by large clearance angles down the sides. Both parting and forming tools have special difficulties associated with small clearance angles, which result in rapid wear at localized positions.

2.7   MILLING

Both grooves and flat surfaces - for example the faces of a car cylinder block - are generated by milling. In this operation the cutting action is achieved by rotating the tool while the work is clamped on a table and the feed action is obtained by moving it under the cutter, *Figure 2.4*.

There is a very large number of different shapes of milling cutters for different applications. Single toothed cutters are possible but typical milling cutters have a number of teeth (cutting edges) which may vary from three to over one hundred. The new surface is generated as each tooth cuts away an arc-shaped segment, the thickness of which is the *feed* or *tooth load*. Feeds are usually light, not often greater than 0.25 mm (0.01 in) per tooth, and frequently less then 0.025 mm (0.001 in) per tooth. However, because of the large number of teeth, the rate of metal removal is often high. The feed often varies through the cutting part of the cycle.

In the orthodox milling operation shown in *Figure 2.4a,* the feed on each tooth is very small at first and reaches a maximum where the tooth breaks contact with the work surface. If the cutter is designed to "climb mill" and rotate in the opposite direction *(Figure 2.4b),* the feed is greatest at the point of initial contact.

An important feature of all milling operations is that the action of each cutting edge is intermittent. Each tooth is cutting during less than half of a revolution of the cutter, and sometimes for only a very small part of the cycle. Each edge is subjected to periodic impacts as it makes contact with the work. Thus, it is stressed and heated during the cutting part of the cycle, followed by a period when it is unstressed and allowed to cool. Frequently cutting times are a small fraction of a second and are repeated several times a second, involving both thermal and mechanical fatigue of the tool. The design of milling cutters is greatly influenced by the problem of getting rid of the chips.

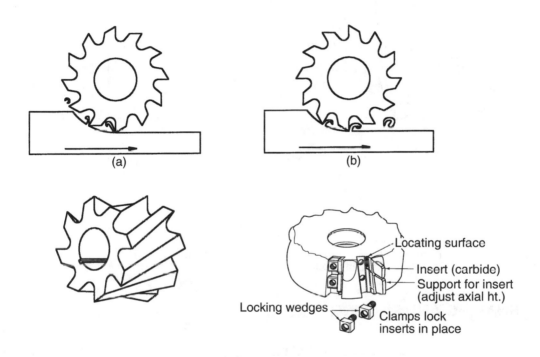

FIGURE 2.4   Milling cutters: a) Up milling, b) Climb milling, c) Edge milling, d) Detail of face milling with indexable inserts - the inserts protrude below the tool holder (Courtesy of J.T. Black)

Milling is used also for the production of curved shapes, while end mills (*Figures 2.5, 2.6,* and *2.7*), which are larger and more robust versions of the dentist's drill, are employed in the production of hollow shapes such as die cavities. The end mill can be used to machine a feature such as a rectangular pocket with (ideally) vertical walls.

In practice it is observed that the sideways pressure on the milling cutter can deflect it away from the vertical surface, as shown in *Figure 2.6b*. Since the slender end-mill is like a cantilever beam the deflection is more pronounced down at its tip. Instead of being vertical, the walls of the pocket begin to have a "ski-slope" cross-section - reasonably vertical at the top, but sloping out near the bottom where the tool is most deflected.[4] Even this is a somewhat simplified view and as shown in the detail of *Figure 2.7*, the non-verticality - the form error - shows an "s-shaped" undulation at the top of the wall, before the ski-slope effect gives rise to the maximum error at the bottom of the wall. Stori[5] has shown that the precise geometry is related to the depth of the pocket versus how many of the flutes shown in *Figure 2.5* are engaged in chip removal.

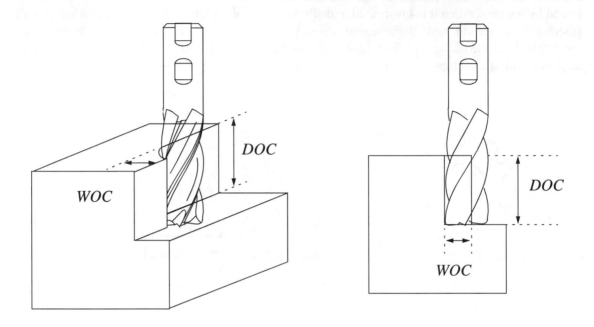

FIGURE 2.5   Ideal cutting conditions of an end mill; WOC = width of cut, DOC = depth of cut (Figure drawings of 2.5 to 2.7, courtesy of Dr. James Stori)

2.8   SHAPING AND PLANING

These are two other methods for generating flat surfaces and can also be used for producing grooves and slots. In the research laboratory, these processes are referred to as orthogonal cutting because the velocity vector of the advancing cutting edge meets the edge of the strip being cut at right-angles.

In shaping, the tool has a reciprocating movement, the cutting taking place on the forward stroke along the whole length of the surface being generated, while the reverse stroke is made with the tool lifted clear, to avoid damage to the tool or the work. The next stroke is made when the work has been moved by the feed distance, which may be either horizontal or vertical.

Planing is similar, but the tool is stationary, and the cutting action is achieved by moving the work. The reciprocating movement, involving periodic reversal of large weights, means that cutting speeds are relatively slow, but fairly high rates of metal removal are achieved by using high feed. The intermittent cutting involves severe impact loading of the cutting edge at every stroke. The cutting times between interruptions are longer than in milling but shorter than in most turning operations.

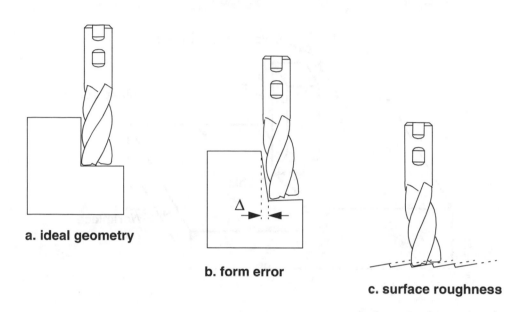

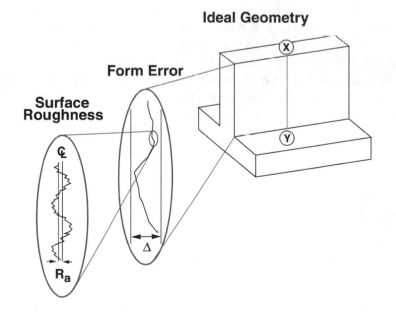

FIGURE 2.6   Effect of tool deflection on form error and surface roughness

FIGURE 2.7   Deviations in form and surface quality

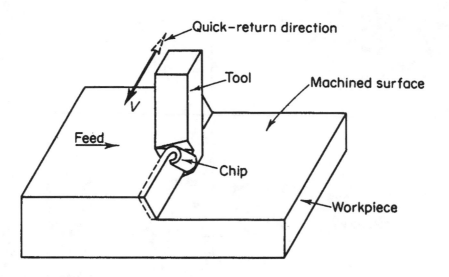

FIGURE 2.8   Shaping (and planing) operation (Courtesy of E.J.A. Armarego)

2.9   BROACHING

Broaching is an operation in which a cutting tool with multiple *transverse* cutting edges (a broach) is pulled or pushed over a surface or through a hole. Each successive cutting edge removes a layer of metal, giving a steady approach to the required final shape. It is an operation designed to produce high-precision forms and the complex tools are expensive. The shapes produced may be flat surfaces, but more often are holes of various forms or grooved components such as fir-tree roots in turbine discs, or the teeth of gears. Cutting speeds and feeds are low in this operation and adequate lubrication is essential.

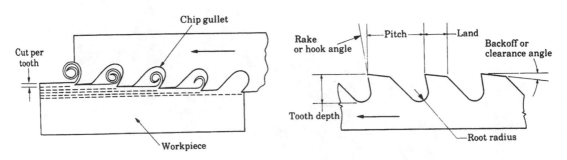

FIGURE 2.9   Broaching operation (Courtesy of S. Kalpakjian)

2.10   CONCLUSION

The operations described are some of the more important ones employed in shaping engineering components, but there are many others. Some of these, like sawing, filing and tapping of threads, are familiar to all of us, but others such as skiving, reaming or the hobbing of gears are the province of the specialist. Each of these operations has its special characteristics and problems, as well as the features which it has in common with the others.

There is one further variable which should be mentioned at this stage and that is the environment of the tool edge. Cutting is often carried out in air, but in many operations the use of a fluid to cool the tool or the workpiece, and/or act as a lubricant, is essential to efficiency. The fluid is usually a liquid, based on water or a mineral oil, but may be a gas such as CO_2. The action of coolants/lubricants will be considered in Chapter 10, but the influence of the cutting lubricant, if any, should never be neglected.

One objective of this very brief discussion of cutting operations is to make a major point which the reader should bear in mind in the subsequent chapters: in spite of the complexity and diversity of machining, analyses are made in the next three chapters of some of the more important features which *all* metal cutting operations have in common. The more theoretical work begins with an analysis of the basic geometry and features of metal cutting in Chapter 3. Stress and heat generation are then considered in Chapters 4 and 5. The work on which these analyses are based has, of necessity, been derived from a study of a limited number of cutting operations - often the simplest - such as uninterrupted turning, and from investigations in the machining of a limited range of work materials. At the same time, the results of these analyses should not be used uncritically, in the complex conditions of machine shops. In considering any problem of metal cutting, the first questions to ask should be - what is the machining operation, and what are the specific features which are critical for this operation?[3] Especially in Chapter 9 on "Machinability", it will be seen that the general pictures of stress and temperature distributions vary quite considerably from one work material alloy to the next.

2.11   REFERENCES

1. United States Cutting Tool Institute, *Metal Cutting Tool Handbook, 7th ed.*
2. Stabler, G.V., *J.I. Prod. E.*, **34**, 264 (1955)
3. *A.S.M. Metals Handbook*, 8th ed., Vol.3. 'Machining' (1967)
4. Kline, W.A., and DeVor, R.E. *Int. J. Mach. Tool Des. Res.*, **23,** (2/3), 123 (1983)
5. Stori, J.A. *Ph.D. Thesis*, University of California, Berkeley, (1998)

2.12   BIBLIOGRAPHY (Also see Chapter 15)

These introductory texts are helpful with diagrams of tool geometry and processes:

1. DeGarmo, E. P., Black, J.T., and Kohser, R. A., *Materials and Processes in Manufacturing*, 8th Edition, Prentice Hall, New York (1997)

2. Groover, M.P., *Fundamentals of Modern Manufacturing*, Prentice Hall, Upper Saddle River, New Jersey (1996)
3. Kalpakjian, S., *Manufacturing Processes for Engineering Materials*, Third Edition, Addison Wesley Longman, Menlo Park, CA (1997)
4. Schey, J.A., *Introduction to Manufacturing Processes*, McGraw Hill, New York (1999)

CHAPTER 3

THE ESSENTIAL FEATURES OF METAL CUTTING

3.1   INTRODUCTION

The processes of cutting most familiar to us in everyday life are those in which very soft bodies are severed by a tool such as a knife, in which the cutting edge is formed by two faces meeting at a very small included angle. The wedge-shaped tool is forced symmetrically into the body being cut, and often, at the same time, is moved parallel with the edge, as when slicing bread. If the tool is sharp, the body may be cut cleanly, with very little force, into two pieces which are gently forced apart by the faces of the tool. A microtome will cut very thin layers from biological specimens with no observable damage to the layer or to the newly-formed surface.

Metal cutting is not like this. Metals and alloys are too hard, so that no known tool materials are strong enough to withstand the stresses which they impose on narrow, knife-like cutting edges. (Very low melting point metals, such as lead and tin, may be cut in this way, but this is exceptional). If both faces forming the tool edge act to force apart the two newly-formed surfaces, very high stresses are imposed, much heat is generated, and both the tool and the work surfaces are damaged. These considerations make it necessary for a metal-cutting tool to *take the form of a large-angled wedge, which is driven asymmetrically into the work material, to remove a thin layer from the thicker work material body (Figure 3.1).* The layer must be sufficiently thin to enable the tool and work to withstand the imposed stress. A clearance angle must be formed on the tool to ensure that the clearance face does not make contact with the new work surface.

In spite of the diversity of the geometries of machining operations (emphasized in Chapter 2), the above restrictions on tool geometry are features of all metal cutting operations, and they provide common ground from which to commence an analysis of machining.

In practical machining, the included angle of the tool edge varies between 55° and 90°, so that the removed layer, the chip, is diverted through an angle of at least 60° as it moves away from the work, across the rake face of the tool. In this process, *the whole volume of metal removed is plastically deformed,* and thus a large amount of energy is required to form the chip and to move it across the tool face. In the process, two new surfaces are formed, the new surface of the work-piece (*OA in Figure 3.1*) and the under surface of the chip (*BC*). The formation of new surfaces requires energy, but in metal cutting, the theoretical minimum energy required to form the new surfaces is an insignificant proportion of that required to deform plastically the whole of the metal removed.

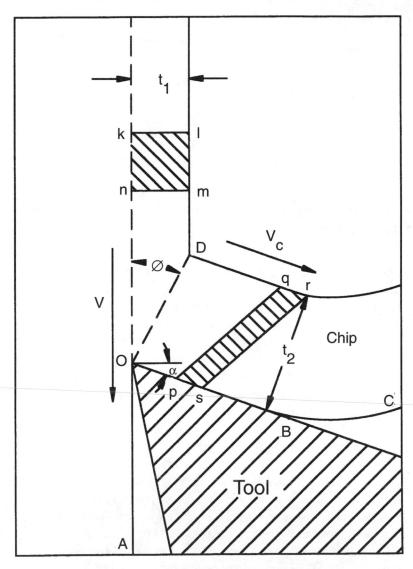

Metal Cutting Diagram

FIGURE 3.1   Metal cutting diagram

3.2   THE CHIP

The chip is enormously variable in shape and size in industrial machining operations; *Figure 3.2* shows some of the forms. The formation of all types of chips involves a shearing of the work material in the region of a plane extending from the tool edge to the position where the upper surface of the chip leaves the work surface (*OD* in *Figure 3.1*). A very large amount of strain takes place in this region in a very short interval of time, and not all metals and alloys can withstand this strain without fracture. Gray cast iron chips, for example, are always fragmented, and the chips of more ductile materials may be produced as segments, particularly at very low cutting speed. This *discontinuous chip* is one of the principal classes of chip form, and has the practical advantage that it is easily cleared from the cutting area. Under a majority of cutting conditions, however, ductile metals and alloys do not fracture on the shear plane and a *continuous chip* is produced (*Figure 3.3*). Continuous chips may adopt many shapes - straight, tangled or with different types of helix. Often they have considerable strength, and control of chip shape is one of the problems confronting machinists and tool designers. Continuous and discontinuous chips are not two sharply defined categories; every shade of gradation between the two types can be observed.

FIGURE 3.2   Chip shapes

The longitudinal shape of continuous chips can be modified by mechanical means, for example by grooves in the tool rake face, which curl the chip into a helix. The cross section of the chips and their thickness are of great importance in the analysis of metal cutting, and are considered later in some detail. For the purpose of studying chip formation in relation to the basic principles of metal cutting, it is useful to start with the simplest possible cutting conditions, consistent with maintaining the essential features common to these operations.

3.3   TECHNIQUES FOR STUDY OF CHIP FORMATION

Before discussing chip shape, the experimental methods used for gathering the information are described. The simplified conditions used in the first stages of laboratory investigations are known as *orthogonal* cutting. In orthogonal cutting the tool edge is straight, it is normal to the direction of cutting, and normal also to the feed direction. On a lathe, these conditions are secured by using a tool with the cutting edge horizontal, on the center line, and at right angles to the axis of rotation of the workpiece. If the workpiece is in the form of a tube whose wall thickness is the depth of cut, only the straight edge of the tool is used. In this method the cutting speed is not quite constant along the cutting edge, being highest at the outside of the tube, but if the tube diameter is reasonably large this is of minor importance. In many cases the work material is not available in tube form, and what is sometimes called *semi-orthogonal* cutting conditions are used, in which the tool cuts a solid bar with a constant depth of cut. In this case, conditions at the nose of the tool are different from those at the outer surface of the bar. If a sharp-nosed tool is used this may result in premature failure, so that it is more usual to have a small nose radius. To avoid too great a departure from orthogonal conditions, the major part of the edge engaged in cutting should be straight. Strictly orthogonal cutting can be carried out on a planing or shaping machine, in which the work material is in the form of a plate, the edge of which is machined. The cutting action on a shaper is, however, intermittent, the time of continuous machining is very short, and speeds are limited. For most test purposes the lathe method is more convenient.

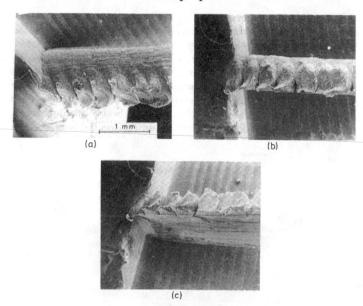

FIGURE 3.3   SEM photographs from three directions of a forming chip of steel cut at 48 m min^{-1} (150 ft/min) and 0.25 mm (0.01 in) per rev feed; image *(a)* from above, image *(b)* is a side view that can be compared with *Figure 3.1* and image *(c)* from below the cutting zone (Courtesy of B.W. Dines)

The study of the formation of chips is difficult, because of the high speed at which it takes place under industrial machining conditions, and the small scale of the phenomena which are to be observed. High speed cine-photography at relatively low magnification has been used. Early employment of this method was confined to study of changes in external shape during chip formation. These observations may mislead if interpreted as demonstrating the cutting action at the center of the chip. This limitation has been partly overcome by a method demonstrated by Tönshoff and his colleagues.[1] A polished and etched work material surface is held against a transparent silica plate and machined in such a way that the role of different phases in the work material (e.g. graphite flakes in cast iron) can be observed during the cutting process. With this method, the constraint of the silica plate provides conditions more like those at the center of the chip. There are still limitations on the magnification at which observations can be made and on the range of cutting speeds. Tabor and colleagues[2,3] have studied the movement of the chip across the rake face of transparent sapphire tools during cutting by observing the interface through the tool. High speed cine-photography was again used. This method is confined to the use of transparent tool materials and it cannot be assumed that the action at the interface is the same as when machining with metallic tools.[3]

No useful information about chip formation can be gained by studying the end of the cutting path after cutting has been stopped in the normal way by disengaging the feed and the drive to the work. By stopping the cutting action suddenly, however, it is possible to retain many of the important details - to "freeze" the action of cutting. Several "quick-stop" mechanisms have been devised for this purpose. One of the most successful involves the use of a humane killer gun to propel a lathe tool away from the cutting position at very high speed in the direction in which the work material is moving.[4] Sometimes the chip adheres to the tool and separates from the bar, but more usually the tool comes away more or less cleanly, leaving the chip attached to the bar. A segment of the bar, with chip attached, can be cut out with a hacksaw and examined in detail at any required magnification. For external examination and photography, the scanning electron microscope (SEM) is particularly valuable because of its great depth of focus. *Figure 3.3* shows an example of chip formation recorded in this way.

Much of the information in this book has been obtained by preparing metallographic sections through "quick-stop" specimens to reveal the internal action of cutting. Because any one specimen illustrates the cutting action at one instant of time, several specimens must be prepared to distinguish those features which have general significance from others which are peculiar to the instant at which cutting stopped.

3.4 CHIP SHAPE

Even with orthogonal cutting, the cross section of the chip is not strictly rectangular. Since it is constrained only by the rake face of the tool, the metal is free to move in all other directions as it is formed into the chip. The chip tends to spread sideways, so that the maximum width is somewhat greater than the original depth of cut. In cutting a tube it can spread in both directions, but in turning a bar it can spread only outward. The chip spread is small with harder alloys, but when cutting soft metals with a small rake angle tool, a chip width more than one and one half times the depth of cut has been observed. Usually the chip *thickness* is greatest near the middle, tapering off somewhat towards the sides.

The upper surface of the chip is always rough, usually with minute corrugations or steps, *Figure 3.3a*. Even with a strong, continuous chip, periodic cracks are often observed, breaking up the outer edge into a series of segments. A complete description of chip form would be very complex, but, for the purposes of analysis of stress and strain in cutting, many details must be ignored and a much simplified model must be assumed, even to deal with such an uncomplicated operation as lathe turning. The making of these simplifications is justified in order to build up a valuable framework of theory, provided it is born in mind that real-life behavior can be completely accounted for only if the complexities, which were ignored first, are reintroduced.

An important simplification is to ignore both the irregular cross section of real chips and the chip spread, and to assume a rectangular cross section, whose width is the original depth of cut, and whose height is the measured mean thickness of the chip. With these assumptions, the formation of chips is considered in terms of the simplified diagram, *Figure 3.1*, an idealized section normal to the cutting edge of a tool used in orthogonal cutting.

3.5   CHIP FORMATION

In practical tests, the mean chip thickness can be obtained by measuring the length, l, and weight, W, of a piece of chip. The mean thickness, t_2, is then

$$t_2 = \frac{W}{\rho w l} \tag{3.1}$$

where ρ = density of work material (assumed unchanged during chip formation) and w = width of chip (depth of cut).

The mean chip thickness is a most important parameter. In practice the chip is never thinner than the feed, which in orthogonal cutting, is equal to the undeformed chip thickness, t_1 (*Figure 3.1*). In most textbooks and papers, it is usual to formally define a term (r), called "*the chip thickness ratio*". The value of r is the ratio of the undeformed chip thickness to the deformed chip thickness, namely, $r = t_1/t_2$, with $r << 1$.

Chip thickness is not constrained by the tooling, and, with many ductile metals, the chip may be five times as thick as the feed, or even more. Thus, it is important to re-emphasize that r will always be less than unity and often in the range 0.2 to 0.5. The chip thickness is related to the tool rake angle, α, and the *shear plane angle* ϕ (*Figure 3.1*).

The latter is the angle formed between the direction of movement of the workpiece *OA* (*Figure 3.1*) and the *shear plane* represented by the line *OD*, from the tool edge to the position where the chip leaves the work surface.

For purposes of simple analysis the chip is assumed to form by shear along the *shear plane*. In fact the shearing action takes place in a zone close to this plane.

It is assumed that the work material is incompressible and no side spread occurs. From the geometry of the cut (see *Figure 3.1*), it can be shown that,

$$OD = \frac{t_1}{\sin\phi} = \frac{t_2}{\cos(\phi - \alpha)}$$

or

$$\frac{t_1}{\sin\phi} = \frac{t_2}{\cos\phi\cos\alpha + \sin\phi\sin\alpha}$$

Hence

$$\tan\phi = \frac{r\cos\alpha}{1 - r\sin\alpha} \qquad (3.2)$$

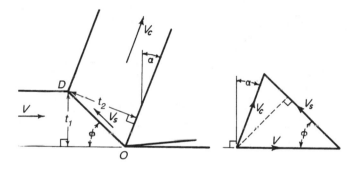

FIGURE 3.4 The relationship between shear angle, chip thickness and velocities. This orientation is the same as *Figures 2.1* and *3.1* but rotated ninety degrees clockwise to show orthogonal planing/shaping.

If the chip thickness ratio is low, the shear plane angle is small and the chip moves away slowly (e.g. for copper), while a large shear plane angle means a thin, high-velocity chip (e.g. for aluminum alloys). As any volume of metal, e.g. *klmn (Figure 3.1)* passes through the shear zone, it is plastically deformed to a new shape - *pqrs*. The amount of plastic deformation (shear strain, γ) is related to the shear plane angle ϕ and the rake angle α by the following equations[5]

From the velocity-vector diagram above,

$$V_c = \frac{\sin\phi}{\cos(\phi - \alpha)}V$$

and

$$V_s = \frac{\cos\alpha}{\cos(\phi - \alpha)}V$$

To derive an expression for the shear strain, the deformation can be idealized as a process of block slip or preferred slip planes, as shown in *Figure 3.5*. The shear strain

$$\gamma = \frac{\Delta S}{\Delta y} = \frac{OA}{CD} = \frac{OD}{CD} + \frac{DA}{CD}$$

i.e.,

$$\gamma = \tan(\phi - \alpha) + \cot\phi$$

or

$$\gamma = \frac{\cos\alpha}{\sin\phi\cos(\phi - \alpha)} \tag{3.3}$$

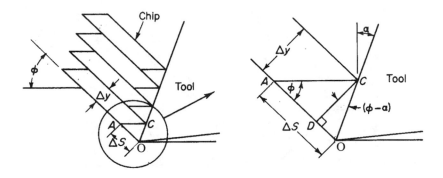

FIGURE 3.5   The shear-strain model. (after Piispanen[5])

The strain may also be expressed in terms of the shearing velocity

$$\gamma = \frac{V_s}{V\sin\phi}$$

The strain rate in cutting is given by

$$\dot{\gamma} = \frac{\Delta S}{\Delta y \Delta t} = \frac{V_s}{\Delta y} = \frac{\cos\alpha}{\cos(\phi - \alpha)} \cdot \frac{V}{\Delta y} \tag{3.4}$$

where Δy is the thickness of the deformation zone, and Δt is the time to achieve the final value of strain.

The meaning of 'shear strain', and of the units in which it is measured, is shown in the inset diagram in *Figure 3.6*. A unit displacement of one face of a unit cube is a shear strain of ($\gamma = 1$).

Figure 3.6 is a graph showing the relationship between the shear strain in cutting and the shear plane angle for three values of the rake angle.

For any rake angle there is a minimum strain when the mean chip thickness is equal to the feed $(t_2 = t_1)$. For zero rake angle, this occurs at $\phi = 45°$. The change of shape of a unit cube after passing through the shear plane for different values of the shear plane angle is shown in the lower diagram of *Figure 3.6* for a tool with a zero rake angle. The minimum strain at $\phi = 45°$ is apparent from the shape change.

At zero rake angle the minimum shear strain is 2. The minimum strain becomes less as the rake angle is increased, and if the rake angle could be made very large, the strain in chip formation could become very small.

In practice the optimum rake angle is determined by experience; too large an angle weakens the tool and leads to fracture. Rake angles higher than $30°$ are seldom used and, in recent years, the tendency has been to decrease the rake angle to make the tools more robust, to enable harder but less tough tool materials to be used (Chapters 7 and 8).

Thus, even under the best cutting conditions, chip formation involves very severe plastic deformation, resulting in considerable work-hardening and structural change. It is not surprising that metals and alloys lacking in ductility are periodically fractured on the shear plane.

3.6   THE CHIP/TOOL INTERFACE

The formation of the chip by shearing action at the shear plane is the aspect of metal cutting which has attracted most attention from those who have attempted analyses of machining. Of at least equal importance for the understanding of machinability and the performance of cutting tools is the movement of the chip and of the work material across the faces and around the edge of the tool.

In most analyses this has been treated as a classical friction situation, in which 'frictional forces' tend to restrain movement across the tool surface, and the forces have been considered in terms of a coefficient of friction (μ) between the tool and work materials. However, detailed studies of the tool/work interface have shown that this approach is inappropriate to most metal cutting conditions. It is necessary, at this stage, to explain why classical friction concepts do not apply and to suggest a more suitable model for analyzing this situation.

The concept of *coefficient of friction* derives from the work of Amontons and Coulomb who demonstrated that, in many common examples of the sliding of one solid surface over another, the force (F) required to initiate or continue sliding is proportional to the force (N) normal to the interface at which sliding is taking place

$$F = \mu N \qquad (3.5)$$

This coefficient of friction μ is dependent only on these forces and is independent of the sliding area of the two surfaces. The work of Bowden and Tabor,[6] Archard, and many others has demonstrated that this proportionality results from the fact that real solid surfaces are never completely flat on a molecular scale, and therefore make contact only at the tops of the hills, while the valleys are separated by a gap.

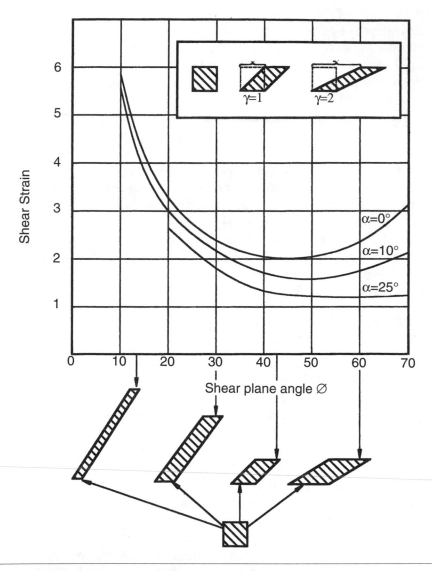

FIGURE 3.6   Strain on shear plane (γ) *vs* shear plane angle (ϕ) for three values of rake angle (α)

Under loading conditions used in engineering sliding mechanisms, the real contact area is very small, often less than one hundredth of the *apparent* area of the sliding surfaces. The mean stress acting on the real contact area supporting the load is equal to the yield stress of the material. If the compressive force normal to the interface is doubled, the real contacts supporting the load are plastically deformed until they double in area, so that the mean stress on them remains constant. In the areas of real contact, the atoms of the two surfaces are brought within range of their very strong attractive forces, i.e. they are atomically bonded. The frictional force is that force required to shear these areas of *real* contact. This friction force is proportional to the *real* contact area and therefore also proportional to the normal force. In engineering sliding mechanisms, the coefficient of friction, *F/N*, is therefore a useful concept - i.e. under conditions where the normal stress on the *apparent* contact area is very small compared with the yield stress of the materials.

When the normal force is increased to such an extent that the real area of contact is a large proportion of the apparent contact area, it is no longer possible for the real contact area to increase proportionately to the load. In the extreme case, where the two surfaces are completely in contact, the real area of contact becomes independent of the normal force, and the frictional force becomes that required to shear the material across the whole interface. When two materials of different strengths are in contact, as in metal cutting, the force required to move one body over the other becomes that *required to shear the weaker of the two materials across the whole area*. This force is almost independent of the normal force, but is directly proportional to apparent area of contact - a relationship directly opposed to that of classical friction concepts!

It is, therefore, important to know what conditions exist at the interface between tool and work material during cutting. This is a very difficult region to investigate. Few significant observations can be made while cutting is in progress, thus the conditions existing must be inferred from studies of the interface after cutting has stopped, and from measurements of stress and temperature. The conclusions presented here are deduced from studies, mainly by optical and electron microscopy, of the interface between work-material and tool after use in a wide variety of cutting conditions. Evidence comes from worn tools, from quick-stop sections and from chips.

The most important conclusion from the observations is that contact between tool and work surfaces is so nearly complete over a large part of the total area of the interface that sliding at the interface is impossible under most cutting conditions.[7] The evidence for this statement is now reviewed.

When cutting is stopped by disengaging the feed and withdrawing the tool, layers of the work material are commonly, but not always, observed on the worn tool surfaces, and micro-sections through these surfaces can preserve details of the interface. Special metallographic techniques are essential to prevent rounding of the edge where the tool is in contact with a much softer, thin layer of work-material.

Figure 3.7 is a photomicrograph (x 1,500) of a section through the cutting edge of a cemented carbide tool used to cut steel. The white area is the residual steel layer, which is attached to the cutting edge (slightly rounded), the tool rake face (horizontal) and down the worn flank. The two surfaces remained firmly attached during the grinding and polishing of the metallographic section. Any gap larger than 0.1 μm (4 micro-inches) would be visible at the magnification in this photomicrograph, but no such gap can be seen. It is unlikely that a gap exists since none of the lubricant used in polishing oozed out afterwards.

Figure 3.8 is an electron micrograph of a replica of a section through a high speed steel tool at the rake surface, where the work material (top) was adherent to the tool. The tool had been used for cutting a very low carbon steel at high cutting speed (200 m min^{-1}; 1,600 ft/min). The steel adhering to the tool had recrystallized and contact between the two surfaces is continuous, in spite of the uneven surface of the tool, which would make sliding impossible. Many investigations have shown the two surfaces to be interlocked, the adhering metal penetrating both major and minor irregularities in the tool surface.

FIGURE 3.7   Section through cutting edge of cemented carbide tool after cutting steel at 84 m min^{-1} (275 ft/min)[7]

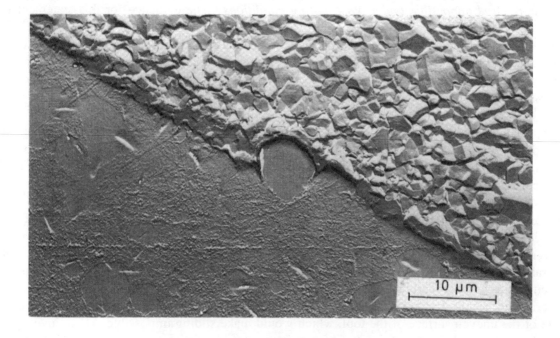

FIGURE 3.8   Section through rake face of steel tool and adhering metal after cutting iron at high speed. Etched in Nital; electron micrograph of replica.

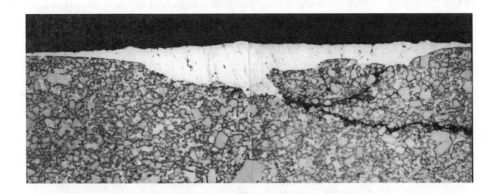

FIGURE 3.9   Section through rake face of cemented carbide tool, after cutting nickel-based alloys, with adhering work material[7]

Figure 3.9 shows a nickel based alloy penetrating deeply into a crack on the rake face of a carbide tool. When cutting gray cast iron, much less adhesion might be expected than when the work material is steel or a nickel based alloy, because of the lower cutting forces, the segmented chips and the presence of graphite in the gray iron. However, micro-sections demonstrate a similar extensive condition of seizure.

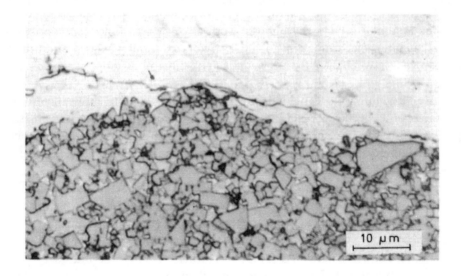

FIGURE 3.10   Section parallel to rake face of cemented carbide tool through the cutting edge, after cutting cast iron; white material is adhering cast iron[7]

Figure 3.10 shows a section parallel to the rake face through the worn flank of a carbide tool used to cut a flake graphite iron at 30 m min^{-1} (100 ft/min). There is a crack through the adhering work material, which may have formed during preparation of the polished section. If there had been gaps at the interface, the crack would have formed there, and the fact that it did not do so is evidence for the continuity and strength of the bond between tool and work material.

FIGURE 3.11   As in *Figure 3.10*, the tool being a steel cutting grade of carbide with two carbide phases[7]

Figure 3.11 is a similar section through the worn edge of a carbide tool containing titanium carbide, as well as tungsten carbide and cobalt. The work material was gray cast iron. Close contact can be observed between the adhering metal and all the grains of both carbide phases present in the structure of this tool.

Naerheim[8] prepared thin foils through the interface of carbide tools with adherent metal on the rake face of the tool after cutting steel at high speed. When examined by transmission electron microscopy (TEM) at the highest possible magnification, no gaps were observed between the carbide and the adherent steel. Any gaps must have been smaller than 5 nm and there was no reason to believe that there were any gaps.

Figure 3.12 is a transmission electron micrograph, produced by R.M. Greenwood of a section through a WC-Co tool used to cut steel. It shows the interface between the carbide tool and adhering steel (*top*). No gaps are visible at the interface. The WC grains are smoothly worn and no structural change was observed at the interface. Similar TEM evidence for high speed steel tools cutting stainless steel has shown continuity of structure across the interface.

The evidence of optical and electron microscopy demonstrates that the surfaces investigated are interlocked or bonded to such a degree that sliding, as normally conceived between surfaces with only the high spots in contact, is not possible. Some degree of metallic bonding is suggested by the frequently observed persistence of contact through all the stages of grinding, lapping and polishing of sections. There is, however, a considerable variation in the strength of bond generated, depending on the tool and work materials and the conditions of cutting. In some cases grinding and lapping causes the work material to break away with the separation occurring along the interface or part of the interface. In other cases, the whole chip adheres to the tool.

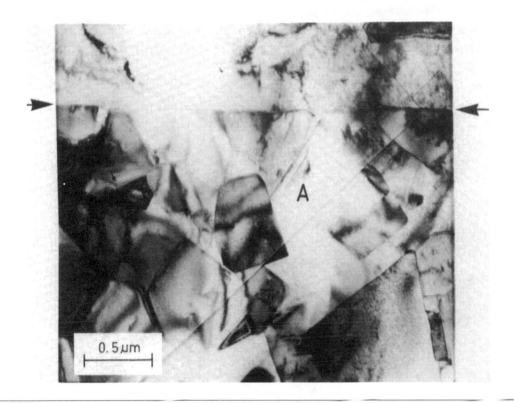

FIGURE 3.12 Transmission electron micrograph of section through rake face of WC-Co tool after cutting steel at high speed; arrows indicate interface between work material (*top*) and carbide tool (Courtesy of R.M. Greenwood)[13]

Further evidence is obtained by use of a quick-stop device enabling cutting to be stopped very rapidly to retain a 'frozen picture' of conditions existing at the instant of stopping. The usual method is to propel the tool from the cutting position by an explosive charge.[4]

Figure 3.13a shows diagrammatically the usual quick-stop direction of separation: the explosive charge reverses the movement of the tool relative to the work piece. It imposes at the interface very rapid change from high compressive stress to high tensile stress. *Figures 3.13b* to *e* show four conditions which commonly occur in quick-stop tests.[14] The tool and work may separate at the interface (*Figure 3.13b*), the tool surface being free from, or showing only traces of, adhering work material. The absence of visible fragments of work material does not disprove atomic bonding at the interface during cutting. It demonstrates that the tensile strength of any interface bond during cutting was lower than that of the work material.

The strength of the bond varies greatly with different tool and work materials and with different cutting speed and feed. The chip may adhere so strongly to the tool that, in quick-stop, it separates by fracture through the chip either near the shear plane (*Figure 3.13c*), the chip remaining attached to the tool, or within the chip, leaving most of the chip attached to the bar (*Figure 3.13d*). When using a cemented carbide tool, the tool may be fractured, leaving a fragment of the tool bonded to the underside of the chip and workpiece (*Figure 3.13e*).

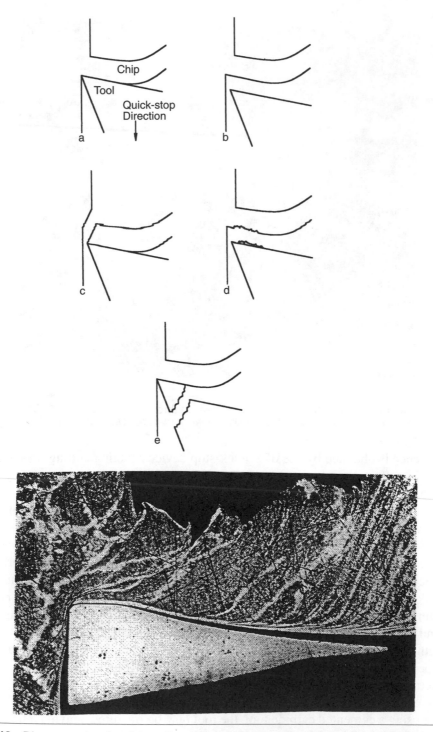

FIGURE 3.13 Diagrams showing (a) mode of action of quick-stop device, (b)-(e) conditions commonly occurring in quick-stop tests, and (f) photograph of condition (e)

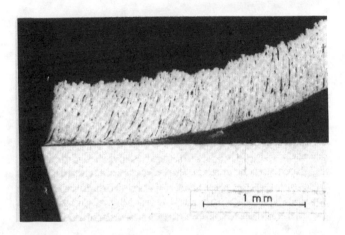

FIGURE 3.14 Section through high speed steel tool, with adhering very low carbon steel chip, after the quick-stop (condition (c) in previous diagram)

Figure 3.14 shows a section through a high-speed steel tool with adhering chip after quick-stop when cutting a very low carbon steel at high speed.

Figure 3.15 shows a layer on the rake face of a tool used to cut a low carbon steel. In this case separation during quick-stop took place by a ductile tensile fracture *within the chip*, close to the tool surface.

FIGURE 3.15 Scanning electron micrograph showing adhering metal (steel) on rake face of tool after quick-stop - condition in *Figure 13.3d* (Note the tool is facing to the right in this photograph)

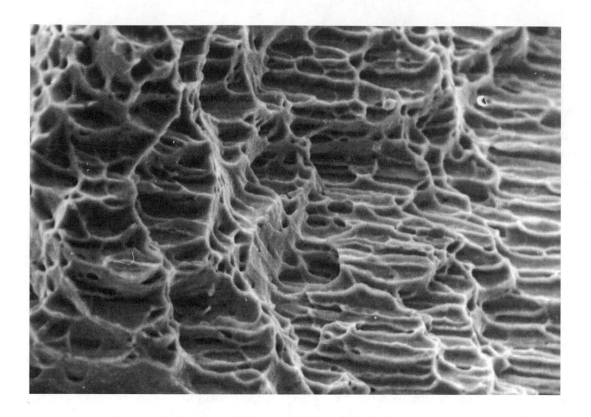

FIGURE 3.16   Fracture surface on adhering metal in *Figure 3.15*

This tensile fracture is seen in *Figure 3.16*, which shows part of the surface in *Figure 3.15* at high magnification. Under the conditions demonstrated by these examples, the high strength of the tool/work interface can have been achieved only by metallic bonding. *Over the areas of bonded contact the tool and work material have effectively become one piece of metallic material.* Under these conditions, the under surface of the chip and the new machined surface on the workpiece must be generated by a process of fracture, which may take place at the interface or within the work material at some distance from the interface.

Solid phase welding is one of the oldest techniques known to craftspeople in metals and, in recent years, considerable research into the mechanisms of solid phase welding has been carried out to facilitate control of industrial operations such as roll bonding, small tool welding and explosive welding.[9] Two factors which have been shown to promote bonding are:

(1) Freedom from contaminants, such as greases, and minimal oxide films on the surfaces.

(2) Plastic deformation of the surfaces.

With respect to these factors the conditions at the interface in metal cutting operations are particularly favorable to metallic bonding. Work material surfaces are being freshly generated and the clean metal flows across the tool surfaces without being exposed to the atmosphere. Freshly generated, cut metallic surfaces are exceptionally active chemically (see also Chapter 10) and bond very readily to other metal surfaces. During cutting the work material is subjected to a level of plastic strain much greater than that encountered in roll bonding.[9] The tool surface is initially contaminated by oxide films and sometimes by lubricants, but the flow of clean metal across the tool surface is unidirectional and often continues for long periods. Contaminants are swept away much more effectively than in processes such as forming, forging or rolling. Thus, in metal cutting, conditions are especially favorable for metallic bonding at the tool/work interface, and the observed bonding should have been predicted.

It is evidence of this character that has demonstrated the mechanically-interlocked and/or metallic-bonded character of the tool-work interface as a normal feature of metal cutting.

Under these conditions the movement of the work material over the tool surface cannot be adequately described using the terms 'sliding' and 'friction' as these are commonly understood. Coefficient of friction is not an appropriate concept for dealing with the relationship between forces in metal cutting for two reasons:

(1) There can be no simple relationship between the forces normal to and parallel to the tool surface.

(2) The force parallel to the tool surface is not independent of the area of contact, but on the contrary, the area of contact between tool and work material is a very important parameter in metal cutting.

The condition where the two surfaces arc interlocked or bonded is referred to here as *conditions of seizure* as opposed to *conditions of sliding* at the interface.

The generalization concerning seizure at the tool/work interface having been stated must now be qualified. The enormous variety of cutting conditions encountered in industrial practice has been discussed and there are some situations where there is sliding contact at the tool surface. It has been demonstrated at very low cutting speed (a few centimeters per minute) and at these speeds sliding is promoted by the use of active lubricants. Sliding at the interface occurs, for example, near the center of a drill where the action of cutting becomes more of a forming operation. Even under seizure conditions, it must be rare for the *whole* of the area of contact between tool and work surfaces to be seized together. This is illustrated diagrammatically in *Figure 3.17* for a lathe cutting tool. Compare the sketch on the right with *Figure 3.1*. Examination of used tools provides positive evidence of seizure on the tool rake face close to the cutting edge, *OEBF* in *Figure 3.17*, the length *OB* being considerably greater than the feed. Beyond the edge of this area there is frequently a region where visual evidence suggests that contact is intermittent, *EHB´FO* in *Figure 3.17*. On the worn flank surface, *OG*, it is uncertain to what extent seizure is continuous and complete, but *Figures 3.7, 3.10* and *3.11* show the sort of evidence demonstrating that seizure occurs on the flank surface also, particularly close to the edge.

Much more research is required into the character of the interface, and into movement within a region a small number of atom spacings from the interface under a variety of cutting conditions. Work by Doyle, Horne, Tabor and Wright[2,3] using sapphire tools has demonstrated that sliding at the interface may sometimes occur during cutting. Movement at the interface was reported to have been observed through transparent tools and recorded using high speed cinematography. Wright[3] concluded that sliding occurs when the interfacial bond is weak - particularly when soft metals such as lead are cut with sapphire tools, or where tools contaminated with a few molecular layers of organic substances are used for short time cutting. Conditions of seizure are encouraged by high cutting speed and long cutting time where difference in hardness between tool and work material is relatively small and bond strength between them is high.

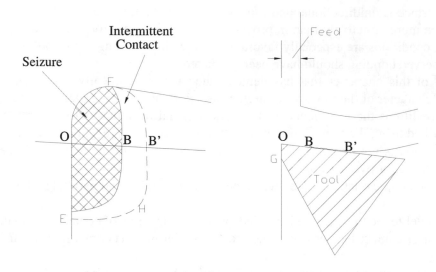

FIGURE 3.17   Areas of seizure on cutting tool in the cross-hatched region up to B

3.7   CHIP FLOW UNDER CONDITIONS OF SEIZURE

Since most published work on machining is based implicitly or explicitly on the classical friction model of conditions at the tool/work interface, the evidence for seizure requires reconsideration of almost all aspects of metal cutting theory. Engineers are accustomed to thinking of seizure as a condition in which relative movement ceases, as when a bearing or a piston in a cylinder is seized. Movement stops because there is insufficient force available to shear the metal at the seized junctions, and if greater force is applied, the normal result is massive, and often catastrophic, fracture at some other part of the system. With metal cutting, however, it is necessary to accustom oneself to the concepts of a system in which relative movement continues under conditions of seizure. This is possible because the area of seizure is small, and sufficient force is applied to shear the work material near the seized interface. Tool materials have high yield stress to avoid destruction under the very severe stresses which seizure conditions impose.

Under sliding conditions, relative movement can be considered to take place at a surface which is the interface between the two bodies. Movement occurs at the interface because the force required to shear the bonds at any areas of real contact is much smaller than that required to shear either of the two bodies. Under seizure conditions it can no longer be assumed that relative movement takes place at the interface, because the force required to overcome the interlocking and bonding is normally higher than that required to shear the adjacent metal. Relative motion under seizure involves shearing in the weaker of the two bodies. In metal cutting this is the work material. This shear strain is not uniformly distributed across the chip.

The bulk of the chip, formed by shear along the shear plane - *OD* in *Figure 3.1* - is not further deformed but moves as a rigid body across the contact area on the tool rake face. The shear strain resulting from seizure is confined to a thin region which may lie immediately adjacent to the interface or at some distance from it.

In sections through chips and in quick-stop sections, zones of intense shear strain near the interface are normally observed, except under conditions where sliding takes place.

Figure 3.18 shows the sheared zone adjacent to the tool surface in the case of steel being cut at high speed. The thickness of these zones is often on the order of 25-50 μm (0.001-0.002 in), and strain within the regions is much more severe than on the shear plane, so that normal structural features of the metal or alloy being cut are greatly altered or completely transformed.

Figure 3.19 shows, at high magnification, the structure in this highly strained region. The pearlite areas, which are elongated but unmistakable in the body of the chip, are so severely deformed in the highly strained region near the interface that they cannot be resolved by optical microscopy. Examination of these regions by transmission electron microscopy (TEM) shows that the main feature of all alloys so far investigated is a structure consisting of equi-axed grains of very small size (e.g. 0.1 to 1.0 μm).

Figure 3.20 shows the structure near the interface of a low carbon steel chip cut with a high speed steel tool. These structures are clearly the result of recovery or recrystallization during the very short time when the material was strained and heated as it passed over the contact area on the tool. This behavior of the work material is, in many ways, more like that of an extremely viscous fluid than that of a normal solid metal. For this reason the term *flow-zone* is used to describe this region. As can be seen from *Figure 3.18*, there is not a sharp line separating the flow-zone from the body of the chip, but a gradual blending in.

There is in fact a *pattern of flow*, in the work material around the cutting edge and across the tool faces, which is characteristic of the metal or alloy being cut and the conditions of cutting. A pattern of flow and a velocity gradient within the work material, with velocity approaching zero at the tool/work interface, are the basis of the model for relative movement under conditions of seizure, to replace the classical friction model of sliding conditions. A flow-zone of this character indicates seizure at the interface.

3.8   THE BUILT-UP EDGE

Seizure at the interface does not always give rise to a flow-zone at the tool surface. An alternative feature, commonly observed, is a *built-up edge*. When cutting many alloys with more than one phase in their structures, strain-hardened work material accumulates, adhering around the cutting edge and on the rake face of the tool, displacing the chip from direct contact with the tool, as shown in *Figure 3.21*. The built-up edge is not observed when cutting pure metals[10] but occurs frequently with alloys used in industry. It can be formed with either a continuous or a discontinuous chip - for example when cutting steel or cast iron. Most commonly it occurs at intermediate cutting speeds - sliding may occur at extremely low speed, a built-up edge at intermediate speed and a flow-zone at high speed. The actual speed range in which a built-up edge exists depends on the alloy being machined and on the feed. This is further discussed when considering the machinability of steel (Chapter 9).

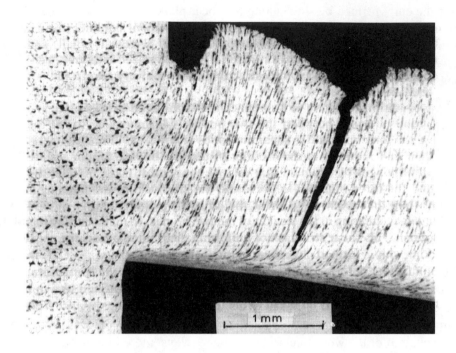

FIGURE 3.18   Section through quick-stop showing flow-zone in 0.1%C carbon steel after cutting at high speed

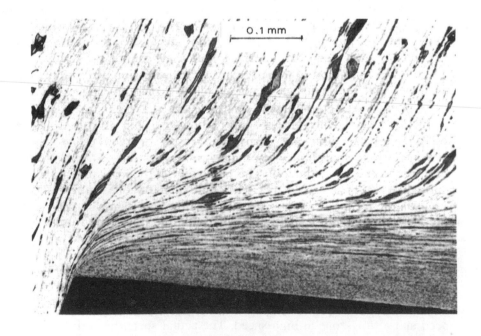

FIGURE 3.19   Detail of *Figure 3.18*

FIGURE 3.20 Transmission electron micrograph (TEM) of the structure of the flow zone in 0.19%C steel machined at 70 m min^{-1}(225 ft/min) (Courtesy of A. Shelbourn)

The built-up edge is not a separate body of metal during the cutting operation. Diagrammatically it should be depicted as in *Figure 3.22*. The new work surface is being formed at O' and the under surface of the chip at B, but between O' and B the built-up edge and the work material are one continuous body of metal, not separated by free surfaces.

The zone of intense and rapid shear strain has been transferred from the tool surface to the top of the built-up edge. This illustrates the principle that, under seizure conditions, relative movement does not necessarily take place immediately adjacent to the interface.

The built-up edge is a dynamic structure, being constructed of successive layers greatly hardened under extreme strain conditions. When cutting steel, for example, many workers have shown the hardness of a built-up edge to be as high as 600 or 650HV, determined by micro-hardness tests. Wallbank[11] studied the structures of the built-up edges formed when cutting steel, using TEM.

Figure 3.23 is a typical micrograph showing the very fine elongated ferrite cell structure. The cementite was so finely dispersed as to be very difficult to resolve and identify at the highest magnification. The material of the built-up edge had thus been very severely strain hardened, but the temperature had not increased to the stage where recrystallization could take place. The built-up edge structures contrast strongly with the equi-axed structures of the flow-zone (*Figure 3.20*). The strain-hardened work material of the built-up edge can support the stress imposed by the cutting operation and it functions as an extension to the cutting tool.

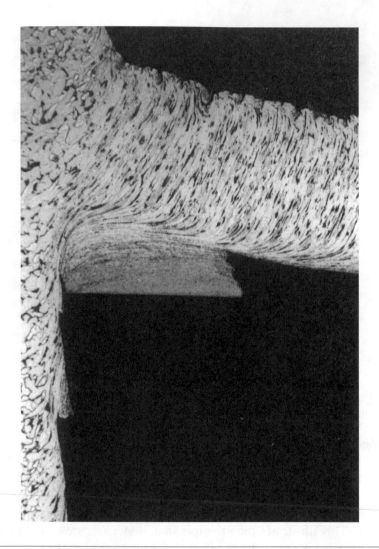

FIGURE 3.21   Section through quick-stop showing built-up edge after cutting 0.15% C steel at low speed in air

The new work surface initiates at O´ *(Figure 3.22)* and the under surfaces of the chip at B are formed by fracture through the work-hardened material. Some severely strained material remains on the two newly formed surfaces. At the top of the built-up edge (between O´ and B in *Figure 3.22*) intense strain-hardening is accompanied by piling up of dislocations at inclusions and other structural discontinuities. In the center of this region, formation of micro-cracks is inhibited by high compressive stress, but as the strain-hardened material flows towards O´ or B, compressive stress is reduced and micro-cracks develop and join together to initiate fracture. The new surfaces follow the general direction of the flow-lines in the deforming structure, often along elongated plastic inclusions, but frequently moving from one line of micro-cracks to another to produce a typically rough surface.[15]

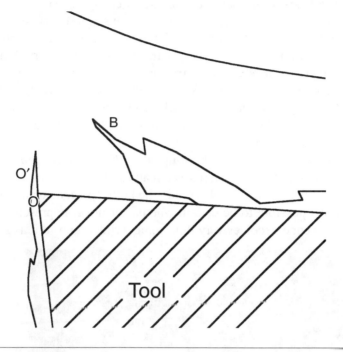

FIGURE 3.22 Form of built-up edge[7]

FIGURE 3.23 TEM from built-up edge of 0.1% C steel (After Wallbank[11])

This mode of fracture leads to an increase in size and change of shape of the built-up edge. This continues until the growing built-up edge becomes unstable in the stress field when fragments are broken away by a different fracture mechanism. Fracture across, rather than along, the flow direction is initiated, probably as a thermoplastic instability, in a very thin shear zone forming a much smoother fracture surface. Such a fracture can be seen on the cut surface in *Figure 3.21*. (The important role of thermoplastic shear bands in metal cutting is further treated in Chapter 5.) This leads to a typical feature of machined surfaces and the under surface of chips when a built-up edge is present. There are smooth, shiny patches on a rough surface as shown on the under side of a chip in *Figure 3.24*.[16]

There is no hard and fast line between a built-up edge (*Figure 3.21*) and a flow-zone (*Figure 3.18*). Seizure between tool and work material is a feature of both situations and every shade of transitional form between the two can be observed. The built-up edge occurs in many shapes and sizes and it is not always possible to be certain whether or not it is present. In the transitional region, when the built-up edge becomes very thin, some writers refer to it as a built-up layer.

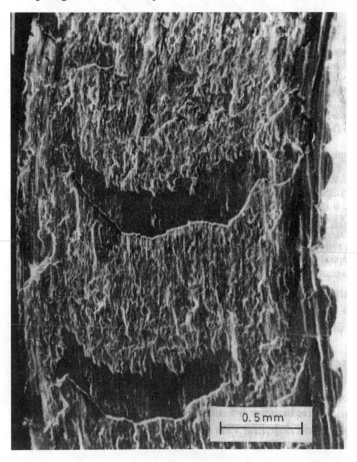

FIGURE 3.24 Scanning electron micrograph of under surface of steel chip, after cutting with a built-up edge. Rough and smooth areas result from fracture along and across lines of flow, respectively (Courtesy of S. Barnes[16])

3.9   MACHINED SURFACES

In conventional language, surfaces formed by cutting are spoken of as different in character from those formed by fracture. *Figures 3.18* and *3.21* demonstrate that in metal cutting the machined surfaces are, in fact, formed by fracture under shearing stress. The new surface rarely originates precisely at the cutting edge of the tool. *Figure 3.21* shows that, in the presence of a built-up edge, the fracture forming the new surface may have its origin above the tool edge. Wallbank[11] has shown that, when a flow-zone is present, the work material wraps itself around a sharp cutting edge and the new surface is formed where the work material breaks contact with the tool flank, a short distance below the edge. With ductile metals and alloys, both sides of a shear fracture are plastically strained, so that *some degree of plastic strain is a guaranteed feature of machined surfaces*. The amount of strain and the depth below the machined surface to which it extends, can vary greatly, depending on the material being cut, the tool geometry, and the cutting conditions, including the presence or absence of a lubricant.[12]

The deformed layer on the machined surface can be thought of as that part of the flow-pattern around the cutting edge which passes off with the work material, so that an understanding of the flow-pattern, and the factors which control it, is important in relation to the character of the machined surface. The presence or absence of seizure on those parts of the tool surface where the new work surface is generated can have a most important influence, as can the presence or absence of a built-up edge and the use of sharp or worn tools. These factors influence not only the plastic deformation, hardness and properties of the machined surface, but also its roughness, its precise configuration and its appearance.

3.10   SUMMARY AND CONCLUSION

The main objective of machining is the shaping of the new work surface. It may seem, therefore, that too much attention is paid in this book to the formation of the chip, which is a waste product. But the consumption of energy occurs mainly in the formation and movement of the chip, and for this reason, the main economic and practical problems concerned with rate of metal removal and tool performance can be understood only by studying the behavior of the work material as it is formed into the chip and moves over the tool.

Detailed knowledge of the chip formation process is also required for the understanding of the accuracy and condition of the machined surface of the desired component. Machined surfaces are inevitably damaged to some degree, since the chip is formed by the shear fracture at high strain (as described in the previous section).

During the "normal machining" of routine consumer products (shown at the top-right of *Figure 1.1.*), the sub-surface damage may not be of great concern. Probably, the in-service performance of such consumer products will not suffer much from some minor amounts of sub-surface strain of the kind that can be seen in the quick-stop sections.

However, in the machining of more critical items - certain aircraft components for example - the sub-surface strain damage might negatively impact on fatigue life. In this regard, it is advisable to approach the final part dimensions with a sequence of carefully planned roughing and finishing cuts. It is important to consider that the finishing cuts might "inherit" damage (*Figure 3.25*) from the previous roughing cuts.[17-18]

To arrive at an acceptable accuracy/tolerance with minimal damage, it may, in certain circumstances, be advisable to use a modest roughing cut followed by two lighter finishing cuts, rather than one more aggressive roughing cut which causes such sub-surface damage to the extent that only one finishing cut cannot correct the surface. A small sacrifice in overall machining time may, in such a case, be a better strategy for minimizing sub-surface damage. Further research is desirable to investigate such strategies for different commercial alloys.

The issues related to sub-surface damage become even more critical in the "precision" and "ultra-precision" ranges in *Figure 1.1*. In such operations, the value of the undeformed chip thickness (t_1) during finishing cuts is very small. It is useful to refer to *Figures 2.1* and *3.1* and relate the value of t_1, the feed rate, to any possible tool wear, blemishes, nicks, or general rounding of the tool's edge during use (at position O in *Figure 3.1*). During roughing cuts, with a large value of t_1, such blemishes or rounding will not be a major influence on the general smoothness of the chip flow process. By contrast, thinking about relative proportions, when t_1 is very small, as in precision finishing, any blemishes or tool-edge rounding by wear will have a more dramatic interference on the general smoothness of chip flow and consequently on the quality of the surface finish of the part being machined.

Thus, when considering the condition of the machined surface, especially for precision machining, the sharpness and integrity of the cutting edge is probably the most important issue to focus on.

During a particular cut as the edge wears and becomes slightly blunt or rounded, the strain imposed on the machined surface increases. Intuitively, it can be imagined that a rounded tool edge is, at high magnification, like a blunt wedge being dragged across the surface. The more blunt and rounded the edge becomes, the more the action resembles a blunt hardness-tester being dragged on the surface causing the familiar strain fields seen in standard plasticity textbooks.[19] As the tool wears even more, the effect can cause shear lamellae to spall-off, visibly creating rough "ears" on the machined surface and on the underside of the chip.

Figures 3.26 to *3.30* have been drawn to summarize the development of such effects over time. Experience has shown that by the time the tool exhibits 0.75mm (0.03 inches) of flank wear, the tool is essentially worn-out because its blunted state causes considerable damage on the machined surface.

Also, at the end of the tool's life the "indenting rather than cutting" effect of its worn edges and clearance face can push-up a "collar formation". This occurs at the outer shoulder of the bar in *Figure 2.1* and is shown schematically for face-milling in *Figure 3.29*.

Clearly this effect is unacceptable for high accuracy machining and the tool should be changed well before the usual 0.75mm (0.03 inches) flank wear limit. However, even at lesser amounts of flank wear, some damage occurs, building up on a gradual basis. *Figures 3.26* to *3.28* have been drawn to show the gradual transition that occurs. A highly skilled machinist can track these events, and also correlate them with aural and visual cues as summarized in *Figure 3.30*.

In conclusion, the characteristics of machined surfaces, especially those of high-quality forming dies and components that demand high-precision, are greatly dependent on the flow patterns, stresses and temperatures at this all-important tool edge, O, and these are the subjects of the next chapters.

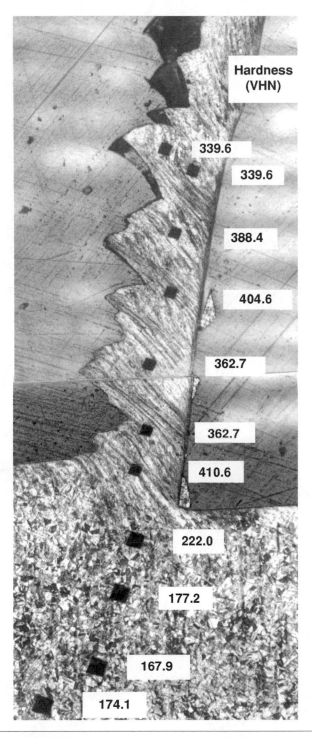

FIGURE 3.25 Microhardness results show damage by work hardening of tool surface. Machining austenitic stainless steel at 100 m min^{-1}

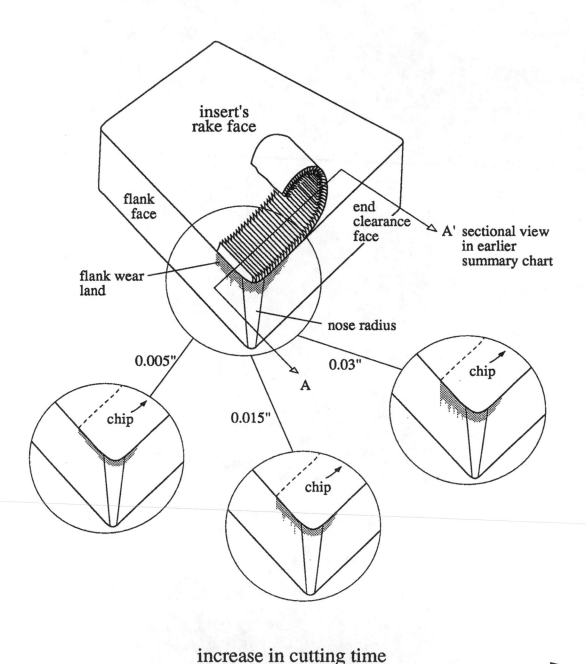

FIGURE 3.26 Expanded detail of insert from *Figure 2.1* showing the development of flank wear from 0.005 inches (~0.15mm) to 0.03 inches (~0.75mm).

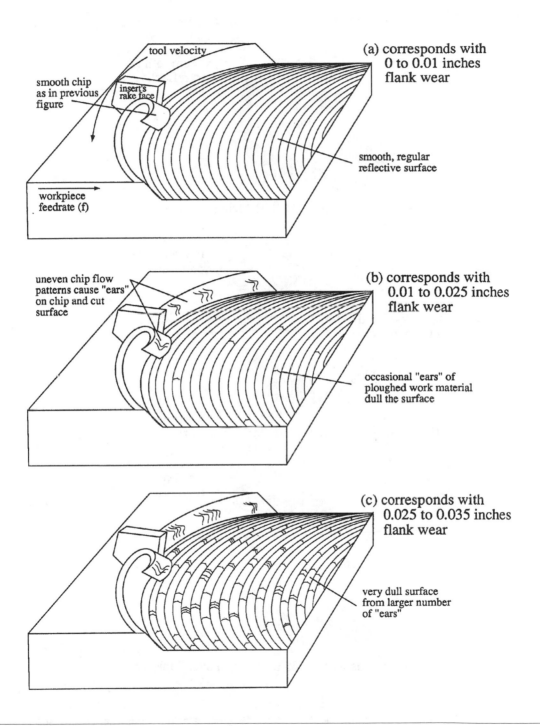

tool velocity

smooth chip as in previous figure

insert's rake face

(a) corresponds with 0 to 0.01 inches flank wear

smooth, regular reflective surface

workpiece feedrate (f)

uneven chip flow patterns cause "ears" on chip and cut surface

(b) corresponds with 0.01 to 0.025 inches flank wear

occasional "ears" of ploughed work material dull the surface

(c) corresponds with 0.025 to 0.035 inches flank wear

very dull surface from larger number of "ears"

FIGURE 3.27   Corresponding changes in surface finish of the part as tool wear progresses

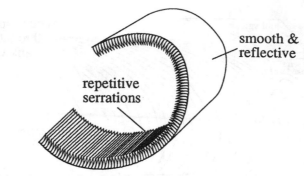

(a) corresponds with 0 to 0.01 inches flank wear

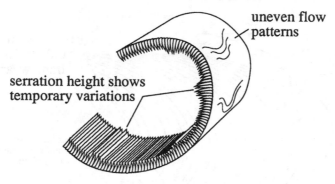

(b) corresponds with 0.01 to 0.025 inches flank wear

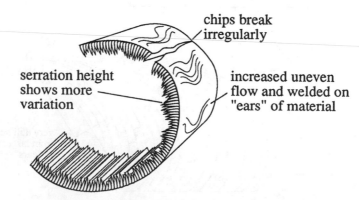

(c) corresponds with 0.025 to 0.035 inches flank wear

FIGURE 3.28   Corresponding changes in the undersurface of the chip as tool wear progresses

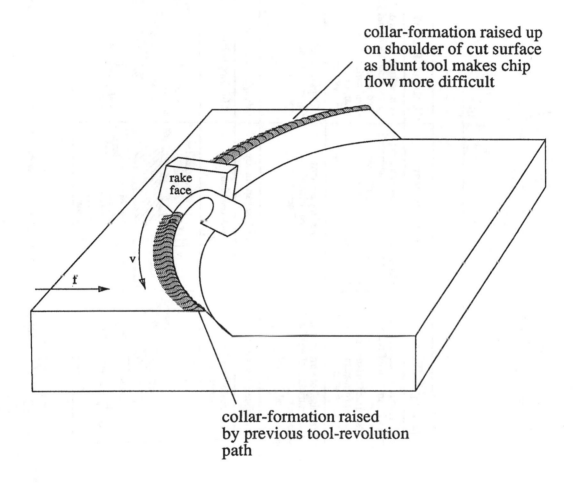

collar-formation raised up
on shoulder of cut surface
as blunt tool makes chip
flow more difficult

rake
face

v

f

collar-formation raised
by previous tool-revolution
path

FIGURE 3.29 Collar formation in face-milling arising from worn tool

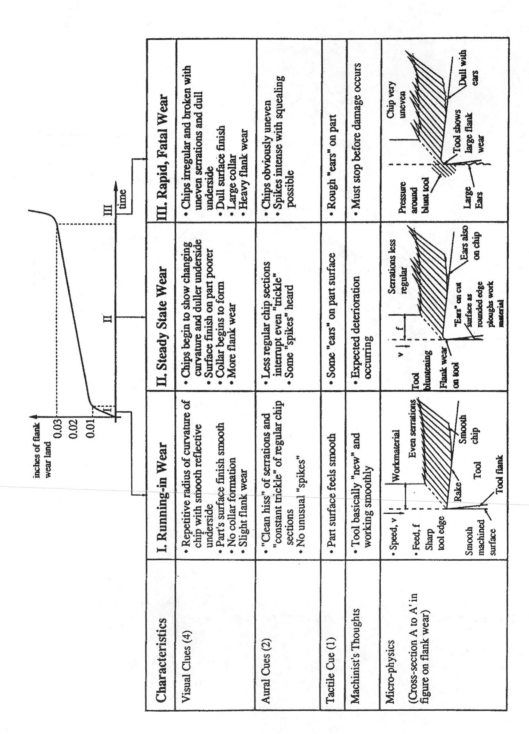

Characteristics	I. Running-in Wear	II. Steady State Wear	III. Rapid, Fatal Wear
Visual Clues (4)	• Repetitive radius of curvature of chip with smooth reflective underside • Part's surface finish smooth • No collar formation • Slight flank wear	• Chips begin to show changing curvature and duller underside • Surface finish on part poorer • Collar begins to form • More flank wear	• Chips irregular and broken with uneven serrations and dull underside • Dull surface finish • Large collar • Heavy flank wear
Aural Cues (2)	• "Clean hiss" of serrations and "constant trickle" of regular chip sections • No unusual "spikes"	• Less regular chip sections interrupt even "trickle" • Some "spikes" heard	• Chips obviously uneven • Spikes intense with squealing possible
Tactile Cue (1)	• Part surface feels smooth	• Some "ears" on part surface	• Rough "ears" on part
Machinist's Thoughts	• Tool basically "new" and working smoothly	• Expected deterioration occurring	• Must stop before damage occurs
Micro-physics (Cross-section A to A' in figure on flank wear)			

FIGURE 3.30 Summary chart that correlates advancing tool wear with cues to the machinist (or possible sensor system)

3.11 REFERENCES

1. Tonshoff, H.K., *et al.* (University of Hanover, FGR) *Micro-cinematograph Investigations of Cutting Process*, Film at Colloquium on Cutting of Metals, St. Etienne, France (Nov. 1979)
2. Doyle, E.D., *et al., Proc. Roy. Soc.* **A366**, 173 (1979)
3. Wright, P.K., *Metals Tech.*, **8,** (4), 150 (1981)
4. Williams, J.E., Smart, E.F. and Milner, D.R., *Metallurgia*, **81**, 6 (1970)
5. Hill, R., *The Mathematical Theory of Plasticity*, 2nd ed., p. 207, Oxford-Clarendon Press (1956). {Also, in 1937, V. Piispanen published "Lastunmoudostimisen Teoriaa" in *Teknillinen Aika Kauslehti*, **27**, page 315. Historians of metal cutting indicate that this was the first paper to represent the shear plane as a "stack of cards" model}
6. Bowden, F.P. and Tabor, D., *Friction and Lubrication of Solids*, Oxford University Press (1954)
7. Trent, E.M., *I.S.I. Special Report*, **94**, (1967)
8. Naerheim, Y. and Trent, E.M., *Metals Tech.*, **4**, (12), 548 (1977)
9. Milner, D.R. and Rowe, G.W., *Metall. Reviews*, **7**, (28), 433 (1962)
10. Williams, J.E. and Rollason, E.C., *J. Inst. Met.*, **98**, 144 (1970)
11. Wallbank, J., *Metals Tech.*, **6**, (4), 145 (1979)
12. Camatini, E., *Proc. 8th Int. Conf. M.T.D.R., Manchester* (1967)
13. Greenwood, R.M., *PhD Thesis*, University of Birmingham, England (1984)
14. Trent, E.M., *Wear*, **128**, 29 (1988)
15. Trent, E.M., *Wear*, **128**, 47 (1988)
16. Barnes, W., *MSc Thesis*, University of Birmngham, England (1986)
17. Dornfeld, D.A., and Wright, P.K., *Proc. 25th North American Research Inst. Conf.*, 359 (1997)
18. Thiele, J.D. and Melkote, S.N., *Proc. 27th North American Research Inst. Conf.*, 135 (1999)
19. Rowe, G.W., *Principles of Industrial Metalworking Processes*, Arnold, London (1977)

CHAPTER 4

FORCES AND STRESSES IN METAL CUTTING

4.1   INTRODUCTION

The forces acting on the tool are an important aspect of machining. For those concerned with the manufacture of machine tools, a knowledge of the forces is needed for estimation of power requirements and for the design of machine tool elements, tool-holders and fixtures, adequately rigid and free from vibration.

The cutting forces vary with the tool angles, and accurate measurement of forces is helpful in optimizing tool design. Scientific analysis of metal cutting also requires knowledge of the forces, and in the last hundred years, many force measurement devices, known as dynamometers, have been developed, capable of measuring tool forces with increasing accuracy.

Detailed accounts of the construction of dynamometers for this work are available in the technical literature. Early methods were based on strain-gage measurement of the elastic deflection of the tool under load.[1] Today, one of the most commonly used dynamometers is a force-platform that incorporates piezo-electric load cells; this is shown schematically in *Figure 2.1*.[2] For a semi-orthogonal cutting operation in lathe turning, the force components can be measured in three directions (*Figure 4.1*), and the force relationships are relatively simple.

The component of the force acting on the rake face of the tool, normal to the cutting edge in the direction YO is called here the *cutting force*, F_c. This is usually the largest of the three force components and acts in the direction of the cutting velocity. The force component acting on the tool in the direction OX, parallel with the direction of feed, is referred to as the *feed force*, F_f. This force acts tangential to the main cutting forces F_c. To maintain consistency with all machining processes the symbol for this tangential force is most often written as F_t rather than F_f.

The third component, acting in the direction OZ, tending to push the tool away from the work in a radial direction, is the smallest of the force components in semi-orthogonal cutting and, for purposes of analysis of cutting forces in simple turning, it is usually ignored.

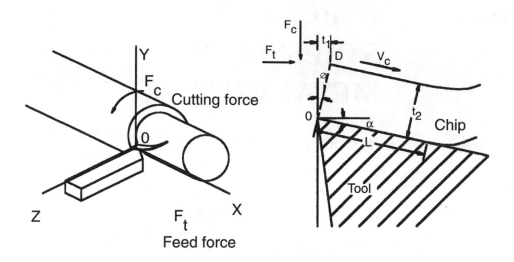

FIGURE 4.1   Forces acting on cutting tool (Note L= OB´ in the figures of Chapter 3)

It is of interest that the *forces* involved in machining are relatively low compared with those in other metal working operations such as forging. Because the layer of metal being removed - the chip - is thin, the forces to be measured are usually not greater than a few tens or hundreds of kilograms. Nevertheless, the small areas of contact make the *stresses* in metal cutting among the highest of all the metal processing operations.

4.2   STRESS ON THE SHEAR PLANE

Measurement of the forces and the chip thickness make it possible to explore the stresses for simple, orthogonal cutting conditions. Where a continuous chip is formed with no built-up edge, the work is sheared in a zone close to the shear plane (*OD* in *Figure 4.1*) and, for the purposes of this simple analysis, it is assumed that shear takes place on this plane to form the chip.

The force acting on the shear plane F_s is calculated from the measured forces and the shear plane angle:

$$F_s = F_c \cos\phi - F_t \sin\phi \qquad (4.1)$$

The shear stress k_s, required to form the chip is

$$k_s = F_s/A_s \qquad (4.2)$$

where A_s = area of shear plane.

The force required to form the chip is dependent on the shear yield strength of the work material under cutting conditions, and on the area of the shear plane. Many calculations of shear yield strength in cutting have been made, using data from dynamometers and chip thickness measurements.

In general the shear strength of metals and alloys in cutting has been found to vary only slightly over a wide range of cutting speeds and feeds. *Table 4.1* shows the values of k_s measured during cutting for a variety of metals and alloys. Provided the shear plane area remains constant, the force required to form the chip, being dependent on the shear yield strength of the metal, is increased by any alloying or heat treatment which raises the yield strength. In practice, however, the area of the shear plane is very variable, and it is this area which exerts the dominant influence on the cutting force, often more than outweighing the effect of the shear strength of the metal being cut.

TABLE 4.1   Material shear yield strength in cutting k_s

	(tonf/in^2)	(MPa)
Iron	24	370
0.13% C steel	31	480
Ni-Cr-V steel	45	690
Austenitic stainless steel	41	630
Nickel	27	420
Copper (annealed)	16	250
Copper (cold worked)	17	270
Brass (70/30)	24	370
Aluminium	6.3	97
Magnesium	8	125
Lead	2.3	36

In orthogonal cutting the area of the shear plane is geometrically related to the undeformed chip thickness t_1 (the feed), to the chip width w (depth of cut) and to the shear plane angle ϕ.

$$A_s = \frac{t_1 w}{\sin \phi} \tag{4.3}$$

The forces increase in direct proportion to increments in the feed and depth of cut, which are two of the major variables under the control of the machine tool operator. The shear plane angle, however, is not directly under the control of the machinist, and in practice it is found to vary greatly under different conditions of cutting. For a zero-degree rake angle tool, the shear plane might vary from a maximum of approximately 45° to a minimum of 5° or even less. *Table 4.2* shows how the chip thickness, t_2, the area of the shear plane, A_s and the shearing force, F_s, for a low carbon steel, vary with the shear plane angle for orthogonal cutting at a feed of 0.5 mm/rev and a depth of cut of 4 mm.

TABLE 4.2   Data for machining low carbon steel

Shear plane angle ϕ	Chip thickness t_2 (mm)	Shear plane area A_s (mm²)	Shearing force F_s (N)
45°	0.05	2.8	1,340
35°	0.71	3.5	1,680
25°	1.07	4.7	2,260
15°	1.85	7.7	3,700
5°	5.75	23.0	11,000

Thus, when the shear plane angle is very small, the *shearing force* may be more than five times that at the minimum where $\phi = 45°$, under conditions where the *shear stress* of the work material remains constant. It is important, therefore, to investigate the factors which regulate the shear plane angle if cutting forces are to be controlled or even predicted. Much of the work done on analysis of machining has been devoted to methods of predicting the shear plane angle.

4.3   FORCES IN THE FLOW ZONE

Before dealing with the factors determining the shear plane angle, consideration must be given to the other main region in which the forces arise - the rake face of the tool. For the simple case where the rake angle is $0°$, the feed force F_t is a measure of the drag which the chip exerts as it flows away from the cutting edge across the rake face. The origin of this resistance to chip flow is discussed in Chapter 3, where conditions of seizure occur over a large part of the interface between the under surface of the chip and the rake face of the tool. Although there are areas where sliding occurs (at the periphery of the seized contact region), the force to cause the chip to move over the tool surface is mainly that required to shear the work material in the flow-zone across the area of seizure.

Under most cutting conditions, the contribution to the feed force made by friction in the non-seized areas is probably relatively small. The feed force, F_t, can, therefore, be considered as the product of the shear strength of the work material at this surface (k_r) and the area of seized contact on the rake face (A_r).

$$F_t = A_r k_r \qquad (4.4)$$

While the feed force can be measured accurately, the same cannot be said for the area of contact, which is usually ill-defined. Observation of this area during cutting is not possible. When the tools are examined after use, worn areas, deposits of work material as smears and small lumps, and discoloration due to oxidation or carbonization of cutting oils are usually seen. The deposits may be on the worn areas, or in adjacent regions and the limits of the worn areas are often difficult to observe. Many of these effects are relatively slight and not easily visible on a

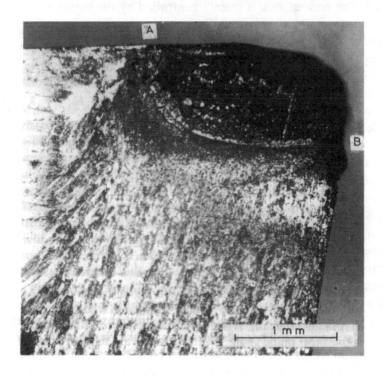

FIGURE 4.2 Contact area on rake face of tool used to cut titanium

ground tool surface. In experimental work on these problems, therefore, the use of tools with polished rake faces is strongly recommended, unless the quality of the tool surface is a parameter being investigated. There is no universal method for arriving at an estimate of the contact area, the problem being different with different work materials, tool materials and cutting conditions. It is necessary to adopt a critical attitude to the criteria defining the area of contact, and a variety of techniques may have to be employed in a detailed study of the used tool.

It may be possible to chemically dissolve the adhering work material to expose the worn tool surface for examination. This technique is used with carbide tools after cutting steel or iron, but it cannot usually be applied with steel tools because reagents effective in removing work material also attack the tool. Quick-stop tests, followed by examination of the tool and the mating chip surfaces, or sections through the tool surfaces, may be required. An example in *Figure 4.2* shows the rake surface of a high speed steel tool used to cut titanium. A quick-stop technique had been used and the area of complete seizure is, in this case, defined by the demarcation line *AB* which encloses an area within which the adhering titanium shows a tensile fracture, where it has separated from the under-surface of the chip. Beyond this line small smears of titanium indicate a region where contact was occasional and short lived.

In orthogonal cutting, the width of the contact region is usually equal to the depth of cut, or only slightly greater, although with very soft metals there may be considerable chip spread. The length of contact (*L* in *Figure 4.1*) is always greater than the undeformed chip thickness t_1, and may be as much as ten times longer; it is usually uneven along the chip width, and a mean value must be estimated. The contact area is mainly controlled by the length of contact L. It is a most important parameter, having a very large influence on cutting forces, on tool life and on many aspects of machinability.

When an estimate can be made of the area of contact, A_r, the mean shear strength of the work material at the rake face k_r, can be calculated (*Equation 4.4*). It is doubtful whether the values of k_r are of much significance, partly because of the inaccuracies in measurement of contact length, but also because of the extreme conditions of shear strain, strain rate, temperature and temperature gradient which exist in this region.

The values of k_r are not likely to be the same as those of the shear stress on the shear plane, k_s, and are usually lower, decreasing with increasing cutting speed. Experimental evidence shows that the feed force also decreases with increased cutting speed. However, the feed force F_f is increased by any changes in composition or structure of the work material which increase k_r. Alloying may either increase or reduce k_r.

When the cutting tool is sharp, the forces related to strain on the shear plane and movement of the chip across the rake face of the tool are the only forces which need to be considered. A force must arise also from pressure of the work material against a small contact area on the clearance face just below the cutting edge. This force is small enough to neglect as long as the tool remains sharp and the area of contact on the clearance face is very small.

If use in cutting results in a wear land on the tool parallel to the cutting direction on the clearance face (for example, *Figure 3.26*), the area of contact is increased. The forces arising from pressure of work material normal to this worn surface and movement of the work parallel to this surface may greatly increase. The increment in force may be used to monitor wear on the tool. For the purpose of a simple analysis of tool-force relationships in cutting, a sharp tool is assumed and forces on the flank are neglected.

4.4   THE SHEAR PLANE AND MINIMUM ENERGY THEORY

4.4.1 Merchant's shear 'plane' analysis

This section of the book now returns to the problem of the shear plane angle. The thickness of the chip is not constrained by the tool, and the question is:

"What *does* determine whether there is a thick chip with a small shear plane angle and high cutting force, or a thin chip with large shear plane angle and minimum cutting force?"

In the last 60 years there have been many attempts to answer this question and to devise equations which will predict quantitatively the behavior of work materials during cutting from a knowledge of their properties. Colloquially speaking, one could even say that the prediction of the shear plane angle has become a sort of "holy grail" preoccupation in the machining research community!

In the pioneering work of Ernst and Merchant,[3] followed by Lee and Shaffer,[4] Kobayashi and Thomsen[5] and others, a model of the cutting process was used in which the shear in chip formation was confined to the shear plane, and movement of the chip over the tool occurred by classic sliding friction, defined by an average friction angle λ.

This approach did not produce equations from which satisfactory predictions could be made of the influence of parameters, such as cutting speed, on the behavior of materials in machining. The inappropriate use of friction relationships relevant only to sliding conditions was probably mainly responsible for the weakness of this analysis. With this model the important area of contact between tool and work was not regarded as significant and no attempt was made to measure or to calculate it.

Nevertheless, Merchant's force circle, shown in *Figure 4.3*, remains an important milestone in metal cutting theory.

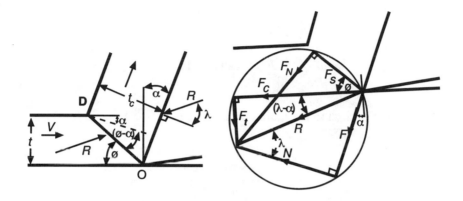

FIGURE 4.3   Merchant's force circle

The forces can be found from *Figure 4.3* as the two equations below:

$$F_c = \frac{t_1 wk \cos(\lambda - \alpha)}{\sin\phi\cos(\phi + \lambda - \alpha)} \tag{4.5}$$

$$F_t = \frac{t_1 wk \sin(\lambda - \alpha)}{\sin\phi\cos(\phi + \lambda - \alpha)} \tag{4.6}$$

Differentiating the first equation with respect to the shear plane angle gives:

$$\frac{dF_c}{d\phi} = \frac{t_1 wk \cos(\lambda - \alpha)\cos(2\phi + \lambda - \alpha)}{\sin^2\phi\cos^2(\phi + \lambda - \alpha)} = 0 \tag{4.7}$$

and the Merchant equation as

$$\phi = \frac{\pi}{4} - \frac{1}{2}(\lambda - \alpha) \tag{4.8}$$

After this expression has been found, the cutting forces can then be written.

For the main cutting force acting in the tool/work direction:

$$F_c = \frac{wt_1 k \cos(\lambda - \alpha)}{\sin\left[(\pi/4) - \frac{1}{2}(\lambda - \alpha)\right]\cos\left[(\pi/4) + \frac{1}{2}(\lambda - \alpha)\right]} \qquad (4.9)$$

and

$$F_c = 2wt_1 k \cot\phi \qquad (4.10)$$

For the feed force acting normal to the main cutting force

$$F_t = \frac{wt_1 k \sin(\lambda - \alpha)}{\sin\left[(\pi/4) - \frac{1}{2}(\lambda - \alpha)\right]\cos\left[(\pi/4) + \frac{1}{2}(\lambda - \alpha)\right]} \qquad (4.11)$$

and

$$F_t = wt_1 k(\cot^2\phi - 1) \qquad (4.12)$$

These forces allow the analysis of shear stresses, temperatures, etc., on the rake face of the tool.

From the force diagram in *Figure 4.3* the forces on the shear plane are given by

$$F_s = shear\ force = F_c \cos\phi - F_t \sin\phi \qquad (4.13)$$

$$F_N = normal\ force = F_c \sin\phi + F_t \cos\phi \qquad (4.14)$$

The shear stress and normal stress may be found from

$$k_s = shear\ stress = \frac{F_s}{A_s} = \frac{[F_c \cos\phi - F_t \sin\phi]\sin\phi}{wt_1} \qquad (4.15)$$

$$\sigma = normal\ stress = \frac{[F_c \sin\phi + F_t \cos\phi]\sin\phi}{wt_1} \qquad (4.16)$$

where A_s is the shear-plane area $wt_1 / \sin\phi$.

The coefficient of friction is defined in usual engineering practice as the ratio of the force in the direction of sliding to the force normal to the sliding interface. Thus, from the geometry shown, *Figure 4.3,* it follows that on the rake face of the tool, the coefficient of friction,

$$\mu = \frac{F}{N} = \frac{F_c \sin\alpha + F_c \cos\alpha}{F_c \cos\alpha - F_t \sin\alpha} \qquad (4.17)$$

$$= \frac{F_c + F_t \tan\alpha}{F_c - F_t \tan\alpha} = \tan\lambda \qquad (4.18)$$

where F is the friction force on the rake face, N is the normal force on the rake face, and λ is the friction angle.

The usual concept of friction implies that F and N are forces which are uniformly distributed over the sliding interface. However, this is not the case in metal cutting and thus this approach to coefficient of friction is too simple to adequately describe a seizure situation.

Although there have been many attempts to update the model, change its boundary conditions and consider new material properties, it still remains a useful, direct way of thinking about the energy used in cutting and how deformation will take place.

4.4.2 Oxley's shear 'zone' analysis

More recently other research workers, notably Oxley and colleagues,[6] have refined the analysis of cutting using more realistic models. Chip formation is considered to take place in a fan-shaped zone at low speed, and a parallel-sided zone at higher speeds, rather than on a single plane. The work-hardening characteristics of the material are taken into account. The frictional conditions are described as shear within a layer of the chip adjacent to the rake face of the tool. Within this layer "near-seizure" conditions are said to exist, with velocity in the work material approaching zero at the interface. To make quantitative predictions from this model, attempts are made to use realistic data for the stress/strain behavior of the work material allowing for strain hardening and the influence of high strain rates and temperature. This model is a considerable advance. For example, the conditions in which a built-up edge will occur during the cutting of steel can be predicted.

In *Figure 4.4* , the directions of maximum shear stress and maximum shear strain-rate are: i) the line OD, near the center of the shear zone, and ii) the tool/chip interface.

The theory analyzes the stress distribution along OD and the tool/chip interface in terms of the shear angle ϕ, and the work material properties.

The key idea is to select ϕ so that the resultant forces transmitted by the shear plane OD and the interface are in equilibrium.

Once ϕ is known then the chip thickness t_2 and the various components of force can be determined.

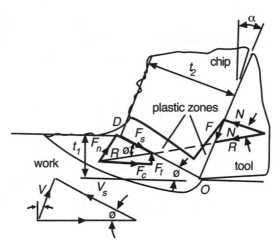

FIGURE 4.4   Model of chip formation used in Oxley's orthogonal analysis.

By starting at the free surface just ahead of the shear plane, and applying the appropriate stress equilibrium equation along the shear plane, it can be shown that for $0 < \phi \le \frac{1}{4}\pi$ the angle θ made by the resultant R with the shear plane is given by

$$\tan\theta = 1 + 2\left(\frac{1}{4}\pi - \phi\right) - Cn \tag{4.19}$$

In this equation, there are two constants:

- First, C is the constant in an empirical strain-rate relation from Stevenson and Oxley's work. C itself came from detailed studies of deformed grids in experimental work and calculations of the primary shear strain rate:

$$\dot{\gamma}_P = CV_P/(s) \tag{4.20}$$

in which $\dot{\gamma}_P$ is the maximum shear strain-rate at the shear plane, V_P is the shear velocity in the shear plane and (s) is the length of the shear plane

- Second, the other constant n, in equation 4.19 is the strain-hardening index in the empirical stress/strain relation

$$\sigma = \sigma_1 \varepsilon^n \tag{4.21}$$

in which σ and ε are the uniaxial (effective) flow stress and strain σ_1 and n are constants. {Note, the constant n here, is quite different from the n in Taylor's equation.} The angle θ can thus be expressed in terms of other angles by the equation

$$\theta = \phi + \lambda - \alpha \tag{4.22}$$

There appear to be two major challenges when using Oxley's new models for the quantitative prediction of behavior in the region where a flow-zone is present at the interface - i.e. in most high speed cutting operations.

Firstly, adequate data on stress/strain relations are not available, particularly for the amounts of strain, extreme strain rates, times and temperatures at which material is deformed in the flow-zone at the chip/tool interface.

Secondly, while the importance of the contact area is recognized, estimations of this area rely on calculations of a mean contact length and the basis for this calculation seems to be inadequate. There appears to be no alternative to experimental measurement of this area which, as has been indicated, presents serious difficulties.

4.4.3 Rowe and Spick's approximate analysis

It is not intended, therefore, to present here a method of making quantitative predictions of cutting behavior from properties of the work material. Instead, a very simple guide is offered to the understanding of certain important features observed when cutting metals and alloys, based on consideration of the energy expended in cutting. This treatment of the subject is a simplified version of the analysis proposed by Rowe and Spick.[7] It is based on a model that assumes shear strain on the shear plane to form the chip and shear strain in a thin layer of the work material adjacent to the tool rake face. The major hypothesis is that, since it is not externally constrained, the shear plane will adopt such a position that the total energy expended in the system (energy on the shear plane plus energy on the rake face) is a minimum; this is similar to Merchant's proposal.

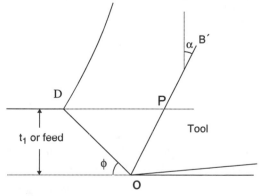

FIGURE 4.5   Rowe-Spick Definitions for Chip Formation (OP = the projected contact length, as opposed to the full contact length OB'. Thus L= OB' = χOP. Also note that the seizure part of the contact length will extend well beyond position P usually to about 80% of OB'-see Figure 3.17)

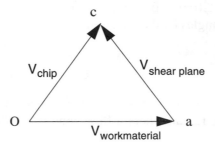

FIGURE 4.6   Corresponding velocity triangle (also called a Hodograph) during chip formation

Consider first the rate of work done on the shear plane. The principal of least action states that the shear plane will adopt such a position as to make the rate of work done in cutting a minimum.

$$\frac{d}{d\phi}\ \text{(work done per sec.)}\ =\ \frac{d}{d\phi}(F_C \cdot V)\ =\ 0.$$

Here, V, is the cutting velocity and F_C is the main cutting force. The approximate technique summarized in *Figure 4.5* and the velocity triangle of *Figure 4.6* with its labels of **o, a** and **c**, lead to the following:

Find the sum of **internal** rate of doing work $\sum \dfrac{dw}{dt} = \sum kus$

Rate of Work done $= \sum$ Primary Shear + Secondary Shear

$$\text{ie.}\frac{d}{d\phi}[\ k \cdot ac \cdot OD + \beta k \cdot oc \cdot OB'\ ] = 0$$

$$= k \cdot ac \cdot OD + \beta k \cdot oc \cdot OB'$$

Then if k is constant this simplifies to

$$\frac{d}{d\phi}[ac \cdot OD + \beta \cdot oc \cdot OB'] = 0$$

The term β arises from the unknown issues around how much seizure versus sliding exists. This is handled by Rowe and Spick by defining two extremes

Perfect Sliding	Complete Sticking
No friction force	All shear within chip's under-surface
Shear stress on rake face = 0	Shear stress = k. (Shear yield strength of material.)

the shear stress k_r on rake face is then a scalar function of the material strength $= \beta k$ (where $0 < \beta < 1$).

Note that $\beta =$ a constant between zero and one, not a friction angle. It has been used in this book to be consistent with the Rowe and Spick analysis. (Note also in other text books on metal processing, that β is often used in a very different way for the contact angle in strip rolling.) The analysis also needs to estimate the contact length OB', which is a function χ of the projected contact length OP. Assume, as shown in *Figure 4.5*, that

$$OB' = \chi OP = \chi \text{projected length}$$

$$OB' = \chi \frac{t_1}{\cos \alpha}$$

The velocities can be found by geometry

$$OD \sin \phi = t_1$$

$$ac = \frac{\cos \alpha}{\cos(\phi - \alpha)}$$

$$oc = \frac{\sin \phi}{\cos(\phi - \alpha)}$$

And

$$\frac{d}{d\phi} \left[\frac{t_1}{\sin \phi} \frac{\cos \alpha}{\cos(\phi - \alpha)} + \frac{t_1 \beta \chi}{\cos \alpha} \frac{\sin \phi}{\cos(\phi - \alpha)} \right] = 0$$

This can be differentiated to give the final equation defining the shear plane angle

$$\cos \alpha \cos(2\phi - \alpha) - \beta \chi \sin^2 \phi = 0 \tag{4.23}$$

The next equation rearranges the final equation:

$$\beta \sin^2 \phi = \frac{1}{\chi} \cos(\alpha) \cos(2\phi - \alpha)$$

This next set of curves plots the left side (A= $\beta\sin^2 \phi$) of the above equation against the shear angle:

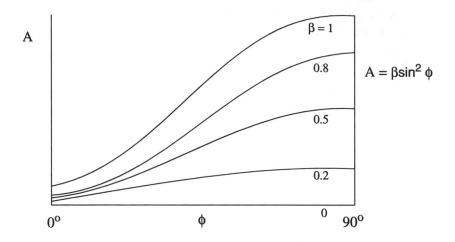

FIGURE 4.7   Effect of ϕ and β on A = $\beta\sin^2 \phi$

This next set is the right side (B= $\cos\alpha\cos(2\phi - \alpha)/\chi$) against the shear angle:

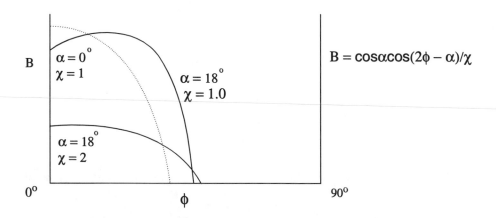

FIGURE 4.8   Effect of ϕ, α, and χ on B = $\cos\alpha\cos(2\phi - \alpha)/\chi$

For different values of rake angle (α) friction condition (β) and contact lengths (χ) the Rowe and Spick equation can be solved graphically for shear angle (ϕ).

This could be computed to get the value of the shear plane angle, but the intersection of the graphs in *Figure 4.9* shows how the physical parameters influence the situation.

If the curves are superimposed, the intersection gives the value of ϕ for different values of χ, β, and α.

For reference, some standard values are:

- χ: for free-machining steel = 1.
- β: for most turning operations, β = 0.8 to 1.
- α: this is a constant set by tooling choices. Most often, it is six degrees.

Since α is a constant and β is unity to a first approximation, a value of the shear plane angle can be obtained by making a reasonable assumption about (χ). It is possible to grind the tool in such a way as to fix (χ). And tools like this are often used in practice to curl the chip deliberately, so this idea is not totally contrived. Once the value of χ, has been fixed, and the value of β is assumed to be unity for free-machining steel, the shear plane angle can be predicted and important parameters such as the cutting forces can be calculated.

As a final check the reader is invited to pin-point the shear plane angle for:

- Rake angle = 18 degrees, full sticking friction and contact length equal to the undeformed chip thickness.

- Rake angle = 0 degrees, β =0.8 and contact length equal to the undeformed chip thickness.

In the simple model, the force along the primary shear plane OD, can now be calculated using the predicted value of the shear plane angle:

$$F_{OD} = \frac{k \cdot w \cdot t_1}{\sin\phi}$$

Also the resolved force along the rake face OB´ can be calculated as $F_{OB'} = k \cdot OB' = \dfrac{k\chi\, t_1}{\cos\alpha}$

These forces can be resolved into the main cutting force directions F_C and F_t as follows:

$$F_c = \frac{F_{OD} - F_t \sin\phi}{\cos\phi}$$

$$F_t = \frac{F_{OB'} - F_c \sin\alpha}{\cos\alpha}$$

Solving these equations simultaneously gives a final equation only in F_C

$$F_c = \frac{1}{(\cos\phi + \tan\alpha\sin\phi)}\left(\frac{kwt_1}{\sin\phi} - \frac{k\chi t_1}{\cos^2\alpha}\right) \tag{4.24}$$

To summarize, if all the cutting conditions are held constant, it is found that the rate of work on the shear plane is proportional to $1/\sin\phi\cos\phi$, which has been shown to be the amount of strain (γ) on the shear plane (*Equation 3.5*) when the rake angle $(\alpha) = 0°$. Curve #1 in *Figure 4.10* shows how the work done on the shear plane varies with the shear plane angle, and this has, of course, the same shape as the curve for zero rake angle in *Figure 3.4*, with a minimum at $\phi = 45°$.

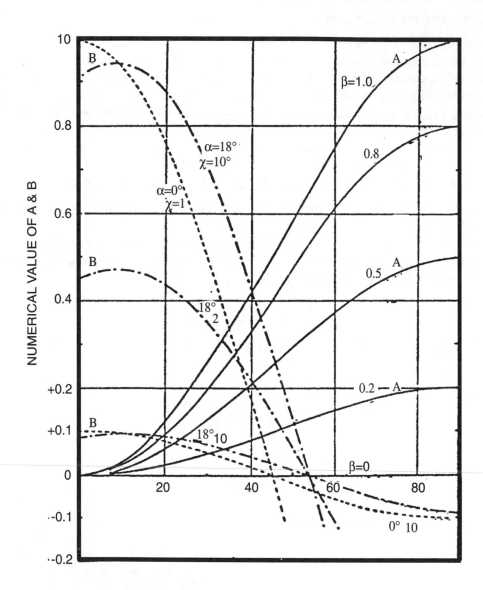

FIGURE 4.9   Intersection of curves gives the value of shear plane angle (from Rowe & Spick 1966)

Thus, if the feed force F_t and the work done on the rake face were so small that they could be neglected, the minimum energy theory proposes that the shear plane angle would be 45°, with the chip thickness t_2 equal to the feed t_1. Where the rake angle is higher than zero, the minimum work is at a shear plane angle higher than 45°, but always at a value where $t_2 = t_1$. In practice, the chip is sometimes approximately equal in thickness to the feed, never thinner, but often much thicker.

From curve 2, *Figure 4.10*, dW_r/dt is seen to increase with the shear plane angle.

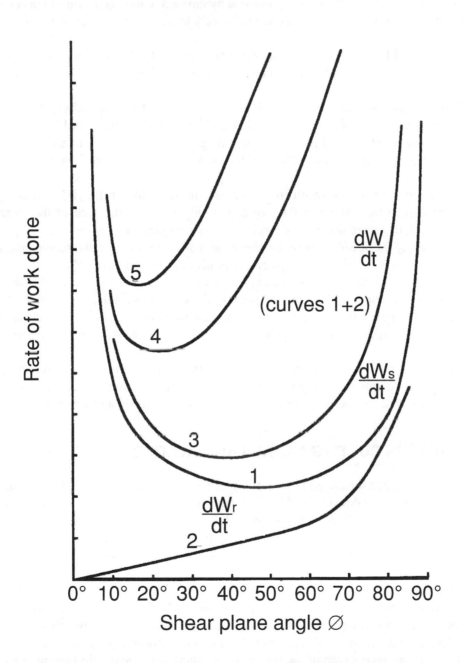

FIGURE 4.10 Rate of work done *vs* shear plane angle ϕ. Curve 1 is the work done in the primary shear zone for a zero rake angle tool; curve 2 is the work done on the rake face for an arbitrarily small value of λ; curve 3 gives the total work done; curves 4 and 5 are for larger values of λ along the rake face, showing how the total work increases *and*, consequently, the value of the shear plane angle decreases (After Rowe and Spick[7])

The total rate of work done, dW/dt is represented by curve 3, which is a sum of curves 1 and 2. In curve *3* the minimum work done occurs at a value of ϕ less than $45°$ and at an increased rate of work.

Curve *2* was plotted for an arbitrary small value of L and of k_r. The contact length L is one of the major variables in cutting; its influence can be demonstrated by plotting a series of curves for different values of L.

Curves *4* and *5* show the total rate of work done for values of L four and eight times that of curve 3. The minimum value of ϕ in curves 1, 3, 4 and 5 is given by Rowe and Spick's analysis.

This family of curves shows that, as the contact length on the rake face of the tool increases, the minimum energy occurs at lower values of the shear plane angle and the rate of work done increases greatly.

A similar reduction in ϕ would result from increases in the value of the yield stress, k_r. It has already been explained that, at low values of ϕ, the chip is thick, the area of the shear plane becomes larger, and, therefore the cutting force, F_c, becomes greater.

The consequence of increasing either the shear yield strength at the rake face or the contact area (length) is to raise not only the feed force F_t but also the cutting force F_c.

The contact area on the tool rake in particular is seen to be a very important region, controlling the mechanics of cutting, and becomes a point of focus for research on machining. Not only the forces, but temperatures, tool wear rates, and the machinability of work materials are closely associated with what happens in this region which receives much attention in the rest of this book.

Thus F_c can be found to a first approximation for the case when $\chi = 1$. The analysis has assumed plane strain conditions, a single shear zone, a constant friction value, and a single value of shear stress, equal to a text-book value - i.e. ignoring the high strain and strain rates that occur.

4.5   FORCES IN CUTTING METALS AND ALLOYS

Cutting forces have been measured in research programs on many metals and alloys, and some of the major trends are now considered.[8,9] When cutting many metals of commercial purity, the forces are high; this is true of iron, nickel, copper and aluminium, among others. With these metals, the area of contact on the rake face is very large, the shear plane angle is small and the very thick, strong chips move away at slow speeds. For these reasons, pure metals are notoriously difficult to machine.

The large contact area is associated with the high ductility of these pure metals, but the reasons are not completely understood. That the high forces *are* related to the large contact area can be simply demonstrated by cutting these pure metals with specially shaped tools on which the contact area is artificially restricted. This is shown diagrammatically in *Figure 4.11*.

In *Table 4.3* an example is given of the forces, shear plane angle and chip thickness when cutting a very low carbon steel (in fact, a commercially pure iron) at a speed of 91.5 m min^{-1} (300 ft/min), a feed of 0.25 mm (0.022 in)/rev and depth of cut of 1.25 mm (0.05 in). In the first column are the results for a normal tool, *Figure 4.11a*, and in the second column those for a tool with contact length restricted to 0.56 mm (0.022 in), *Figure 4.11b*.

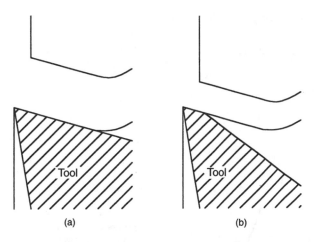

FIGURE 4.11 (a) Normal tool (b) Tool with restricted contact on rake face

TABLE 4.3 **Significant reduction in forces occurs by cutting a low carbon steel with controlled contact tools.**

	Normal tool	Tool with restricted contacts
Cutting force F_c	1,400 N	670 N
	315 lbf	150 lbf
Feed force F_t	1,310 N	254 N
	295 lbf	51 lbf
Shear plane angle ϕ	8°	22°
Chip thickness t_2	1.83 mm	0.66 mm
	0.072 in	0.026 in

Reduction in forces by restriction of contact on the rake face may be a useful technique in some conditions, but in many cases it is not practical because it weakens the tool. Not all pure metals form such large contact areas when subjected to high forces.

For example, when cutting commercially pure magnesium, titanium and zirconium, the forces and contact areas are much smaller and the chips are thin. It is common experience when cutting most metals and alloys that the chip becomes thinner and forces *decrease* as the cutting speed is raised.

Figure 4.12 shows force/cutting speed curves for iron, copper and titanium at a feed of 0.25 mm rev^{-1} (0.010 in/rev) and a depth of cut of 1.25 mm (0.050 in). The decrease in both F_c and F_t with cutting speed is most marked in the low speed range. This drop in forces is partly caused by a decrease in contact area and partly by a drop in shear strength (k_r) in the flow-zone as its temperature rises with increasing speed. This is discussed in Chapter 5.

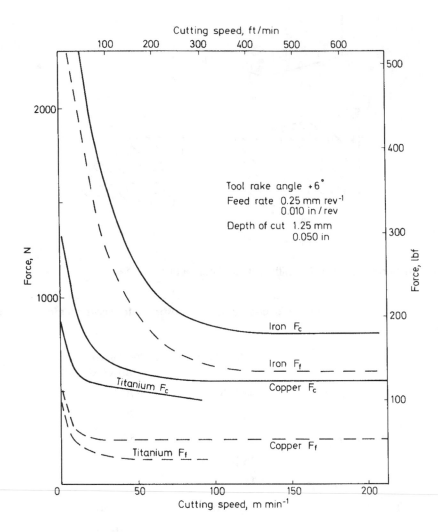

FIGURE 4.12 Cutting force *vs* cutting speed for iron, titanium and copper (From data of Williams, Smart and Milner[8])

Alloying of a pure metal normally increases its yield strength but often reduces the tool forces because the contact length on the rake face becomes shorter. For example, *Figure 4.13* compares force/cutting speed curves for iron and a medium carbon steel, as well as for copper and 70/30 brass.[8] In each case the tool forces are lower for the alloy than for the pure metal over the whole speed range, the difference being greatest at low speeds.

The kink in the curve for the carbon steel in the medium speed range illustrates the effect of a built-up edge. With steels, a built-up edge forms at fairly low speeds and disappears when the speed is raised. Where it is present the forces are usually abnormally low because the built-up edge acts like a restricted contact tool, effectively reducing contact on the rake face *(see Figure 3.21).*

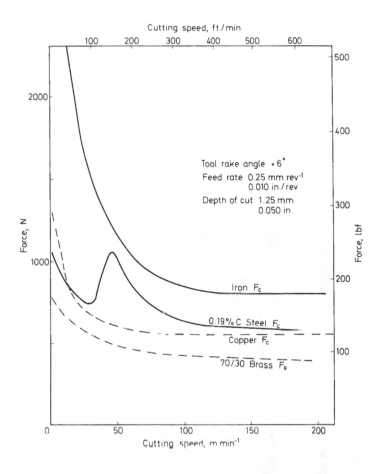

FIGURE 4.13 Cutting force *vs* cutting speed for iron, steel, copper and brass (From data of Williams, Smart and Milner[8])

The tool forces are also influenced by tool geometry, the most important parameter being the rake angle. Increase in the rake angle lowers both the cutting force and feed force *(Table 4.4)*, but reduces the strength of the tool edge and may lead to fracture.

The strongest tool edge is achieved with negative rake angle tools, and these are frequently used for the harder grades of carbide and for ceramic tools which lack toughness. The high forces make negative rake angle tools unsuitable for machining slender shapes which may be deflected or distorted by the high stresses imposed on them. Tool forces usually rise as the tool is worn, the clearance angle is destroyed, and the area of contact on the clearance face is increased by flank wear. The forces acting on the tool are one of the factors which must be taken into consideration in the design of cutting tools - a very complex and important part of machining.

The tool material can also influence the tool forces. When one major type of tool material is substituted for another, the forces may be altered considerably, even if the conditions of cutting and the tool geometry are kept constant. This is caused mainly by changes in the area of seized contact.

TABLE 4.4   Typical cutting forces for low carbon free-cutting steel; cutting speed 27 m min⁻¹ (90 ft/min)

Rake angle α	Cutting force F_c, feed =				Feed Force F_b feed =			
	0.10 mm rev⁻¹ (0.004 in/rev)		0.20 mm rev⁻¹ (0.008 in/rev)		0.10 mm rev⁻¹ (0.004 in/rev)		0.20 mm rev⁻¹ (0.008 in/rev)	
	(N)	(lbf)	(N)	(lbf)	(N)	(lbf)	(N)	(lbf)
+5°	913	205	-	-	392	88	-	-
+10°	840	189	1520	342	289	65	520	117
+15°	743	167	1328	298	200	45	320	72
+20°	716	161	1210	272	151	34	222	50
+25°	627	141	1158	2660	80	18	116	26
+30°	600	135	1090	245	49	11	45	10

Finally, the contact length and tool forces may be greatly influenced by cutting lubricants. When cutting at very low speed the lubricant may act to prevent seizure between tool and work and thus greatly reduce the forces. In the speed range used in most machine shop operations it is not possible to *prevent* seizure near the edge, but liquid or gaseous lubricants, by penetrating from the periphery, can restrict the area of seizure to a small region.

The action of lubricants is discussed in Chapter 10, but in relation to forces, it is important to understand that lubricants can act to reduce the seized contact area and thus the forces acting on the tool. They are most effective in doing this at low cutting speeds and become largely ineffective in the high speed range.[10]

4.6   STRESSES IN THE TOOL

With the complex tool configurations and cutting conditions of industrial machining operations, accurate estimation of the localized stresses acting on the tool near the cutting edge defies the analytical methods available. In fact cutting tools are rarely, if ever, designed on the basis of stress calculations; trial and error methods and accumulated experience form the basis for tool design. However, in trying to understand the properties required of tool materials, it is useful to have some knowledge of the general character of these stresses.

In a simple turning operation, two stresses of major importance act on the tool:

(1) The cutting force acting on a tool with a small rake angle imposes a stress on the contact area on the rake face which is largely compressive in character. The mean value of this stress is determined by dividing the cutting force, F_c, by the contact area. Since the contact area is usually not known accurately, there is considerable error in estimation of the mean compressive stress, but the values can be very high when cutting materials of high strength. Examples of estimated values for the mean compressive stress are shown in *Table 4.5* to give an idea of the stresses involved. Unlike the cutting *force*, the compressive *stress* acting on the tool is related to the shear strength of the work material. High *forces* when cutting a pure metal are an indirect result of the large contact area, and the mean stress on the tool is relatively low, compared to that imposed when cutting an alloy of the metal.

(2) The feed force, F_t, imposes a shearing stress on the tool over the area of contact on the rake face. The mean value of this stress is equal to the force, F_t, divided by the contact area. Since F_t is normally smaller than F_c, the shear stress is lower than the compressive stress acting on the same area. Frequently the mean shear stress is between 30% and 60% of the value of the mean compressive stress.

When a worn surface is generated on the clearance face of a tool ('flank wear') both compressive and shear stresses act on this surface. Although the contact area on the flank is sometimes clearly defined, it is very difficult to arrive at values for the forces acting on it, and there are no reliable estimates for the stress on the worn flank surface.

Other stresses acting on the body of the tool are related to the general construction of the tool and to the rigid connection where the tool is fastened to the machine tool. In a lathe, where the tool acts as a cantilever, there are bending stresses giving tension on the upper surface between the contact area and the tool holder. In a twist drill the stresses are mainly torsional. Tools must be strong enough and rigid enough to resist fracture and to give minimum deflection under load.

This is an important area of tool design, but is not further discussed because stress is being considered here in relation to tool life and chip formation.

TABLE 4.5   Mean compressive stress on the tool

Work material	Cutting force F_c		Contact area		Mean compressive stress	
	(N)	*(lbf)*	*(mm²)*	*(in²)*	*(MPa)*	*(tonf/in²)*
Steel (medium carbon)	490	110	0.65	0.001	770	50
Titanium	455	102	0.77	0.0012	570	37
Iron	1,070	240	3.1	0.0048	340	22
Copper	800	180	3.96	0.021	202	13
70/30 brass	500	111.5	12.2	0.019	42	2.6
Lead	323	72.5	22.5	0.035	14	0.9

4.7   STRESS DISTRIBUTION

4.7.1 Introduction

Since the maximum stress acting on any part of the tool/work interface is the most critical stress determining the requirements for the tool material, it is essential to know the distribution of the compressive and shear stresses. Both calculation and experimental determination of stress distribution have been attempted.

To calculate stress distribution, Zorev[11] assumed a model (*Figure 4.14*) in which, in the absence of a built-up edge, sticking or seizure occurs at the interface near the tool edge and sliding takes place beyond the sticking region.

Zorev's assumed distribution of compressive stress in *Figure 4.14* is based on a simple hypothesis that this stress (σ_c) at any position must be represented by the expression:

$$\sigma_c = qx^y \tag{4.25}$$

where x = distance from the point B where the chip breaks contact with the tool, and q and y are both constants.

The essential feature is the gradient of compressive stress, with the maximum at the cutting edge, falling to zero where the chip breaks contact with the tool. The shear stress shows a lower maximum and a more uniform distribution across the surface. The real situation at the tool/work interface is much more complex than that of the simple models adopted for calculation, and experimental determination of stress distribution is therefore of more interest. The results of four methods will now be mentioned.

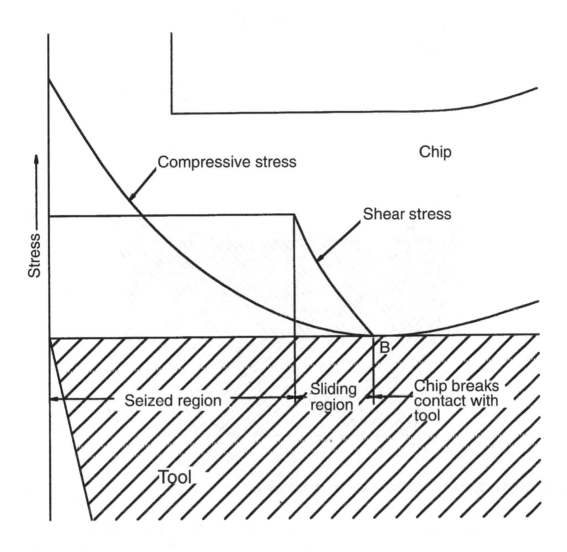

FIGURE 4.14 Model of stress distribution on tool during cutting (After Zorev[11])

4.7.2 Experiments with split tools

A first method is to employ a split steel or carbide tool as shown diagrammatically in *Figure 4.15*. The split is parallel to the cutting edge and the design is such that the length *OC* in *Figure 4.15* is much shorter than the contact length of the chip, *OB*. The deflection of the outer part of the tool *EOCD* can be measured using wire strain gauges or a piezoelectric sensor to determine the force acting on the part of the contact area near the tool edge. Since the area corresponding to the length *OC* can be measured, the stress acting upon it can be determined accurately. The length *OC* can be varied from a practical minimum of about 0.12 mm (0.005 inch) to the full contact length, and a series of measurements give the stress distribution on the tool rake face.

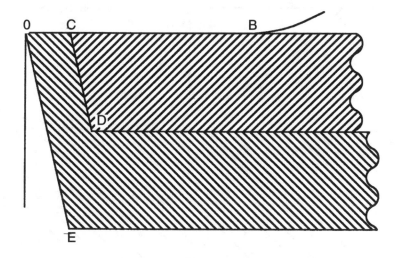

FIGURE 4.15 Split tool used for measurement of stress distribution

This method was described by Loladze in his book *Wear of Cutting Tools*.[12] He gives the distribution of compressive stress on the rake face of tools used to cut several steels at different speeds, feeds and tool rake angles. In general, the compressive stress was a maximum at or very near to the tool edge. Values for the maximum stress varied from 900 MPa to 1600 MPa (60 to 100 tonf/in^2) for different steels. In each case the maximum stress near the tool edge was higher than the yield strength of the work material by a factor greater than 2, so that a stress greater than 1100 MPa (75 tonf/in^2) is normal when cutting steel. The maximum stress is strongly related to the yield stress of the work material; it seems to increase to a small extent with cutting speed and feed, and decreases as the rake angle of the tool is increased.

In the split tool method of Kato *et al*.[13] tools of two designs were used to measure the distribution of both compressive stress and stress parallel to the tool rake face. *Figure 4.16* shows the results obtained when high speed steel tools were used to cut aluminium, copper and zinc at a speed of 50 m min^{-1}. *Table 4.6* gives the values for the maximum compressive stress and compares these with the values of yield flow stress of the same materials measured in compression at a natural strain of 0.2. The maximum stress acting on the tool is commensurate with the yield stress of the work material at a rather high level of plastic strain.

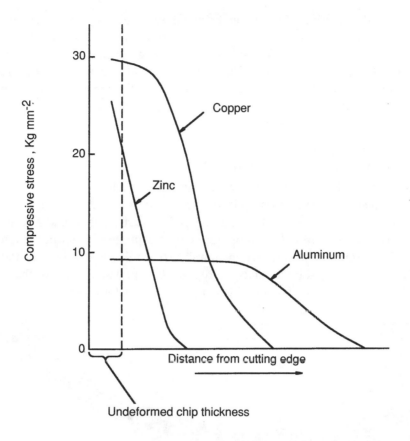

FIGURE 4.16 Distribution of compressive stress on rake face of tools used to cut three metals (After Kato *et al.*[13])

TABLE 4.6 Work material strength (Data after Kato *et al.*[13])

Work material	Maximum compressive stress on tool (MPa)	Compressive yield strength of work material at $\varepsilon = 0.2$ (MPa)
Aluminium	83	93
Copper	284	294
Zinc	245	304

4.7.3 Experiments with tools of different strengths

A second method to estimate distribution of compressive stress is by use of a series of tools of different hardness levels and known yield strength. A tool with adequate yield strength is not deformed plastically during cutting, but if the yield strength is too low, the tool edge is permanently deformed downward.

An estimate of the distribution of stress can be made by cutting the workpiece with a series of tools of decreasing yield strength and measuring the permanent deformation of each edge. This method was explored by Rowe and Wilcox[14] at low cutting speeds to avoid any heating of the tool which would alter its yield strength. The published results agree with the other methods, indicating a maximum compressive stress near the tool edge. For use over a wide range of speed, temperature measurement in the tool would be required as well as knowledge of the variation of yield strength of the tool with temperature.

4.7.4 Experiments with photoelastic polymers

A photo-elastic method uses a tool made of a polymer such as PVC. Because of the low strength of the polymer, and the rapid drop in strength with temperature, these tools can be used for the cutting only of soft metals of low melting points such as lead and tin, and the cutting speed must be kept low. *Figure 4.17* is a photograph of a photo-elastic tool taken with monochromatic light, while cutting lead at 3.1 m min^{-1} (10 ft/min) and a feed of 0.46 mm rev^{-1} (0.0185 in/rev).[15] The dark and light bands are regions of equal strain within the tool and, from these, the distribution of compressive and shear stress can be determined.

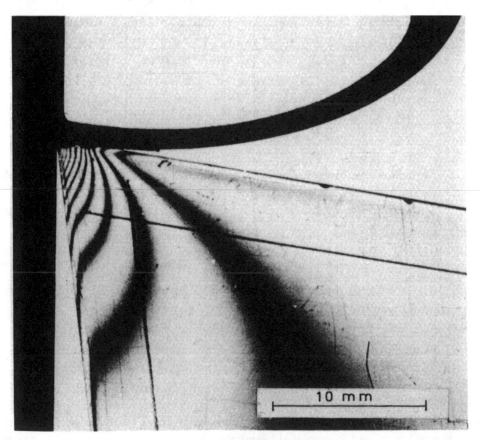

FIGURE 4.17 Stress distribution in photo-elastic model tool when cutting lead (Courtesy of E. Amini[15])

Figure 4.18 is an example of the type of stress distribution found.

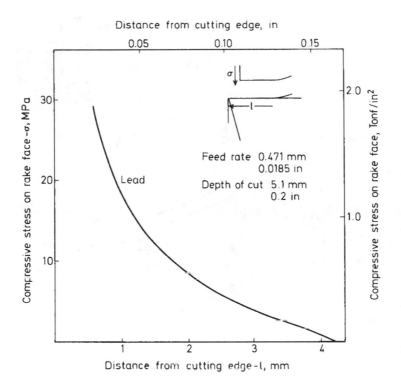

FIGURE 4.18 Compressive stress on rake face calculated from photo-elastic tool, as in *Figure 4.17* (From data of E. Amini[15])

4.7.5 Experiments with photoelastic sapphire

Stress birefringence in sapphire tools can also be used to determine the stress boundary conditions in machining. The stress birefringence effect in sapphire is relatively weak and, of course, sapphire is inherently brittle. Despite these difficulties, Bagchi and Wright[16] were able to machine steel and brass specimens at speeds of up to 75 m min^{-1} and at a maximum feed rate of 0.381 mm per revolution to study the effect of speed and feed rate on stress distributions.

As shown in *Figure 4.19*, the tool holder was set up on a photoelastic bench on top of a three-axis piezoelectric dynamometer, which measured the *x, y* and *z* components of the force on the tool during machining. The *x*-direction on the dynamometer was parallel to the cutting during machining, the *y*-direction normal to the cutting edge and on a horizontal plane containing the edge, and *z*-direction orthogonal to the *x* and *y* plane. The photoelastic bench, sapphire insert and tool holder were mounted on the carriage of the lathe. AISI 1020 and 12L14 steels, and 360 brass in the form of tubular sections made from round bars were used as work materials. Each of these materials exhibited similar machining characteristics and produced continuous chips for the cutting conditions used.

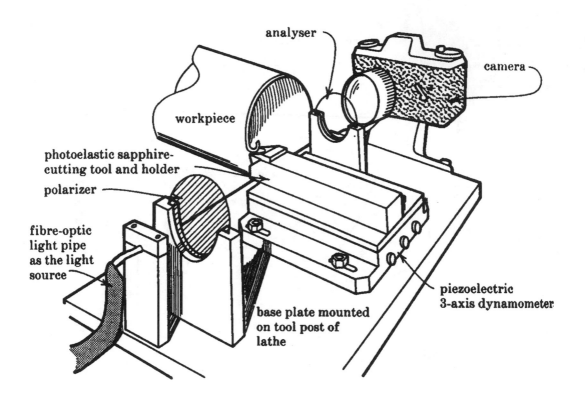

FIGURE 4.19 View of the experimental arrangement used for photoelastic stress analysis in machining.

The transparent, rectangular, parallelepiped photoelastic sapphire tool (7 x 10 x 15 mm^3) was mounted in a tool holder with a -5° rake and a 5° clearance angle, which supported the insert only on its back and bottom faces. This ensured that the sides of the crystal, that is, the c-faces, were stress free, thus conforming to plane stress conditions for the analysis. A suitably shaped window in the tool holder permitted the observation and recording of the photoelastic fringes during machining.

Fast-frame photography allowed a series of images to be collected. These had a similar appearance to *Figure 4.17*. A first set of color images provided "isochromatic fringes". These are "contours of light extinction" that correspond to positions where the difference in the principal stress values is equal to an integer value of wavelength.[†] A second set of black and white images

---

[†] An easy to read review of stress analysis with photoelasticity may be found in "Advanced Strength and Applied Stress Analysis" by R.G. Budynas, 1977, McGraw Hill, Pages 375 et seq. The basic photoelastic equation is $(\sigma_1 - \sigma_2) = N\lambda/f_a t$ where N = 0,1,2,3,4... In this equation, σ is a principal stress value, N is zero or an integer, λ is traditionally used for the wavelength of the light, f_a is the stress-optical coefficient for the material being used and t is the thickness of the model being studied, in this case the cutting tool thickness in Figure 4.19.

provided the "isoclinic fringes" (θ). These occur when either of the principal stresses are aligned with the polarizer. Each photoelasticity experiment, with a fixed orientation of the crossed polarizer-analyser combination, can produce only a single isoclinic. Therefore, each experiment was repeated (with identical speed and feed rate) for six orientations of polarizer-analyser settings to obtain six isoclinics.

The three unknown quantities in the stress field are σ_z, σ_x and τ_{xy}. Because only two independent parameters are known, it is not possible to determine the three unknowns uniquely from these two parameters. A third independent relationship or parameter is necessary for separating the stresses. The shear-difference method was thus selected and used by Bagchi and Wright to calculate the normal and shear stresses from isochromatics and isoclinics obtained experimentally. The composite isochromatic and isoclinic contours for the three cutting conditions are shown in *Figures 4.20 to 4.22*. The rake surface stress distribution for these three machining conditions - from the experimentally obtained isochromatics and isoclinics, and the shear difference method - are shown in *Figures 4.23 to 4.25*.

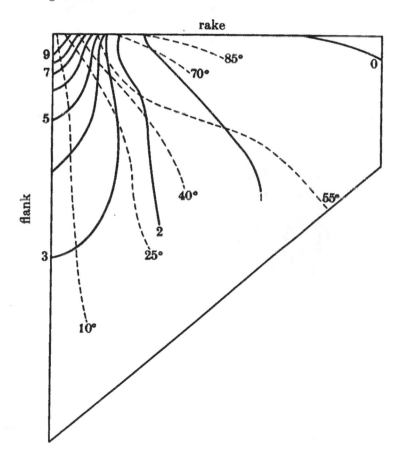

FIGURE 4.20 Composite isochromatics and isoclinics obtained during the machining of 1020 steel at 10 m min⁻¹ and an uncut chip thickness of 0.132mm. Solid lines; isochromatics, dashed lines isoclinics.

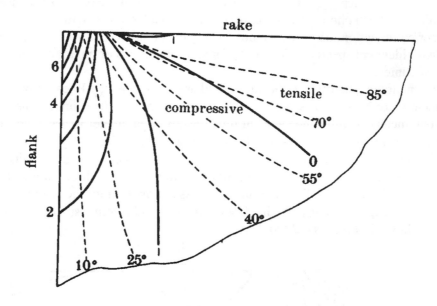

FIGURE 4.21 Composite isochromatics and isoclinics obtained during the machining of 12L14 steel at 75 m min⁻¹ and an uncut chip thickness of 0.132mm. Solid lines, isochromatics; dashed lines, isoclinics.

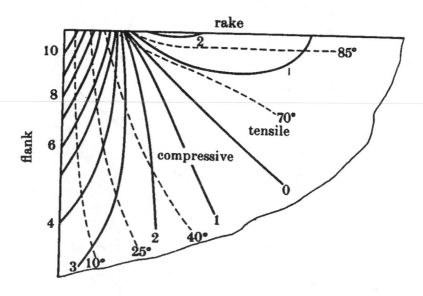

FIGURE 4.22 Composite isochromatics and isoclinics obtained during the machining of 12L14 steel at 75 m min⁻¹ and an uncut chip thickness of 0.381mm. Solid lines, isochromatics; dashed lines, isoclinics.

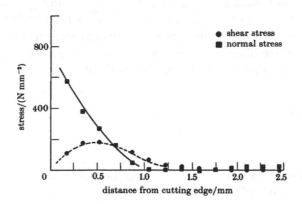

FIGURE 4.23 Normal and shear stress distributions along the rake face surface during the machining of 1020 steel at 10 m min[-1] and an uncut chip thickness of 0.132 mm.

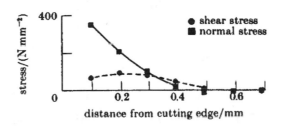

FIGURE 4.24 Normal and shear stress distributions along the rake face surface during the machining of 12L14 steel at 75 m min[-1] and an uncut chip thickness of 0.132 mm.

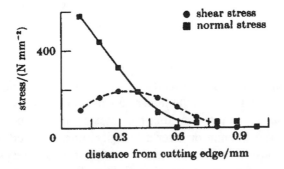

FIGURE 4.25 Normal and shear stress distributions along the rake face surface during the machining of 12L14 steel at 75 m min[-1] and an uncut chip thickness of 0.381mm.

These rake-face normal and shear-stress distributions for the cutting conditions shown, qualitatively resemble those obtained by Chandrasekaran and Kapoor[17] who machined commercially pure lead with a -10 degree rake angle epoxy resin tool. In the recent set of experiments with sapphire tools it was found that the boundary stress distributions varied with the speed and feed rate as expected, although the variation was more pronounced with respect to feed rate.

4.7.6 Normal stress distribution: typical results

4.7.6.1 Influence of speed

TABLE 4.7   Comparisons of the Rake-face Stress Distributions from Photoelastic Analysis for Different Machining Conditions

work material	speed	thickness (uncut chip)	width (uncut chip)	Cutting Forces for F_f and F_c newtons		average normal stress	peak normal stress
	m min^{-1}	mm	mm			MPa	MPa
360 brass	60	0.132	5.00	947	1373	---	600
1020 steel	10	0.132	4.50	1225	1778	---	565
12L14 steel	10	0.132	5.00	558	1307	286	580
12L14 steel	25	0.132	4.84	294	904	326	490
12L14 steel	75	0.132	4.93	201	821	339	530
12L14 steel	25	0.381	5.00	847	2633	498	800
12L14 steel	75	0.381	4.99	450	2140	495	840

It can be seen in *Table 4.7* that the maximum normal stress at the edge decreased by only 10% for 650% increase in speed (from 10 to 75 m min^{-1}) at 0.132 mm per revolution feed rate. At a feed rate of 0.381 mm per revolution, although the speed was increased by 200% (from 25 to 75 m min^{-1}) the decrease in peak stress was not noticeable. The variation in the stresses calculated from experimental data closest to the cutting edge also showed the same trend. Such small changes were not considered to be indicative of any variation of the stresses with speed.

4.7.7 Influence of feed rate

The variation of the normal stresses with feed rate was more noticeable. The rake-face normal stresses increased by a factor of 1.5 when the feed rate was increased near three times from 0.132 to 0.381 mm per revolution. It is thus clear that the interfacial stresses increase with the uncut chip thickness. However experimentation over a broader range of feed rates will be necessary before any functional relationship between stresses and feed rate can be established.

The experimental data was also approximately validated by comparing the total normal force measured by the 3 axis dynamometer with that obtained by integrating the normal stress distribution from photoelastic stress analysis (*Figures 4.23 to 4.25*) over the entire contact length. *Table 4.8* shows the comparison between the rake face normal forces calculated from photoelastic stress analysis and those measured by the dynamometer. It can be seen that the maximum error is 32% but the mean absolute error is 14.3% and the mean error is 10.5%. Owing to the variabilities inherent in chip formation a difference of 10-15% between the maximum and minimum measured force can be expected within a sequence of experiments.

TABLE 4.8   Comparison of the Rake-face Normal Cutting Forces from Photoelastic Analysis with Dynamometer Measurements

work material	cutting speed m min^{-1}	uncut chip thickness/ mm	normal force (dynamometer) in Newtons	normal force (photoelastic) in Newtons	difference between dynamometer and photoelastic forces (%)
brass	60	0.132	1450	1462	-0.8
1020	10	0.132	1877	2165	-15
12L14	10	0.132	1350	1787	-32
12L14	25	0.132	926	1150	-24
12L14	75	0.132	835	961	-15
12L14	25	0.381	2696	2582	+4
12L14	75	0.381	2171	1968	+9

4.7.8   Shear stress distribution: typical results

It is noteworthy that the shear stress in *Figures 4.23-4.25* is low immediately behind the cutting edge, increasing to a maximum over the remainder of the contact length before decreasing, as might be expected, at the rear of contact as the chip leaves the tool face. This was also observed in movie-films taken through the back-face of the transparent sapphire tools.

In Doyle *et al*'s work[18] copper coatings were evaporated on to sapphire tools before cutting lead. It was found that the coating was sheared away in the main part of the contact area (a region they termed zone 1*b*) but not in the region immediately behind the cutting edge (termed zone 1*a*).

Similarly, Wright[19] demonstrated the same partition into zones 1a and 1b by spraying enamel paint onto the sapphire tools before cutting soft workmaterials. Once again the region close to the edge remained covered in paint indicating a lower imposed shear stress in that region. After cutting a wide range of materials at both low and high speeds, low magnification photography of the rake faces of the tools showed that there is generally less transfer near the cutting edge, particularly for soft work materials (*Figures 4.26a and b*).

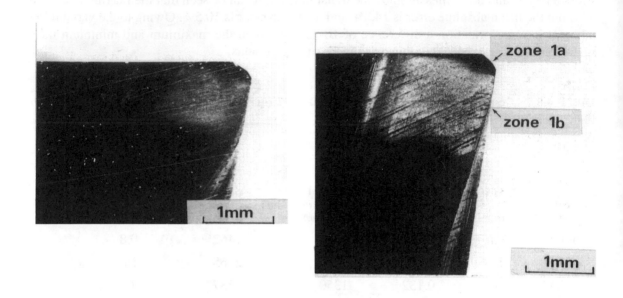

FIGURE 4.26 Rake face photographs at x25 magnification confirming the reduced transfer and hence shear stress in Zone 1a.[18,19] From machining copper at 120 m min^{-1} for a) only 5 seconds and b) 50 seconds. Longer cutting times "clean-up" any organic films on rake face, creating increased amounts of the seizure type contact as time progresses.

Photomicographs of sections through partially formed chips show that the zone 1a feature is associated with the way in which material flows around the cutting edge. Metallography has been carried out and *Figure 4.27* shows a section through a copper quick-stop produced at 25 mm s^{-1}. In this case the zone 1a region varied in length from 0.2 to 0.3mm. These lengths are shown by *OP* and *OQ* in *Figure 4.28*. It is emphasized that the total length of zones 1a and 1b was 3.1mm and that the first 1.2 mm only is shown in *Figures 4.27 and 4.28*.

The work material is initially sheared in the primary zone and the equiaxed grains of copper are deformed to such an extent that the grain boundaries appear as single lines *AB*. Such lines then show the deformation flow pattern. In the chip, all these deformation lines show a point of inflection. This is marked along *EF* and may first be detected in the line *CD*. At *C*, material continues to flow into the machined bar whereas at *D*, material is turning into the chip.

As the chip is formed, such deformation lines bend towards the rake face as shown in the progression from *CD* to *EF*. The lines *GH* and *IJ* are further development of this process, but in these cases the deformation line terminates at the rake face rather than in the machined surface. Finally, the line *KLMN* contacts the rake face as the tangent at M. Beyond M all the lines curve in towards the tool to meet the rake face at ~0°. In contrast, between *0* and *M* there is a region, approximately bounded by *KLM* and the rake face, where the flow pattern is substantially different from that beyond *M*.

FIGURE 4.27 Quick-stop section through copper chip machined at a slow speed of 1.5 m min⁻¹ for 300 seconds.

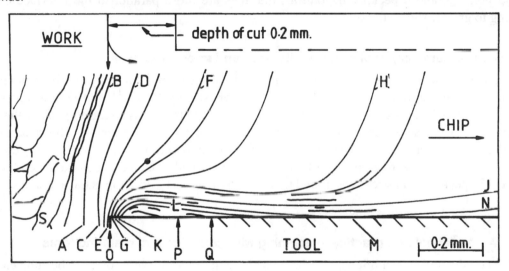

FIGURE 4.28 Tracing of flow lines above. The Zone 1a region was 0.2 to 0.3 mm (OP to OQ). The total contact length was 3.1 mm. ON is approximately 1.2mm.

Other quick-stop sections that were prepared showed that when cutting speed was increased, the height and length of the area bounded by *KLM* decreased. At the highest speed used of 120 m min⁻¹ such a region could not be resolved in the quick-stop sections. However, there continued to be differences in the flow pattern along the contact length. Immediately behind the edge (zone 1a) the deformation lines resembled *GH* and met the rake face at an acute angle; further back (zone 1b) the lines resembled *NM* meeting the tool at ~ 0°. Examination of the used tools from these experiments at higher speeds and longer cutting times still showed the zone 1a effect as indicated in *Figure 4.26*.

The deformation lines in *Figure 4.28* exhibit the inflection points because material continuity must be maintained between metal flowing into the bar and into the chip. Schey[20] has presented micrographs of metal flow in plane strain forging which show similar patterns. If metal flow tangential to the die is restrained at the interface then the test sample "barrels", showing flow patterns similar to the line *IJ*. In Schey's specimens, regions such as *KLM* may also be observed on either side of the vertical axis of the forging. There are also some similarities between the edge region *KLM* and the "stagnation point" that occurs in the centre of a fluid jet impacting on a flat plate (*see* Batchelor[21]). In using this term stagnation it is not the intention to imply that the region *KLM* is a built-up edge and therefore stationary. There was no evidence for this on the used tools nor in the movie-films of the transparent tool machining copper. In addition, built-up edges are not expected with commercially pure metals.[8] Rather, the intent is to illustrate that the metal flow over *OM* is different from the remainder of the contact length. In the region *KLM* the deformation lines turn back on to the rake face. In particular, near *K*, *P* and *Q* the lines give the impression that a small 'cap' of metal exists immediately behind the cutting edge in zone 1*a*. Referring only to the chip material immediately adjacent to the tool, these deformation lines confirm that the shear strain tangential to the rake face is much less over *OPQ* than beyond *Q*. Beyond point *Q*, and especially beyond *M*, the lines are more parallel to the interface corresponding to greater shear stresses.

4.7.9 General summary of photoelasticity work and stress analyses

The salient features of the stress analysis in machining commercial alloys with photoelastic sapphire tools are summarized in *Table 4.7*. As proposed by Zorev,[11] and demonstrated by Amini[15] and by Chandrasekaran and Kapoor,[17] the normal stress was found to peak at the cutting edge and decrease exponentially to zero at the end of the contact length. The shear stress was either zero or very small at the edge, increasing to a maximum in the middle of the contact length and reducing to zero at the end of the contact length and reducing to zero at the end of the contact. The peak normal stress was found to be approximately 1.5 to 2 times the average normal stress.

4.7.10 A special note on practical machining with ceramics and diamond tools

Under certain cutting conditions, the photoelastic sapphire tools were found to be subjected to tensile stresses near the stress-free rake face beyond the rear of the contact length. The top right side of *Figures 4.21 and 4.22* show this effect developing. Some discussion on this observation is also found in Amini[15] and in Chandrasekaran and Kapoor.[17] In practice, tensile stresses should be avoided particularly when more brittle tool materials are being used. Users of such tool materials often experience catastrophic failures of the whole corner of an insert, if the production staff have not set the speeds and feeds at appropriate values. The results above suggest that feed is much more damaging than speed in this regard and that adjustments should be made on this basis. Today, with the increasing use of ceramic and diamond tools, new experiments with these photoelastic sapphire materials may well be the best way of determining the specific feed/speed combinations at which such unfavorable tensile stresses begin to appear at the rake face. Modeling of such conditions is also described in Chapter 11, based on the work of Kistler and colleagues.

4.8   CONCLUSION

The very high normal stress levels account for the conditions of seizure on the rake face. Comparison can be made with the process of friction welding, in which the joint is made by rotating one surface against another under pressure. With steel, for example, complete welding can be accomplished where the relative speed of the two surfaces is 16-50 m min^{-1} (50-150 ft/min) and the pressure is 45-75 MPa (3-5 tons/in^2). The stress on the rake face in cutting steel may be five to ten times as high as this near the tool edge. Seizure between the two surfaces is, therefore, a normal condition to be expected during cutting.

The existing knowledge of stress and stress distribution at the tool-work interface is far too scanty to enable tools to be designed on the basis of the localized stresses encountered. Even for the simplest type of tooling only a few estimates have been made, using a two-dimensional model, where the influence of a tool nose does not have to be considered. However, the present level of knowledge is useful in relation to analyses of tool wear and failure, the properties required of tool materials and the influence of tool geometry on performance.

4.9   REFERENCES

1. Boothyrod, G., *Fundamentals of Metal Machining.*, Marcel Dekker (1990)
2. Kistler, Piezo-electric Force Platforms, Product Literature may be found at <http://www.kistler. com>.
3. Merchant, M.E., J. *Appl. Phys.*, **16,** (5), 267 (1945)
4. Lee, H. and Shaffer, B.W., *J. Appl. Mech., Trans. A.S.M.E.*, **73**, 405 (1951)
5. Kobayashi, S. and Thomsen, E.G., *Trans. A.S.M.E.*, **B84**, 71 (1962)
6. Palmer W.B., and Oxley, P.L.B., *Proc. Instn. Mech. Engrs.*, 173, 623 (1959)
    The above is based on Oxley's Ph.D. thesis. Also see:
        Oxley, P.L.B. *Int. J. of Machine Tool Des. and Res.* 2, 219 (1962)
        Fenton, R.G., and Oxley, P.L.B., *Proc. Instn. Mech. Engrs.,* **183**, (Part 1) 417 (1968-69)
        Stevenson, M.G., and Oxley, P.L.B., *Proc. Inst. Mech. Engrs.*, **185**, 741, (1970)
        Roth, R.N., and Oxley, P.L.B., *J. Mech. Eng. Sci.*, **14**, (2) 85 (1972)
        Hastings, W.F., Matthew, P. and Oxley, P.L.B., *Proc. Roy. Soc., Lond.*, **A371**, 569 (1980)
        Manyindo, B.M., and Oxley, *P.L.B., Proc. Inst. Mech. Engrs.*, **200**, (C5), 349 (1986)
        Oxley, *P.L.B., Proceedings of the CIRP International Workshop on Modeling of Machining Operations,* held in Atlanta, GA., Published by the University of Kentucky Lexington, KY, 40506-0108. p.35, (1998)
7. Rowe, G.W. and Spick, P.T., *Trans. A.S.M.E.*, **89B**, 530 (1967) A more detailed treatment for material properties can be found in Wright, P.K., *Trans. ASME, Journal of Engineering for Industry*, **104**, (3), 285 (1982).
8. Williams, J.E., Smart, E.F. and Milner, DR., *Metallurgia*, **81**, (3), 51, 89 (1970)
9. Eggleston, D.M., Herzog, R. and Thomson, E.G., *J. of Engineering for Industry*, August, 263 (1959)
10. Rowe, G.W. and Smart, E.F., *Proc. 3rd Lubrication Conv., London*, p. 83, (1965)
11. Zorev. N.N., *International Research in Production Engineering*, Pittsburgh, p. 42 (1963)
12. Loladze, T.N., *Wear of Cutting Tools*, Mashqiz, Moscow (1958)

13. Kato, S. *et al., Trans. A.S.M.E.*, **B94**, 683 (1972)
14. Rowe, G.W. and Wilcox, A.B., *J.I.S.I.*, **209**, 231 (1971)
15. Amini, E., J. *Strain Analysis*, 3, 206 (1968)
16. Bagchi, A. and Wright, P.K., *Proc. Royal Society of London*, A **409**, 99 (1987)
17. Chandrasekaran, H., and Kapoor, D.V., *J. of Engineering for Industry*, **87**, 495 (1965)
18. Doyle, E.D., et al., *Proc. Royal Society of London*, A 366, **173** (1979)
19. Wright, P.K., *Metals Tech.*, **8,** (4), 150 (1981)
20. Schey, J.A., Friction and Lubrication in Metalworking, (eds F.F. Ling et al), American Society for Mechanical Engineers, New York, 20 (1966)
21. Batchelor, G.K., *An Introduction to Fluid Dynamics*, plate 7, Cambridge University Press (1967)

CHAPTER 5

HEAT IN METAL CUTTING

5.1 INTRODUCTION

The power consumed in metal cutting is largely converted into heat near the cutting edge of the tool, and many of the economic and technical problems of machining are caused directly or indirectly by this heating action. The cost of machining is very strongly dependent on the rate of metal removal, and costs may be reduced by increasing the cutting speed and/or the feed rate, but there are limits to the speed and feed above which the life of the tool is shortened excessively. This may not be a major constraint when machining aluminium and magnesium and certain of their alloys, in the cutting of which other problems, such as the ability to handle large quantities of fast moving chips, may limit the rate of metal removal. The bulk of cutting, however, is carried out on steel and cast iron, and it is in the cutting of these, together with the nickel-based alloys, that the most serious technical and economic problems occur. With these higher melting point metals and alloys, the tools are heated to high temperatures as metal removal rate increases and, above certain critical speeds, the tools tend to collapse after a very short cutting time under the influence of stress and temperature.

That heat plays a part in machining was clearly recognized by 1907 when F. W. Taylor, in his paper 'On the art of cutting metals', surveyed the steps which had led to the development of the new high speed steels.[1] These, by their ability to cut steel and iron with the tool running at a much higher temperature, raised the permissible rates of metal removal by a factor of four. The limitations imposed by cutting temperatures have been the spur to the tool materials development of the last 90 years. Problems remain however, and even with present day tool materials, cutting speeds may be limited to 30 m min^{-1} (100 ft/min) or less when cutting certain creep resistant alloys.

It is, therefore, important to understand the factors which influence the generation of heat, the flow of heat, and the temperature distribution in the tool and work material near the tool edge. The importance of the heat distribution, not just the amount of heat, was clear to J. T. Nicolson who studied metal cutting in Manchester, England around the turn of the century. In *The Engineer* (1904) he stated "There is little doubt that when the laws of variation of the temperature of the shaving and tool with different cutting angles, sizes and shapes of cut, and of the rate of abrasion... are definitely determined, it will be possible to indicate how a tool should be ground in order to meet with the best efficiency... and the various conditions to be found in practice."[2] However, determination of temperatures and temperature distribution in the vitally important region near the cutting edge is technically difficult, and progress has been slow in the more than 100 years since the problem was clearly stated. Recent research is clarifying some of the principles, but the work done so far is only the beginning of the fundamental survey that is required.

5.2   HEAT IN THE PRIMARY SHEAR ZONE

The work done in (1) deforming the bar to form the chip and (2) moving the chip and freshly cut work surface over the tool is nearly all converted into heat. Because of the very large amount of plastic strain, it is unlikely that more than 1% of the work done is stored as elastic energy, the remaining 99% going to heat the chip, the tool and the work material.

Under most normal cutting conditions, the largest part of the work is done in forming the chip at the shear plane. Boothroyd describes the method for calculating the approximate mean temperature rise in the body of the chip from measurement of tool forces, knowledge of the cutting parameters and data for the thermal properties of the work material.[3]

Using an upper bound type analysis the rate of doing work in a discrete parallel sided primary zone is

$$\frac{dW}{dt} = k_p S_p V_s \tag{5.1}$$

where k_p is the average shar strength in the primary zone, S_p is the area of the shear plane and V_s the shear velocity. Assume first that a fraction of the heat, β, is conducted back into the bar. This leaves a fraction $(1-\beta)$ to be absorbed by the volume of material $(S_p V_n)$, flowing across the zone. This is converted into heat giving the temperature rise for the primary zone as

$$\Delta T_p = \frac{k_p V_s}{\rho c V_n}(1 - \beta) \tag{5.2}$$

where ρ and c are the density and specific heat and V_n is the velocity normal to the shear plane. V_s and V_n may be calculated from the simple hodograph for machining. Alternatively, expressing equation 5.2 in terms of measurable quantities

$$\Delta T_p = \frac{(F_c \cos\phi - F_f \sin\phi)\cos\alpha}{\rho c w_1 t_1 \cos(\phi - \alpha)}(1 - \beta) \tag{5.3}$$

where F_c and F_f are the cutting forces, ϕ is the shear plane angle, α is the tool rake angle and $w_1 t_1$ is the product of the undeformed chip width and thickness. As indicated above, most of the heat generated on the shear plane passes into the chip, but a proportion is conducted into the work material. Theoretical calculations and experimentally determined values show that this proportion (β) may be as high as 50% for very low rates of metal removal, materials with high conductivity and small shear plane angles. However, for high rates of metal removal and machining steel (where temperature effects become a serious problem) it is of the order of 10-15%. To calculate the temperature rise in the chip the value of β should be determined using Boothroyd's experimental graph based on the thermal number for machining (*Figure 5.1*).[3] The thermal number is given by

$$R_T = \frac{\rho c V_w t_1}{k_t} \tag{5.4}$$

where V_w is the work material velocity and k_t the thermal conductivity.

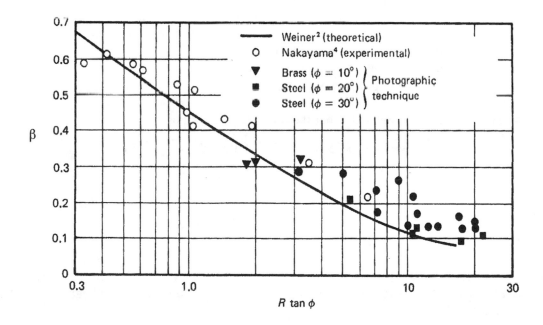

FIGURE 5.1   Boothroyd's expermental curve for the proportion of heat conducted back into the workmaterial

The temperature of the body of the chip (in the absence of a built-up edge) is not greatly influenced by the cutting speed. The value of β decreases as the cutting speed is raised, and there is an increase in temperature because less heat flows back into the work material, but the chip body temperature tends to become constant at high speeds. Calculation of chip body temperature to higher accuracy requires data for the variation with temperature of the specific heat and thermal conductivity of the work material.

The most obvious indication of the temperature of steel chips is their color. When steel is machined at high speed without the use of a coolant, the chip is seen to change color, usually to a brown or blue, a few seconds after leaving the tool. These 'temper colors' are caused by a thin layer of oxide on the steel surface and indicate a temperature on the order of 250 to 350° C. At very low speeds the chip does not change color, indicating a lower temperature, usually associated with a built-up edge. Under very exceptional conditions, when cutting fully hardened steel or certain nickel alloys at high speed, chips have been seen to leave the tool red hot - i.e. a temperature over 650° C - but in most operations when cutting steel the chip body reaches a temperature in the range 200 - 350° C.

The temperature of the chip can affect the performance of the tool only as long as the chip remains in contact; the heat remaining in the chip after it breaks contact is carried out of the system. Any small element of the *body* of the chip, after being heated in passing through the shear zone, is not further deformed and heated as it passes over the rake face, and the time required to pass over the area of contact is very short. For example, at a cutting speed of 50 m min^{-1} (150 ft/min), if the chip thickness ratio is 2, the chip velocity is 25 m min^{-1} (75 ft/min). If the contact on the rake face (*L*) is 1 mm long, a small element of the chip will pass over this area in just over 2 milliseconds. Very little of the heat can be lost from the chip body in this short time interval by radiation or convection to the air, or by conduction into the tool.

In one investigation, the temperature of the top surface of the chip was measured by a radiation pyrometer.[4]

A low carbon steel was being cut under the following conditions:

cutting speed	160m min^{-1}(500 ft/min)
feed	0.32 mm rev^{-1} (0.013 in/rev)
depth of cut	3 mm (0.125 in)

The temperature of the top surface of the chip was shown to be 335°C and a very small temperature *increase*, 2°C, was recorded as it passed over the contact area, due to heat generated on the rake face.

While heat may also be lost from the body of the chip by conduction into the tool through the contact area, it will be shown that, under most conditions of cutting, particularly at high rate of metal removal, the work done by the feed force, F_f, heats the flow-zone at the under surface of the chip to a temperature higher than that of the body of the chip. More heat then tends to flow into the body of the chip from the flow-zone, and no heat is lost from the body of the chip into the tool by conduction.

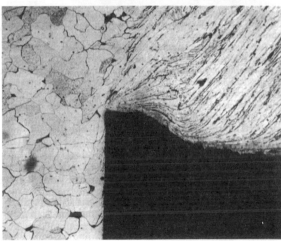

FIGURE 5.2 Metal flow around edge of tool used to cut very low carbon steel (as Figure 3.18); (a) using normal tool (b) using tool with restricted contact on rake face (see Figure 4.11)

The amount of heat conducted into the workpiece is often higher than that calculated using Boothroyd's idealized model in which heat is generated *on the shear plane (OD in Figure 4.1).*[3] In reality the strain is not confined to a precise shear plane, but takes place in a finite volume of metal, the shape of which varies with the material being cut and the cutting conditions. Quick-stop sections provide evidence that the strained region does not terminate at the cutting edge. *Figure 3.18* shows the flow around the edge of a tool used to cut a low carbon steel.

Figure 5.2a shows the region near the cutting edge when machining a commercially pure iron at 16 m min^{-1} (50 ft/min) at a feed of 0.25 mm (0.01 in per rev). With both materials the deformed region visibly extends into the work material at the cutting edge. Thus a zone of strained and heated material remains on the new workpiece surface. In many operations, such as turning, much of the metal heated during one revolution of the workpiece is removed on the next revolution, and this portion of the heat is also fed into the chip. However, some of the heat from the deformed layer of work material is conducted back into the workpiece and goes to raise the temperature of the machined part. It is sometimes necessary to remove this heat with a liquid coolant to maintain dimensional accuracy.

The thickness of the deformed layer on the workpiece is very variable. For example, *Figure 5.2b* shows a quick-stop section through the same work material as *Figure 5.2a* but the cutting speed was 110 m min^{-1} (350 ft/min) and the tool was shaped to give reduced contact length. The visibly deformed zone at the work surface below the cutting edge extends to a depth of not more than 20 μm (compared with more than 100 μm in *Figure 5.2a*) and heating of the surface must have been correspondingly reduced.

Low cutting speeds, low rake angles and other factors which give a small shear plane angle tend to increase the heat flow into the workpiece. Alloying and treatments which reduce the ductility of the work material will usually reduce the residual strain in the workpiece.

To sum up, most of the heat resulting from the work done on the shear plane to form the chip remains in the chip and is carried away with it, while a small but variable percentage is conducted into the workpiece and raises its temperature. This part of the work done in cutting makes a relatively unimportant contribution to the heating of the cutting tool.

5.3 HEAT AT THE TOOL/WORK INTERFACE

The heat generated at the tool/work interface is of major importance in relation to tool performance, and is particularly significant in limiting the rates of metal removal when cutting iron, steel and other metals and alloys of high melting point. In most publications the generation of heat in this region is treated on the basis of classical friction theory; here the subject is reconsidered in the light of the evidence that *seizure* is a normal condition at the tool/work interface. First it is necessary to discuss in more detail the pattern of strain in the built-up edge or in the flow-zone, which constitute the main heat sources raising the tool temperature.

5.3.1 Interface temperatures with a built-up edge

From photomicrographs of built-up edges (*Figures 3.21* and *5.3*) the amount of strain in the main part of these structures is seen to be many times greater than in the body of the chip. Consequently, much more energy is expended per unit volume in the built-up edge, and higher temperatures are generated, than in the body of the chip. The lower part of the built-up edge is formed during the first few seconds cutting. For most of the duration of cutting this part remains stationary and unstrained.

It is the region near the boundary between chip and built-up edge (*O'* to *B* in *Figure 3.22*) that is subjected to continuous strain. The heat generated in this region raises its temperature above that of the chip. Heat is conducted both into the chip, to be carried out of the system, and through the body of the built-up edge into the tool. This is the main heat source raising the tool temperature. The highest temperature is in this region of heat generation and the tool/work interface temperatures are somewhat lower. The distance from the top of the built-up edge to the tool face is usually a few tenths of a millimeter and the tool interface temperature is probably only a few degrees lower.

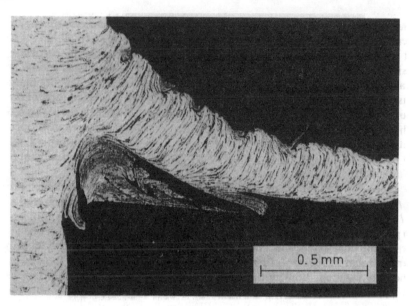

FIGURE 5.3   Section through quick-stop specimen after cutting steel, showing a built-up edge with fragments being sheared away

In *Figures 3.24* and *5.3* fragments of strained material are seen on the new work surface and on the under side of the chip. These were removed by shear fracture across the flow direction, probably on a thermoplastic shear band where the material was locally weakened by higher temperatures. Although these local temperatures may be much higher than those in the surrounding material, the shear zones are extremely thin - e.g. ~ 1 μm - and their duration is on the order of 1 ms. The high temperature 'flashes' are rapidly dispersed by conduction and have insignificant effects on the tool/work interface temperature.

The rate of heat generation at the top of the built-up edge increases as the cutting speed is raised and its temperature rises as cutting speed is increased. The calculation of temperature in relation to cutting speed or feed would be extremely difficult because the shape and size of the built-up edge change. It is unlikely that a method of accurate calculation for these conditions will be achieved.

5.3.2   Interface temperatures with a flow-zone

As the rate of metal removal is increased by raising cutting speed or feed, the built-up edge disappears and in its place a flow-zone is observed, also strongly bonded to the tool rake face, as described in Chapter 3. *Figures 3.14* and *3.18* show chips formed with a flow-zone at the interface. *Figures 3.19* and *5.4* show enlargements of parts of these.

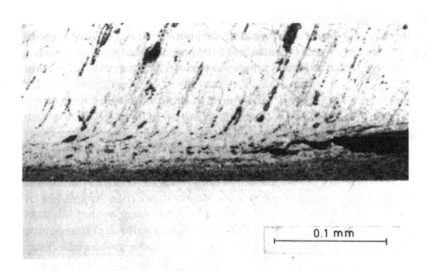

0.1 mm

FIGURE 5.4   Flow-zone at rake face of tool used to cut very low carbon steel at high speed (Detail of *Figure 3.14*) Cutting speed was 153 m min^{-1}

Cutting with a flow-zone is the condition which exists in high speed machining operations in industry, and influences many operations where productivity is limited by tool life problems.

Behavior of the work material in the flow-zone will now be considered, using the example of the tool illustrated in *Figures 3.14* and *5.4*. The cutting speed was 153 m min^{-1} (500 ft/min) and the chip thickness ratio was 4:1. The body of the chip was therefore moving over the rake face at 38 m min^{-1} (125 ft/min), while, at the tool face, the two surfaces were not in relative movement. Shear strain (γ) and the units in which it is measured are discussed briefly in Chapter 3 and illustrated diagrammatically in *Figure 3.6* in relation to strain on the shear plane. Values of 2 to 4 for γ on the shear plane are commonly found, and in the present example γ was approximately 4.

In *Figure 5.4*, the flow zone was, on the average, 0.075 mm (0.003 in) thick over most of the seized contact which was 1.5 mm (0.06 in) long. The flow-zone material was therefore subjected to a mean shear strain of 20 as the chip moved across the contact area, i.e. five times the strain on the shear plane. This was not the total extent of strain in the flow-zone, however. Since the bottom of the flow-zone remained anchored to the tool surface, the material in that part very close to the tool surface continued to be subjected to strain indefinitely. Thus the amount of strain in the flow-zone is very large.

There is no certain knowledge as to the distribution of strain throughout the flow-zone. The usual methods of measuring strain such as making a grid on the surface and measuring its shape change after deformation can be used to map the plastic strain at the shear plane, but cannot be applied to the flow-zone because of its small size and the extremes of strain.

"Natural markers," structural features such as grain boundaries and plastic inclusions, either disappear in the flow-zone or are drawn out so nearly parallel to the tool surface that their angle of inclination is too small to be measured in the part of the flow-zone near the tool surface. The strain pattern observed in photo-micrographs of quick-stop sections is usually such as would be expected from the rigidity of the chip body and seizure at the interface (*Figures 3.17, 5.2a* and

5.4). Since this is a very rare condition, not covered in engineering treatment of the behavior of solid materials, a simple model is proposed below which indicates some of the consequences of this condition.

In *Figure 5.5* a flow-zone is part of the chip, the body of which is moving away from the cutting edge. It is seized to the tool rake face from the cutting edge O to the position Y where the chip separates from the tool. (For clarity the flow-zone thickness relative to the contact length is exaggerated.) The simplifying assumptions are made that the flow-zone is of uniform thickness, Oa, that the chip body is rigid, and that the shear strain in the flow-zone is uniform from the tool surface to the chip body/flow-zone interface.

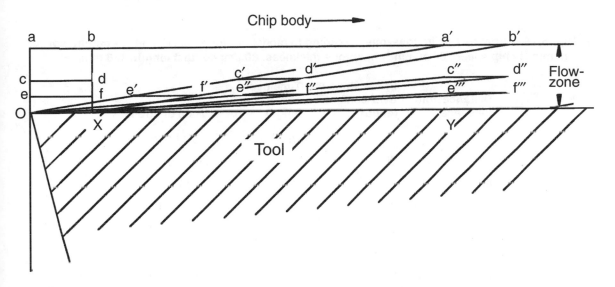

FIGURE 5.5 Idealized model of flow-zone on rake face of tool

Consider a unit body of work material, $OabX$, at the cutting edge of the tool, each side of which is equal to the flow-zone thickness. When the chip moves across the seized contact area, the upper surface of this body, ab moves to $a'b'$ to produce the uniformly strained body $Oa'b'X$. The material at the center line of the unit body cd has been subjected to the same strain as the material at $a'b'$ but has moved only to $c'd'$, half way along the seized length OY. The material at $a'b'$ continues to move at the same rate after leaving the tool at position Y. When the material at $c'd'$ has reached $c''d''$ (above the position Y) it has been subjected to twice the shear strain as ab at $a'b'$. Correspondingly, the material at ef, where Oe is one-quarter of Oa, is subjected to four times the strain when it reaches $e'''f''''$ compared to $a'b'$.

The amount of shear strain in the flow-zone is inversely proportional to the distance from the tool rake face. Using this model, *Table 5.1* gives an indication of the shear strain across the flow-zone at the end of the seized contact, taking as an example the cutting of steel at high speed. *Table 5.1* was prepared from a low carbon iron "quick stop" after cutting at 180 m min^{-1}. Note in *Figure 5.4*, the speed was slightly lower at 153 m min^{-1} but in this range, the physics and the temperature do not change much with such small increments of speed.

Theoretically, the amount of shear strain would become infinite at the tool surface if complete seizure persisted, as in the ideal model. The surfaces of real tools, however, are never perfectly smooth and the mean surface roughness is usually of the order of a few microns. Therefore uniform laminar flow cannot be considered as persisting closer to the tool surface than a few microns. In *Table 5.1*, the shear strain γ over the contact length (the flow zone being 80 μm thick) is 20 at a distance of 80μm from the rake face, and 640 at a distance of 2.5μm. With this model, the rate of strain is constant and the time required for any small element of work material to traverse the contact area becomes longer as the tool surface is approached. At 80μm from the tool surface it is 1.6 milliseconds (ms) and at a distance of 2.5μm it is 51.2 ms.

TABLE 5.1   **Shear strain in flow-zone according to model in *Figure 5.5*. Example - cutting speed: 180 m min⁻¹; chip speed: 60 m min⁻¹; flow zone thickness: 80 μm; contact length: 1. 6 mm.**

Distance from rake face	Shear strain γ over contact length	Time over contact length	Rate of strain
(μm)		(ms)	(s⁻¹)
80	20	1.6	1.25×10^4
40	40	3.2	1.25×10^4
20	80	6.4	1.25×10^4
10	160	12.8	1.25×10^4
5	320	25.6	1.25×10^4
2.5	640	51.2	1.25×10^4

It is clear that the amounts of strain in the flow-zone are normally several orders of magnitude greater than on the shear plane. The amounts of strain are far outside the range encountered in normal laboratory mechanical testing, where fracture occurs at very much lower strains. The ability of metals and alloys to withstand such enormous shear strains in the flow-zone without fracture, must be attributed to the very high compressive stresses in this region which inhibit the initiation of cracks, and cause the re-welding of such small cracks as may be started or already existed in the work material before machining. For example, the holes in highly porous powder metal components are often completely sealed on the under surface of the chip and on the machined surface where these areas have passed through the flow-zone. It has been shown (Section 4.8) that the compressive stress at the rake face decreases as the chip moves away from the edge, and, when the compressive component of the stress can no longer inhibit the formation of cracks, the chip separates from the tool, its under-surface being formed by fracture either at the tool rake face, or at points of weakness within the flow-zone.

In real cutting operations, the strain across the flow-zone is not so uniform as in the simple model (*Figure 5.5*). It may be lower in the transition region between chip body and flow-zone and there may be a dead metal region at the tool surface close to the tool edge (*Figures 3.22 and*

5.2a). In general, however, the observed metallurgical structures in the flow-zone are such as would result from the strain pattern of this model. In particular the work material within a few microns of the tool surface becomes almost featureless at the highest magnifications, except for rigid inclusions and for grain boundaries which must have resulted from recrystallization after the chip separated from the tool (*Figure 3.20*).

From *Figure 5.5*, the material in any part of the flow-zone is strained continuously as it moves from the cutting edge to the position where it breaks contact with the tool - for example $cd \rightarrow c'd' \rightarrow c''d''$. The flow zone material is therefore continuously heated as it passes over the contact area, so that an increase in temperature can be expected away from the tool edge. This is different from the body of the chip, which is heated only on the shear plane and not further heated as it passes over the contact area.

Metals and alloys commonly machined are strengthened by plastic deformation, the yield flow stress usually conforming to the empirical relation

$$\sigma = \sigma_1 \varepsilon^n \qquad (5.5)$$

where n is a strain hardening coefficient which increases with the strain rate. This relationship clearly cannot hold for the extreme strain and strain rate conditions in the flow-zone, since the flow strength would be so high that strain would be transferred into the body of the chip where the yield stress is much lower. (This is what happens with poly-phase alloys at low cutting speeds and results in the formation of a built-up edge.)

The yield stress in the flow zone, where strains are on the order of 100 or greater and the strain rate is on the order of 10^4 s^{-1}, cannot be predicted by extrapolating data for *Equation 5.5*, which are usually obtained from laboratory tests where strain is less than 1. It is known that, over certain limits, the yield stress is lowered by further increments of strain and strain rate.[5] This can be explained by two factors:

(1) By adiabatic heating which raises the temperature to high values at which yield stress is reduced and structure is modified by recovery processes

(2) By structural changes in the work piece, brought about by extreme strain as well as by the temperature increase

At the cutting speeds typical of industrial practice, the typical structure in the flow-zone comprises very small (0.1 to 1 µm) equi-axed grains with few dislocations, as shown in electron micrographs (*Figure 3.23*). The result of this weakening of the work material at extreme strain and strain rate is the concentration of strain into the thin flow-zone - an essential feature of the cutting of many metals and alloys where seizure is normal at the tool/work interface during high speed cutting. The thickness of the flow-zone and the stress required to cause flow are characteristic of the particular metal or alloy being machined. They must depend on the temperatures achieved and the change in flow stress brought about by the very high strains and strain rates. This is specific for each work material and adequate data are not available for prediction of these stresses.

The concentration of strain in a narrow flow-zone in cutting is akin to phenomena described for other metal-working processes. Cottrell argues that "if the rate of plastic flow is sufficiently high, there may not be time to conduct away the heat produced from the plastic working and the temperature may rise sufficiently to soften the deforming material. A plastic instability is then possible in which intense, rapid flow becomes concentrated in the first zones to become seriously weakened by this effect... This adiabatic softening effect (has been demonstrated) in the perforation of steel plates with a flat ended punch. Slow indentations produced widespread plastic deformation with a large absorption of energy, but punching by impact enabled the plate to be perforated easily along a thin surface of intense shear deformation."[6]

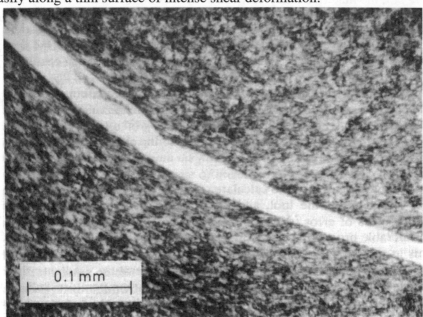

FIGURE 5.6   Thermo-plastic shear band in high tensile steel rope wire crushed by hammer blow [7]

An example of thermoplastic shear bands was observed by Trent 60 years ago when working on high tensile steel wire for wire ropes.[7] When flattened by a hammer on an anvil, a section through this wire showed shear bands at $45°$ to the striking direction, the plane of maximum shear stress. Through the shear band was a thin layer of martensite, 15 μm thick (*Figure 5.6*).

The heat generated by shear strain at a high strain rate, concentrated in the shear band, raised the temperature above $720°C$ (the transformation temperature of the steel) and the fine pearlitic structure was transformed to austenite. Because the very thin shear band was continuous with the body of the wire, this acted as a heat sink, cooling the shear zone so rapidly that the austenite was transformed to martensite. This experiment demonstrates characteristic features of thermoplastic shear bands, which are observed in industrial operations such as punching and blanking of sheet and plate. Very high temperatures are generated and the temperature cycle is very short, usually measured in milliseconds. The life cycle of thermoplastic shear bands is usually very short because the rapid strain in the band relieves the local stress, which inhibits further strain at this position.

The flow-zone at the tool/work interface in high speed metal cutting is a thermoplastic shear band. It is not, however, a transient structure, persisting for milliseconds, as those encountered in other metal working operations, and observed in the built-up edge in metal cutting. It is a body of metal maintaining its integrity for the duration of a cutting operation. High speed metal cutting is a very effective method of trapping a thermoplastic shear band at the tool surface. Many of the essential features of metal cutting depend on the way each work material behaves in such shear bands. Little is known of the behavior of metals and alloys and their constituent phases under these conditions, but high speed metal cutting provides a simple and very effective mechanism which can be used to study the influence of changes in composition and structure of metals and alloys when deformed in plastic instabilities of this type.

From the analysis based on the model in *Figure 5.5* an increase in temperature in the material in the flow-zone as it moves away from the cutting edge can be predicted. The temperature increase depends on the amount of work done and on the quantity of metal passing through the flow-zone. The thickness of the flow-zone provides some measure of the latter, and the thinner the flow-zone the higher the temperature would be for the same amount of work done.

The thickness varies considerably with the material being cut from more than 100 μm (0.004 in) to less than 12 μm (0.0005 in). It tends to be thicker at low speeds, but does not vary greatly with the feed. In general the flow-zone is very thin compared with the body of the chip - commonly of the order of 5% of the chip thickness. Since the work done at the tool rake face is frequently about 20% of the work done on the shear plane, much higher temperatures are found in the flow-zone than in the chip body, particularly at high cutting speed. There is very little knowledge at present of the influence of factors such as work material, tool geometry, or tool material on the flow-zone thickness, and this is an area in which research is required.

The temperature in the flow-zone is also influenced by heat loss through conduction. The heat is generated in a very thin layer of metal which has a large area of metallic contact both with the body of the chip and with the tool. Since the temperature is higher than that of the chip body, particularly in the region well back from the cutting edge, the maximum temperature in the flow-zone is reduced by heat loss into the chip. After the chip leaves the tool surface, that part of the flow-zone which passes off on the under surface of the chip, cools very rapidly to the temperature of the chip body, since cooling by metallic conduction is very efficient. The increase in temperature of the chip body due to heat from the flow-zone is slight because of the relatively large volume of the chip body.

The conditions of loss of heat from the flow-zone into the tool are different from those at the flow-zone/chip body interface because heat flows continuously into the same small volume of tool material. It has been demonstrated (Chapter 3) that the bond at this interface is often completely metallic in character, and where this is true, the tool will be effectively at the same temperature as the flow-zone material at the surface of contact. The tool acts as a heat sink into which heat flows from the flow-zone and a stable temperature gradient is built up within the tool. The amount of heat lost from the flow-zone into the tool depends on the thermal conductivity of the tool, the tool shape, and any cooling method used to lower its temperature.

The heat flowing into the tool from the flow-zone raises its temperature and this is the most important factor limiting the rate of metal removal when cutting the higher melting point metals.

5.3.3   Calculation of interface temperatures with a flow-zone

Many methods have been used to calculate temperatures and temperature gradients on the rake face of the tool.[8-13] Recent work using finite element analysis indicates considerable progress in calculating interface temperatures and temperature gradients in two dimensional sections through cutting tools. Comparisons of calculated temperature distribution with measured temperature gradients show considerable promise.[12]

The amount of heat generated at the tool/work interface can be estimated from force and chip thickness measurements (*Equation 5.3*). The temperature of the *body of the chip* can be calculated from knowledge of heat generated on the shear plane because there is little error in assuming an even distribution of strain across the shear plane and neglecting heat loss during the short time interval involved.

Boothroyd[3] has considered an element in the *bulk* of the chip through which heat is conducted from the primary and secondary regions and through which materials flow. Weiner's[13] assumption that the conduction of heat in the chip direction (x-direction in *Figure 5.7*) is small in comparison to the transfer due to motion is then made so that in the body of the chip:

$$V_c \frac{\delta T}{\delta x} = K \frac{\delta^2 T}{\delta y^2}$$

(5.6)

Figure 5.7 shows the boundary conditions for the analysis, K is the thermal diffusivity, $\delta T / \delta x$ is the rate of change of temperature in the direction of chip flow, the y direction is perpendicular to the rake face and V_c is the bulk chip speed. (Note that $K = k_t / \rho c$ where k_t is thermal conductivity, ρ is density and c is specific heat).

This equation also applies to an element in the secondary shear zone but there is an additional heat generation term for this region of plastic deformation. Expressed in terms of the secondary shear stress, k_s, the strain rate in secondary shear, $\dot{\gamma}_s$, the rate of doing work on an element of length Δx and unit cross section is $(k_s \dot{\gamma} \Delta x)$. The average temperature rise is therefore proportional to $(k_s \dot{\gamma} \Delta x)/V_A$ where V_A is the average velocity through the element. Considering the total thickness, h, of the zone the velocity varies from zero at the interface (due to the seizure) to V_c the bulk chip speed. The simplification is therefore made that material travels through the centre of the zone $(y = h/2)$ at a speed of $V_c/2$. If all the plastic work were converted into heat, the rate of change of temperature along the seizure contact would be

$$\frac{\delta T}{\delta x} = \frac{2 k_s \dot{\gamma}_s}{\rho c V_c} + \frac{K}{V_c} \frac{\delta^2 T}{\delta y^2}$$

(5.7)

In fact this energy balance equation has also been used by Ramalingam and Black[14] for the calculation of temperature in shear lamellae in the primary zone. Although $k_s \dot{\gamma}_s$ and h will vary along the length of the secondary shear zone, a first approximation to the solution of the above equation may be made by making these constant and assuming that *all* the material in the secondary shear zone travels at $V_c/2$ to make $\delta T/\delta x$ constant between $y = 0$ and h.

The equation may be integrated once over the thin secondary shear layer and then the simplification made that conduction of heat from the center of the zone is uniform and symmetrical so that $(\delta T)/\delta y_{y\,=\,0} = -(\delta T)/(\delta y)_{y\,=\,h}$ and

$$h\frac{\delta T}{\delta x} = \frac{2k_s\dot{\gamma}_s h}{\rho c V_c} + \frac{2K}{V_c}\left(\frac{\delta T}{\delta y}\right)_{y\,=\,h} \tag{5.8}$$

This equation has been solved by assuming that the temperature T, and temperature gradient, $\delta T/\delta y_{y\,=\,h}$, arising along the secondary shear zone, are boundary values for the bulk of the chip at $y \approx 0$. It can be shown that the approximate temperature along the chip-tool interface, **in the region of seizure**, is given by the equation below. Further refinements allow the heat source in the sliding area to be considered.[15]

$$T(x, 0) = \frac{2k_s\dot{\gamma}_s h}{\rho c}\left(\frac{x}{\pi K V_c}\right)^{\frac{1}{2}} \tag{5.9}$$

This equation can be used to calculate $T(x, 0)$ as a function of distance from the cutting edge, and is added to the primary zone temperature. Results that compare the calculation with experimental results from the techniques in Section 5.6.4 are in *Figure 5.8*.

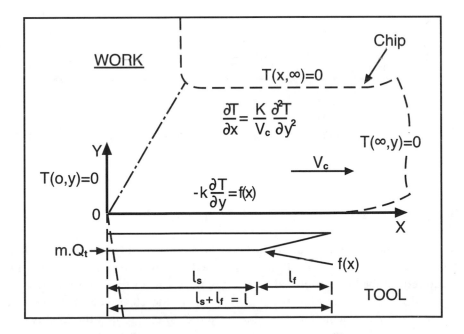

FIGURE 5.7   Summary of the appropriate boundary conditions, equations and the nature of the heat source, f(x), for the heat transfer problem taking into account the secondary shear zone

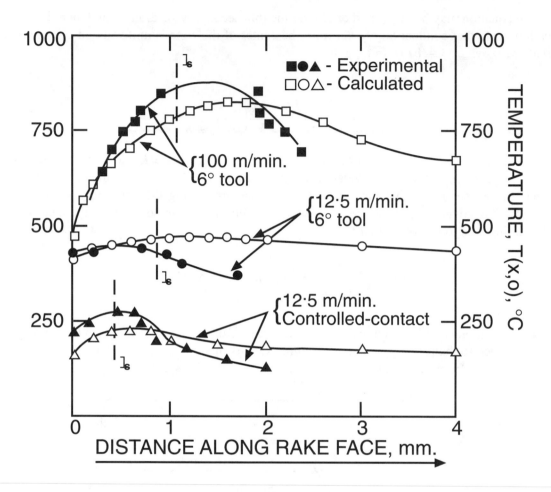

FIGURE 5.8 Comparisons of experimental and theoretical temperature distributions along the chip-tool interface. Results for flat faced tools at two different speeds and results for controlled contact tools are shown. The experimental sticking friction contact length is also shown (l_s). Machining annealed low carbon iron (0.07% carbon)

5.4 HEAT FLOW AT THE TOOL CLEARANCE FACE

The flow-zone on the rake face is an important heat source because the chip is relatively flexible and the compressive stress forces it into contact with the tool over a long path. In some cases it is possible to reduce the heat generated by altering the shape of the tool to restrict the length of contact, *Figure 4.11*. The same objective is achieved on the clearance face of the tool by the clearance angle, *Figure 2.2*, which must be large enough to ensure that the freshly cut surface is separated from the tool face and does not rub against it. The heat generated by formation of the new surface, *Figure 5.1*, is dissipated by conduction into the workpiece and has little heating effect on the tool.

As mentioned in Chapter 3, Wallbank obtained evidence that, even with a sharp tool, when cutting under conditions where a flow-zone is present, the work material may be in contact with the clearance face for a distance of the order of 0.2 mm below the cutting edge of a tool with a conventional clearance angle of about 6°. The workpiece is much more rigid than the chip and the feed force is too low to deflect it to maintain contact with the clearance face for a length greater than this. This contact length is too short for the generation of high temperatures in a flow-zone on the clearance face, as long as the tool remains sharp and the clearance angle is large.

With certain types of tooling, for example form tools or parting-off tools, a large clearance angle would seriously weaken the tool or make it too expensive, and clearance angles as low as 1° are employed. With such small clearance angles there is a risk of creating a long contact path on the clearance face which becomes a third heat source, similar in character to the flow-zone on the rake face. Even with normal clearance angles, prolonged cutting results in flank wear, in which a new surface is generated on the tool more or less parallel to the direction of cutting. The work material is often seized to this 'wear land' as to the rake face of the tool, *Figure 3.7*, and when the worn surface is long enough, the flow-zone in this region becomes a serious heat source.

Temperatures generated at the worn surface may be higher than on the rake surface of the same tool because the work material moves across this surface at the cutting speed of the operation, while the chip speed over the rake face may be a half or a third of this speed, or lower. Generation of high temperatures in this region is usually followed immediately by collapse of the tool. Thus, these preceding discussions have practical importance which should be considered by tool designers, especially for parting-off and form tools.

5.5 HEAT IN AREAS OF SLIDING

When cutting at very low speeds there may be sliding rather than seizure at the interface. At higher cutting speeds there may be sliding or intermittent interfacial contact between work and tool materials at peripheral regions of the contact area (*Figure 3.15*). Where the conditions at the tool/work interface are those of sliding contact, the mode of heat generation is very different. The shearing of very small, isolated metallic junctions provides numerous very short-lived temperature surges at the interface. This is likely to give rise to a very different temperature pattern at the tool/work interface compared with that of the seized flow-zone.

It seems probable that the mean temperatures will be lower, because a much smaller amount of work is required to shear the isolated junctions, but localized high temperatures could also occur at any part of the interface. There is likely to be more even temperature distribution over the contact area than under conditions of seizure. The direct effects of heat generated at sliding contact surfaces have to be taken into consideration in a complete account of metal cutting. However, when considering the range of cutting conditions under which problems of industrial machining arise, the effects of heat from areas of sliding contact can probably be neglected without seriously distorting the understanding of the machining process.

5.6   METHODS OF TOOL TEMPERATURE MEASUREMENT

The difficulty of calculating temperatures and temperature gradients near the cutting edge, even for very simple cutting conditions, gives emphasis to the importance of methods for measuring temperature. This has been an important objective of research and some of the experimental methods explored are now discussed.

5.6.1   Tool/work thermocouple

The most extensively used method of tool temperature measurement has employed the tool and the work material as the two elements of a thermocouple.[16] The thermoelectric e.m.f. generated between the tool and workpiece during cutting is measured using a sensitive millivoltmeter. The hot junction is the contact area at the cutting edge, while an electrical connection to a cold part of the tool forms one cold junction. The tool is electrically insulated from the machine tool (usually a lathe). The electrical connection forming the cold junction with the rotating workpiece is more difficult to make, and various methods have been adopted, a form of slip-ring often being used.

Care must be taken to avoid secondary e.m.f. sources such as may arise with tipped tools, or short circuits which may occur if the chip makes a second contact with the tool, for example on a chip breaker. The e.m.f. can be measured and recorded during cutting, and, to convert these readings to temperatures, the tool and work materials, used as a thermocouple, must be calibrated against a standard couple such as chromel-alumel. Each different type of tool and work material used must be calibrated.

There are several sources of error in the use of this method. The tool and work materials are not ideal elements of a thermocouple: the e.m.f. tends to be low, and the shape of the e.m.f.-temperature curve is far from a straight line. It is doubtful whether the thermo-e.m.f. from a stationary couple, used in calibration, corresponds exactly to that of the same couple during cutting when the work material is being severely strained. The tool/work thermo-couple method has nevertheless been used successfully by many workers to investigate specific areas of metal cutting - for example, to compare the machinability of different work materials, the effectiveness of coolants and lubricants or the performance of different tool materials.

As an example, Kurimoto and Barrow cut a low-alloy engineering steel using a steel cutting grade of cemented carbide tool over a range of speeds and feeds in air and using different cutting lubricants.[17] *Figure 5.9* shows the temperature determined, plotted against cutting speed for three different values of feed, and a comparison of cutting in air and with water as a cutting fluid. The results, the mean of 3 to 12 measurements, are consistent. They indicate an increase in temperature with increments in speed and feed, and a small lowering of temperature when water was used as a coolant. Temperatures over 1000°C are recorded for speeds of 120 m min^{-1}(400ft/min) and above - conditions commonly used in machine shops. This method is useful in demonstrating the influence of such variable parameters, but the significance to attach to the numerical values for temperature which it provides is uncertain.

It is demonstrated later that there are very steep temperature gradients across the contact area when machining steel at high speed, with temperature differences of 300°C or greater at different positions. It is uncertain whether the tool/work thermo-couple measures the lowest temperature at the interface or a mean value. It is unlikely that it records the highest temperature in the contact area.

While this method has been used effectively to study the influence of parameters in specific areas of machining, there does not seem to have been any attempt to correlate the results to give a unified picture of interface temperature in relation to speed, feed, tool design and cutting conditions for a wide range of work materials.

Errors arising from uncertain calibration of the thermocouple can be partially eliminated by using two different tool materials to cut the same bar of work material, simultaneously, under the same conditions.[18] The e.m.f. between the two tools is measured. If it can be assumed that the temperatures at the interfaces of the two tools are the same, then the temperature can be determined from the measured e.m.f. by calibration of a thermocouple of the two tool materials. Few results have been reported using this method and, at best, a mean temperature at the contact area can be determined.

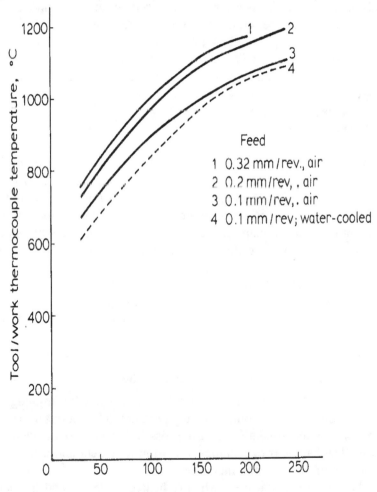

FIGURE 5.9   Tool temperature (tool/work thermocouple); work material is a low alloy engineering steel and the tool is a cemented carbide (Data from Kurimoto and Barrow[17])

5.6.2   Inserted thermocouples

Measurement of the *distribution* of temperature in the tool has been the objective of much experimental work. A simple but very tedious method is to make a hole in the tool and insert a thermocouple in a precisely determined position close to the cutting edge.[19] This must be repeated many times with holes in different positions to map the temperature gradients. The main error in this method arises because the temperature gradients near the edge are very steep, and holes large enough to take the thermocouple overlap a considerable range of temperature. If the hole is positioned very precisely, this may be a satisfactory method for comparing the tool temperature when cutting different alloys, but the method is not likely to be satisfactory for determining temperature distribution.

5.6.3   Radiation methods

Several methods have been developed for the measurement of radiation from the heated areas of the tool. In some machining operations, the end clearance face is accessible for observation, for example in planing a narrow plate. The end clearance face may be photographed, using film sensitive to infra-red radiation. The heat image of the tool and chip on the film is scanned using a micro-photometer, and from the intensity of the image the temperature gradient on the end face of the tool is plotted.[20] The results show that the temperature on this face is at a maximum at the rake surface at some distance back from the cutting edge, while the temperature is lower near the cutting edge itself. This method gives information only about the temperature on exposed surfaces of the tool.

A more refined but difficult technique using radiant heat measurement involves making holes, either in the work material or in the tool to act as windows through which a small area (e.g. 0.2 mm dia.) on the rake surface of the tool can be viewed.[4] The image of the hot spot is focussed onto a PbS photo-resistor which can be calibrated to measure the temperature. By using tools with holes at different positions, a map can be constructed showing the temperature distribution on the rake face. A few results using this method have been published showing temperatures when cutting steel at high speeds using carbide tools. These show very high temperatures (up to 1200°C) and steep gradients with the highest temperature well back from the cutting edge on the rake face. This method requires very exacting techniques to achieve a single temperature gradient, and it has been used to estimate the temperature on the flank of carbide tools used to cut steel.[21]

5.6.4   Changes in hardness and microstructure in steel tools

Much more information about temperature distribution near the cutting edge of tools may be obtained by using the tool itself to monitor the temperature. The room temperature hardness of hardened steel decreases after re-heating, and the loss in hardness depends on the temperature and time of heating. The hardness of a hardened carbon tool steel and a fully heat-treated high speed steel is a function of the re-heat temperature.

Carbon tool steels start to lose hardness when re-heated to 250°C and the hardness is greatly reduced after heating to 600°C. Fully heat-treated high-speed steel tools are not softened appreciably until 600°C is exceeded. Between 600 and about 850°C the hardness falls rapidly, but it rises again at still higher re-heat temperatures if the steel is rapidly cooled, as it is in the heat affected zone of the tools. By calibrating the hardness against the temperature and time of heating, a family of curves can be obtained for any tool steel. Then, if the hardness in any heat-affected region is measured and the time of heating is known, the temperature to which it was heated can be determined by comparing the data in the worn tool areas with the standardized calibration charts.[22-26]

This method has been used to estimate the temperature in critical parts of gas turbines, piston engines and other structures where thermocouples cannot be readily employed. Plugs of a calibrated tool steel are inserted and after use are removed for hardness testing.[22] For temperature measurement in cutting tools, clamped tool tips are used and hardness tests are carried out after machining for a known time. The heat-affected region occupies a few cubic millimeters near the cutting-edge.

The tool may be sectioned through the cutting edge, polished and a series of hardness tests made to cover the heat-affected region (*Figure 5.10*) or the rake surface may be polished and tested. Since the heated area is small and temperature gradients may be steep, the indentations must be closely-spaced, e.g. 0.1 mm apart. The indentations must therefore be small and an accurate micro-hardness machine must be used. The hardness tests in *Figure 5.10a* were made using a Vickers pyramidal diamond at 300 gms load. Using this method the temperature at any position can be estimated within $25°C$ within the temperature range where the tools are softened.

Carbon steel tools are useful for determining tool temperatures when cutting non-ferrous metals of low melting point, such as copper and its alloys, but are of little use for temperature estimation when cutting steel. High-speed steel tools have been used for determining temperatures when cutting steel, nickel alloys, titanium alloys and other high melting point materials.

Temperatures of about $600°C$ occur frequently when cutting steel with high speed steel tools. Temperatures as low as this cannot be determined using a fully heat-treated high-speed steel tool but, with tools tempered at a low temperature after hardening (e.g. $400°C$), the hardness first increases up to $600°C$ and then decreases at higher temperatures.

Micro-hardness measurements on tools heat treated in this way can be used to estimate temperatures in the tool edge region which are often about $500-650°C$. *Figure 5.10b* shows the temperature contours derived from the hardness indentations in *Figure 5.10a*. In this example, a medium carbon steel had been cut at 27 m min^{-1} (90 ft/min). The maximum temperature on the rake face was 700°C at a position about 0.75 mm from the cutting edge.

The hardness decrease after heating hardened steel tools is the result of changes in microstructure. The structural changes can be observed by optical and electron microscopy. With carbon steels and some high speed steels these changes are usually too gradual to permit positive identification of a structure characteristic of a narrow temperature range, but Wright demonstrated that, with certain high speed steels (e.g. cobalt containing steels such as M34 or M42), distinct modifications to structure occur at approximately 50°C intervals between 600 and 900°C, which permits measurement of temperature with an accuracy of approximately $25°C$ within the heat affected region.[24]

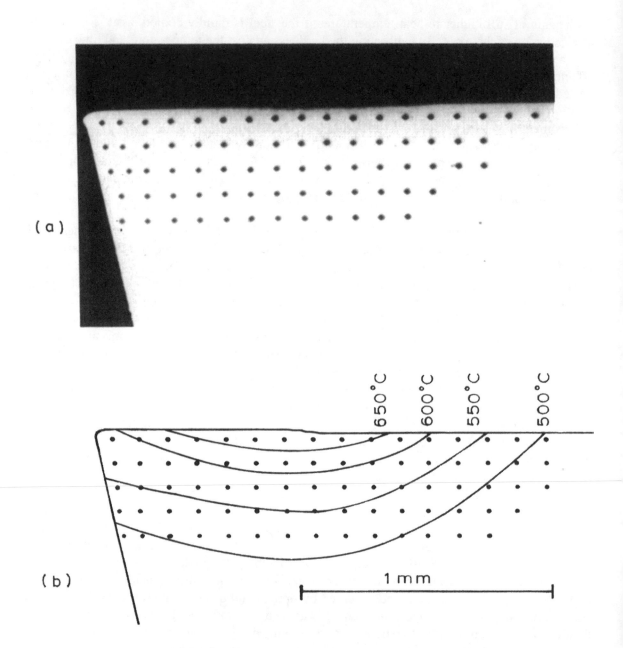

FIGURE 5.10 Section through tool used to cut medium carbon steel at 27 m min^{-1}(90 ft/min) and 0.25 mm/rev feed (a) Photomicrograph showing micro-hardness indentations (b) Isotherms determined by hardness measurements

Figure 5.11a is a photomicrograph of an etched section through the cutting edge of a Type M34 high-speed steel tool used to cut a very low carbon steel at 183 m min^{-1} (600 ft/min) at a feed rate of 0.25 mm/rev for a time of 30 s. The polished section was etched in 2% Nital (a solution of HNO$_3$ in alcohol). The darkened area forming a crescent below the rake face defines the volume of the tool heated above 650°C during cutting.

The structures within this region are shown at higher magnification in *Figure 5.12* with the temperature corresponding to each structure. The light area in the center of the heat-affected region had been above 900°C. The structure became austenitic and this small volume of the tool was re-hardened when cutting stopped and it cooled very rapidly to room temperature. The tool rake surface can be re-polished after use and etched to determine temperature distribution in three dimensions in the tool.

Figure 5.13a shows the structural changes just below the rake face of a tool used under the same conditions of cutting. *Figure 5.11a* is thus a section along the line O-B´ in *Figure 3.17* and *5.13a*. *Figures 5.11b* and *5.13b* show the temperature contours derived from the structural changes.

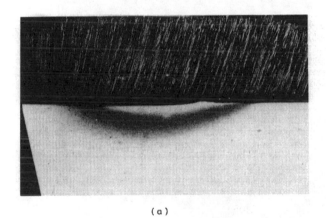

(a)

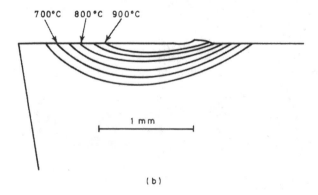

(b)

FIGURE 5.11 (a) Section through tool used to cut very low carbon steel at high speed, etched in Nital to show heat-affected region (b) Temperature contours derived from structural changes in (a)[24]

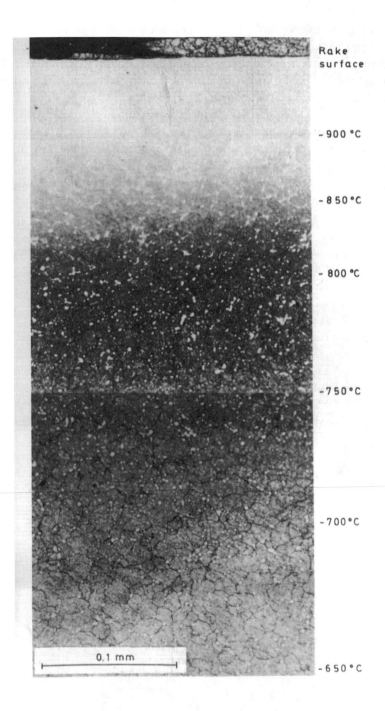

FIGURE 5.12 Structural changes and corresponding temperatures in high speed steel tool[24]

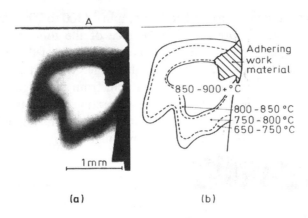

FIGURE 5.13 (a) Rake face of tool used to cut very low carbon steel at high speed, etched to show heat-affected region; (b) temperature contours derived from structural changes in (a)[26]

The micro-hardness method is time consuming and requires very accurate hardness measurement. The structural change method requires experience in interpretation of structures but, where it can be used, it is as accurate and much more rapid. In conjunction with metallographic studies of the interface, important information can be gained which is lost using micro-hardness alone.

The main limitations of both methods are that they can be used only within the cutting speed limitations of steel tools and where relatively high temperatures are generated. Commercial cemented carbide tools can be used at much higher cutting speeds. When heated at temperatures up to their melting point (about 1300°C) and cooled, normal cobalt-bonded carbides do not undergo observable changes in hardness or structure. Certain iron-bonded cemented carbides, however, do undergo structural change at about 800°C.

Dearnley has given metallographic evidence of a sharp structural change in the heat-affected regions of iron-bonded carbide tools, which can be used as a temperature contour.[25] This opens the possibility of temperature measurements in tools used to cut high melting-point materials at much higher speeds. Knowing the character of the heat source, it should be possible, starting from the estimated temperature distribution at the tool rake face, to treat the subject as a heat-flow problem and greatly extend the existing knowledge of temperature distribution near the edge of cutting tools.

5.7 MEASURED TEMPERATURE DISTRIBUTION IN TOOLS

Using the micro-hardness and metallographic methods, a number of investigations of temperature distribution in cutting tools have been made.[23-26] Results of both theoretical importance and practical interest have been demonstrated. Temperature distribution in tools used to cut different work materials is dealt with in relation to machinability in Chapter 9. Here, the influence of cutting speed and feed is discussed on the basis of tests using one work material - a very low carbon steel (0.04% C). *Figure 5.14* shows sections through tools used to cut steel at a feed of 0.25 mm

(0.010 in) per rev, at cutting speeds from 91 to 213 m min^{-1} (300 to 700 ft/min) for a cutting time of 30 s. The maximum temperature on the rake face of the tool rose as the cutting speed increased, while the hot spot stayed in the same position. Even at the maximum speed a cool zone (below 650°C) extended for 0.2 mm from the edge, while the maximum temperature, approximately 1000°C, is at a position just over 1 mm from the edge. This demonstrates the presence of very high temperatures and very steep temperature gradients which can be present in the tool at the rake face when machining high melting-point metals and alloys under conditions where the heat source is a thin flow-zone on the chip's underside.

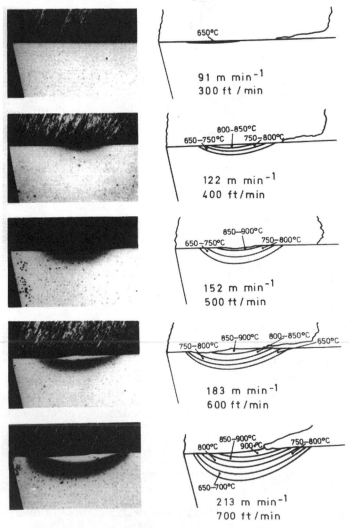

FIGURE 5.14 Temperature contours in tools used to cut very low carbon steel at a feed of 0.25 mm (0.010 in) per rev, and at speeds shown for a cutting time of 30 s[26]

As has already been argued, where the flow-zone is metallurgically bonded to the tool surface, there is no barrier to heat conduction, and the temperature at the rake surface of the tool is essentially the same as that in the flow-zone at any position on the interface. The temperature in the flow-zone cannot readily be estimated from structural changes in the work material because these are confused by the enormous strains to which it is subjected, but both Loladze[27] and Hau-Bracamonte[28] have given evidence that austenite may be formed in the flow-zone when cutting steel - i.e. the temperature may exceed 720°C.

Since the flow-zone is the heat source, the temperature in the hottest position within this zone must be slightly higher than the highest temperature in the tool, but the difference is probably small. The cool edge of the tool is of great importance in permitting the tool to support the compressive stress acting on the rake face. There is no experimental method by which the stress distribution can be measured under the conditions used in these tests, but the general character of stress distribution, with a maximum at the cutting edge, has already been discussed (*Figure 4.14*).

As shown diagrammatically in *Figure 5.15*, the maximum compressive stress is supported by that part of the tool which remains relatively cool under the conditions of these tests.

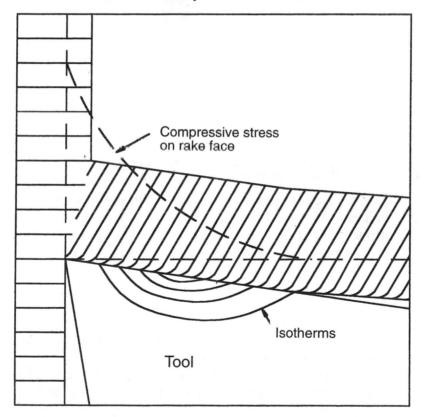

FIGURE 5.15 Temperature and stress distribution in tool used to cut steel at high speed

Heat flows within the tool from the hot spot towards the cutting edge, but this region is cooled by the continual feeding in of new work material. In this series of tests, when the cutting speed was raised above 213 m min^{-1} (700 ft/min), the heated zone approached closer to the edge and the tool failed, the tool material near the edge deforming and collapsing as it was weakened.

As the tool deformed, *Figure 5.16a*, the clearance angle near the edge was eliminated, and contact was established between tool and work material down the flank, forming a new heat source, *Figure 5.16b*. Very high temperatures were quickly generated at the flank, the tool was heated from both rake and flank surfaces, and tool failure was sudden and catastrophic.

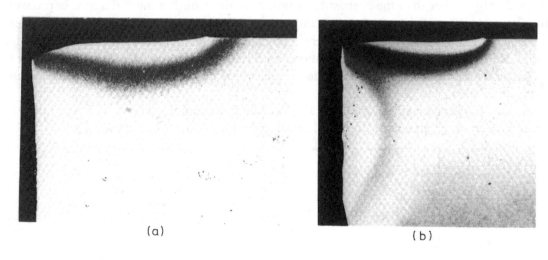

(a)   (b)

FIGURE 5.16 Stages in tool failure (a) High temperature region spreads to cutting edge; (b) second high temperature region generated at worn clearance face

The wear on the clearance face of cutting tools often takes the form of a more or less flat surface - the 'flank wear land'- over which the clearance angle is eliminated (e.g. *Figure 3.7*). A flow-zone at this interface can become a heat source, causing catastrophic failure, if the wear land is allowed to become too large, and it is often recommended that tools be re-ground or replaced when the wear land on the tool reaches some maximum depth, e.g. 0.75 mm or 1.5 mm (0.03 or 0.06 in).

When cutting this low carbon steel, the temperature gradients were established quickly. *Figure 5.14* shows the gradients after 30 s cutting time and they were not greatly different when cutting time was reduced to 10 s. Further experimental work is required to determine temperatures after very short times, such as those encountered in milling.

A few investigations have been made on the influence of feed on temperature gradients. *Figure 5.17* shows temperature maps in tools used to cut the same low carbon steel at three different feeds: 0.125, 0.25, and 0.5 mm (0.005, 0.010, and 0.020 in) per rev. The maximum temperature increased as the feed was raised at any cutting speed. The influence of both speed and feed on the temperature gradients is summarized in *Figure 5.18*, in which the temperature on the rake face of the tool is plotted against the distance from the cutting edge. The conditions are for three different speeds at the same feed, and for three feeds at the same speed.

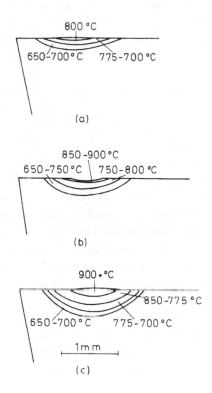

FIGURE 5.17 Influence of feed on temperatures in tools used to cut iron at feeds of (a) 0.125 mm (0.005in); (b)0.25mm (0.010in); and (c) 0.5 mm (0.020 in) per rev

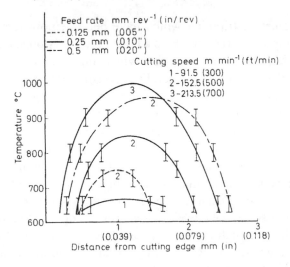

FIGURE 5.18 Temperature distribution on rake face of tools used to cut very low carbon steel at different speeds and feeds[26]

The main effect of increasing feed appears to be an increase in length of contact between chip and tool, with extension of the heated area further from the edge and deeper below the rake face, accompanied by an increase in the maximum temperature. The heat flow back toward the cutting edge was increased, and this back flow of heat gradually raised the temperature of the edge, as the feed was increased.

There was, however, only a relatively small change in the distance from the edge to the point of maximum temperature when the feed was doubled, and the temperature at the edge was only slightly higher. The highly stressed region near the cutting edge must extend further from the edge as the feed is raised, and this, together with the small increase in edge temperature, sets a limit to the maximum feed which can be employed in practice.

To understand tool behavior, the temperature distribution in three dimensions must be considered. This is illustrated by *Figure 5.13* which shows the temperature map of the rake face of a tool used to cut the same steel. The low temperature region is seen to extend along the whole of the main cutting edge, including the nose radius. The light area in the center of the heated region is the part heated above 900°C, while the sharply defined dark area extending from the end clearance face is a layer of work material filling a deformed hollow on the tool surface.

The high temperature region is displaced from the center line of the chip towards the end clearance face. This is because the tool acts as a heat sink into which heat is conducted from the flow-zone. Higher temperatures are reached at the end clearance because the heat sink is missing in this region. With the design of tool used for this test, the only visible surface of the tool heated above red heat was thus at the end clearance face just behind the nose radius. Tool failure had begun at this position because the compressive stress was high where an unsupported edge was weakened by being heated to nearly 900°C.

5.8   RELATIONSHIP OF TOOL TEMPERATURE TO SPEED

The patterns of temperature distribution which have been demonstrated for the low carbon steel, both in the low cutting speed region where a built-up edge is present, and at higher speeds where a flow-zone exists, have been found to occur when cutting all steels so far tested including carbon, low alloy and stainless steels. The general relationship between temperature distribution at the tool/work interface and cutting speed can be summarized in a graph.

Figure 5.19 shows temperature at two positions on the interface - at the cutting edge and about 1 mm from the edge on the rake face - as a function of cutting speed.[29] At lower speeds, in the built-up edge region, where plastic deformation of the chip is dominated by a mechanism of dislocation movement, the temperature rises rapidly as cutting speed is increased and the highest temperature is at the cutting edge. At a critical speed range - A in *Figure 5.19* - the mode of shear strain near the tool/work interface changes. A thermo-plastic shear band becomes established at the tool rake face. Within this band, the mode of shear strain is dominated by recovery processes.

With this mode of plastic deformation strain hardening ceases and the temperature *at the cutting edge* increases much more slowly with cutting speed. The temperature *away from the tool edge* in the direction of chip flow, continues to increase rapidly as shown in *Figure 5.19*. This transition takes place in a critical temperature range - B in *Figure 5.19* which is characteristic of the work material. It is the temperature at which strain hardening is replaced by dynamic recovery where the material is subjected to extreme strain at the high strain rates encountered in this region.

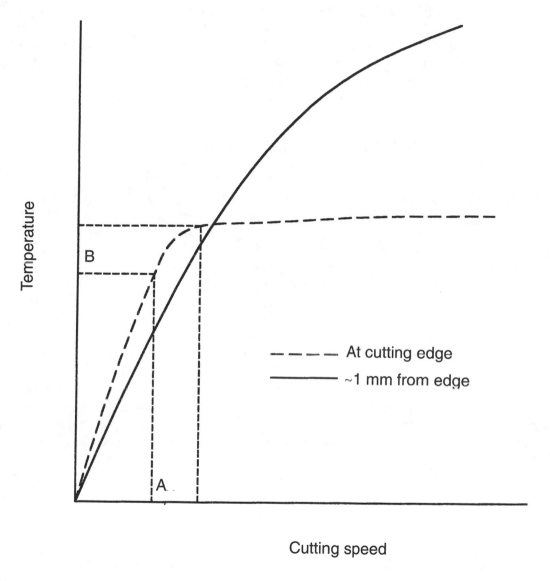

FIGURE 5.19 Diagram showing typical temperature distribution in tools as a function of cutting speed, when cutting steel and other metals and alloys

Many different steels in different heat treatment conditions have been subjected to this test program. It is remarkable that, when machining all of these, the critical temperature, B, has been nearly the same - in the range 600-675°C. Since this temperature is related to recovery processes in the strained work material, it is probably a function of the melting point of the metal iron, being the major element in all steels. The temperature range 600-675°C is a factor of 0.48 to

0.52 of the melting point of iron in Kelvin. While the critical temperature - B in *Figure 5.19* - is nearly the same in machining all steels, the critical cutting speed, A, at which this temperature is reached, varies greatly depending on the composition and heat treatment of the steel. This is discussed in Chapter 9.

The relationship in *Figure 5.19* has been formulated as a result of many cutting tests on several types of steel. If cutting conditions are carefully controlled, the results of temperature measurement tests are very consistent. The relationships appear to hold for machining of other metals and their alloys and much more testing will be valuable. This also is discussed in Chapter 9.

5.9   RELATIONSHIP OF TOOL TEMPERATURE TO TOOL DESIGN

The temperature distribution in tools is influenced by the geometry of the tool, for example by the rake angle, by the angle between end and side clearance faces or by the nose radius. An understanding of temperature distribution makes possible more rational design of tools, and this is an area of continuing investigation.[12,15]

It is evident from *Figure 4.11* that the artificial reduction of the chip-tool contact length substantially reduces the power consumed by the cut (given by F_cV). Boothroyd[3] and Chao and Trigger[9] have presented similar results. As a result, the heat generated is reduced and consequently it was found that none of the reduced-contact length M34 tools from the cutting force tests showed the distinctive tempered structures shown in *Figure 5.11a*. Thus the temperature contours for the reduced-contact tools were determined by machining with 01 and D2S steel inserts that "soften-off" at much lower temperatures and hence cutting speeds.

The experimental data in *Figure 5.20* show that over the speed range 10-30 m min^{-1} the artificial reduction of the chip tool contact length reduced the maximum temperatures by ~30 per cent (e.g. ~150°C at 20 m min^{-1}). The curves obtained have similar gradients over the limited range investigated. However, since no tempering effects were found at any speed in the M34 controlled-contact tools this curve must decrease in slope to give T < 650-700°C in the range 100-175 m min^{-1}. (This is suggested in the broken line extrapolation in *Figure 5.20*). Again, this is equivalent to at least a 30 per cent reduction in maximum temperature. As a result of the reduced temperatures it was found that for a particular tool material type and cutting speed, the reduced contact inserts exhibited a longer tool life than the conventional 6 degree rake tools. For example, when machining at 175 m min^{-1} with the M34, 6 deg tools the cutting edge collapsed after 60 s of cutting. By contrast the reduced-contact length M34 tools continued to cut for 10 minutes at 175 m min^{-1}.

Although no data have yet been obtained that would lead to Taylor equations[1] for the two tool types it is clear that the reduction of the contact length reduces the local stresses and temperature at the edge leading to an improved tool life (*Figure 5.21*).

It is emphasized that this result is for the steady state turning of a ductile low carbon iron and that similar conclusions may not hold for interrupted cuts. For other work materials there is likely to be an optimum side rake face design which can keep cutting temperatures to a minimum and maintain a strong cutting edge.

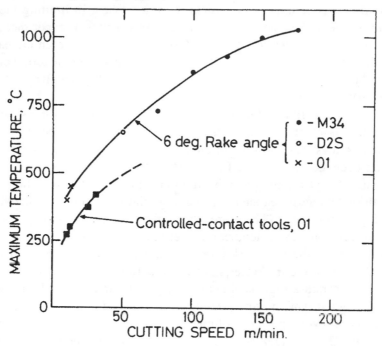

FIGURE 5.20 Maximum rake face temperatures for two tool geometries. The tool materials used are also noted.

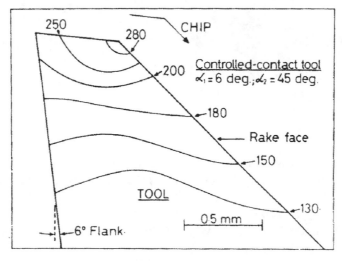

FIGURE 5.21 Temperature contours in a controlled contact tool after machining low carbon iron at 12.5 m min^{-1}

5.10 CONCLUSION

A major objective of this chapter is to explain the role of heat in limiting the rate of metal removal when cutting the higher melting point metals. In Chapter 4, experimental evidence is given demonstrating that the *forces* acting on the tool *decrease*, rather than increase, as the cutting speed is raised, and there is no reason to think that the *stresses* on the tool increase with cutting speed (unless the chip-tool contact area becomes considerably smaller, and even then, the evidence indicates that forces continue to fall with increasing speed).

Temperatures at the tool/work interface do, however, increase with cutting speed and it is this rise in temperature which sets the ultimate limit to the practical cutting speed for higher melting point metals and their alloys. The most important heat source responsible for raising the temperature of the tool has been identified as the flow-zone where the chip is seized to the rake face of the tool. The amount of heat required to raise the temperature of the very thin flow-zone is a small fraction of the total energy expended in cutting, and the volume of metal heated in the flow-zone may vary considerably. Therefore, there is no direct relationship between cutting forces or power consumption and the temperature near the cutting edge.

Very high temperatures at the tool/work interface have been demonstrated. The existence of temperatures over $1000^{\circ}C$ at the interface is not obvious to the observer of the machining process, since the high temperature regions are completely concealed, and it is rare to see any part of the tool even glowing red. The thermal assault on the rolls and dies used in the hot working of steel appears much more severe, but the tool materials used for these processes are quite inadequate for metal cutting. The cutting of steel in particular has stimulated development of the most advanced tool materials because it subjects the critical cutting edge of the tools to the high stresses which characterize cold-working operations, such as cold forming or wire drawing, and simultaneously to the high temperatures imposed by hot-working processes.

5.11 REFERENCES

1. Taylor, F.W., *Trans. A.S.M.E.*, **28**, 31 (1907)
2. Nicolson, J.T., *The Engineer*, **99**, 385 (1905)
3. Boothroyd, G., *Fundamentals of Metal Machining and Machine Tools*, Marcel Dekker (1990)
4. Lenz, E., S.M.E. *Ist International Cemented Carbide Conference, Dearborn*, Paper No. MR 71-905 (1971)
5. Holzer, A.J. and Wright, P.K., *Mat. Sci. Eng.*, **51**, 81 (1981)
6. Cottrell, A.H., *Conf. on Props. of Materials at High Rates of Strain*, p.3, Inst. Mech. Eng., London (1957)
7. Trent, E.M., *J.I.S.I.*, **1**, 401 (1941)
8. Loewen, E.G. and Shaw, M.C., *Trans. A.S.M.E.*, **76**, 217 (1954)
9. Chao, B.T. and Trigger, K.J., *Trans. A.S.M.E.*, **80**, 311 (1958)
10. Child, T.H.C., Maekawa, K. and Maulik, P., *Mat. Scit. & Tec.*, **4**, 1005 (1988)
11. Usui, E., Shirakashi, T and Kigawa, T., *Wear*, **100**, 129 (1984)

12. Stevenson, M.G., Wright, P.K., and Chow, J.G., *J. of Eng. for Industry* (ASME), **105**, 149 (1983). For the original paper on the FEA work see Tay, A.O., Stevenson, M.G., de Vahl Davis, G., *Proc. Inst. of Mech. Engrs.*, **188**, 627, (1974)
13. Weiner, J.H., *Trans., ASME.*, **77**, 1331, (1955)
14. Ramalingam, S. and Black, J.T., *Metallurgical Trans.*, **4**, 1103, (1973)
15. Wright, P.K., McCormick, S.P., and Miller T.R., *J. of Eng. for Industry* (ASME), **102**, (2), 123 (1980)
16. Braiden, P.M., *Proc. Inst. Mech. Eng.*, **182** (3G), 68 (1968)
17. Kurimoto, T. and Barrow, G., *Annals of CIRP*, **31**, (1), 19 (1982)
18. Pesante, M., *Proceedings of Seminar on Metal Cutting O.E.C.D.*, Paris 1966, 127 (1967)
19. Kuidsters, K.J., *Industrie Anzeiger*, **89**, 1337 (1956)
20. Boothroyd, G., *Proc. Inst. Mech. Eng.*, **177**, 789 (1963)
21. Chao, B.T., Li, H.L. and Trigger, K.J., *Trans. A.S.M.E.*, **83**, 496 (1961)
22. Belcher, P.R. and Wilson, R.W., *The Engineer*, **221**, 305 (1966)
23. Trent, E.M. and Smart, E.F., *Materiaux et Techniques*, Aug - Sept, 291 (1981)
24. Wright, P.K. and Trent, E.M., *J.I.S.I.*, **211**, 364 (1973)
25. Dearnley, P. and Trent, E.M., *Metals Tech.* **9**, (2), 60 (1982)
26. Smart, E.F. and Trent, E.M., *Int. J. Prod. Res.*, **13**, (3), 265 (1975)
27. Loladze, T.N., *Wear of Cutting Tools,* Mashgiz, Moscow (1958)
28. Hau-Bracamonte, J.L., *Metals Tech.*, **8**, (11), 447 (1981)
29. Trent, E,M., *Wear,* **128**, 65 (1988)

CHAPTER 6

CUTTING TOOL MATERIALS I: HIGH SPEED STEELS

6.1   INTRODUCTION AND SHORT HISTORY

The development of tool materials for cutting applications has been accomplished very largely by practical craftspeople. It has been of an evolutionary character and parallels can be drawn with biological evolution. Millions of people are daily subjecting metal cutting tools to a tremendous range of environments. The pressures of technological change and economic competition have imposed demands of increasing severity. To meet these requirements, new tool materials have been sought and a very large number of different materials have been tried. The novel tool materials (corresponding to genetic mutations) which have been proved by trials, are the products of the persistent effort of thousands of craftspeople, inventors, technologists and scientists, blacksmiths, engineers, metallurgists and chemists. The tool materials which have survived and are commercially available today, are those which have proved fittest to satisfy the demands put upon them in terms of the life of the tool, the rate of metal removal, the surface finish produced, the ability to give satisfactory performance in a variety of applications, and the cost of tools made from them. The agents of this "natural selection" are the machinists, tool room supervisors, tooling specialists and buyers in the engineering factories, who effectively decide which of all the potential tool materials shall survive.

To reconstruct the whole history of these materials is not possible because so many of the unsuccessful 'mutants' have disappeared without trace. Patents and the back numbers of engineering journals contain records of some of these. For example *The Engineer* for April 13, 1883, records a discussion at the Institution of Mechanical Engineers in England where Mr. W.F. Smith described experiments with chilled cast iron tools, which had shown some success in competition with carbon steel. Fuller accounts are available of the work which led to the development of those tool materials which have been an evolutionary success. It is clear that these innovations were made by people who had very little to guide them in the way of basic understanding of the conditions which tools were required to resist at the cutting edge. The simplest concepts of the requirements such as: that the tool material should be hard, able to resist heat, and tough enough to withstand impact without fracture, were all that was available to those engaged in this work.

It is only in retrospect that a reasoned, logical structure is beginning to emerge to explain the performance of successful tool materials and the failure of other contenders. In this chapter the performance of one of the main groups of present-day cutting tool materials is discussed. As far as possible the properties of the tool materials are related, on the one hand to their composition and structure and, on the other, to their ability to resist the temperatures and stresses discussed in the previous chapters, and the interactions with the work material which wear the tool. The objective is an improved understanding of the performance of existing tools in order to provide a more useful framework of knowledge to guide the continuing evolution of tool materials and the design of tools.

One consequence of this evolutionary form of development has been that the user is confronted with an embarrassingly large number of tool materials from which to select the most efficient.[1,2] A smell of the magical activities of the legendary Wayland Smith lingers in such tool steel trade names as *Super Hydra* or *Spear Mermaid*. Many competing commercial alloys are essentially the same, but there is a very large number of varieties which can be distinguished by their composition, properties or performance as cutting tools. These varieties can be grouped into species and groups of species can be related to one another, the groups being the "genera" of the cutting tool world.

In present-day machine shop practice, the vast majority of tools come from two of these 'genera' - high speed steels and cemented carbides. The other main groups of cutting tool materials are carbon (and lower alloy) steels, cast cobalt-based alloys, ceramics and diamond. The procedure adopted here is to consider first, in this chapter, tool steels and factors governing their performance. Cemented carbides are then dealt with similarly in Chapter 7, and these two classes of tool material are used to demonstrate general features of the behavior of tools subjected to the stresses and temperatures of cutting. Most of this discussion relates to the behavior of tools when cutting cast iron and steel, because these work materials have been much more thoroughly explored than others, but, where relevant, the performance of tools when cutting other materials is described. The present period is one in which tool materials are being developed very rapidly. New developments in steel and carbide tools are dealt with in this chapter and Chapter 7. Recent progress with ceramic type, diamond and other tool materials is considered in Chapter 8.

6.2 CARBON STEEL TOOLS

The only tool material for metal cutting from the beginning of the Industrial Revolution until the 1860s was carbon tool steel. This consists essentially of iron alloyed with 0.8 to about 2% carbon, the other alloying elements present - manganese, silicon, sulfur and phosphorus - being impurities or additions to facilitate steelmaking. Prolonged industrial experience was the guide to the selection of optimum carbon content for particular cutting operations. Carbon tool steel is hardened by heating to a temperature between 750 and 835°C ('cherry-red heat') followed by very rapid cooling to room temperature, usually by quenching in water. This operation was carried out by a skilled smith who was also responsible for shaping the tools. A slowly cooled tool steel has a hardness of less than 200 HV (Vickers Hardness); after quenching the hardness is increased to a maximum of 950 HV. If the quenching temperature is raised above 'cherry-red' there is no further increase in hardness, but the steel becomes more brittle: the tool edge fractures readily under impact. The necessity of controlling hardening temperature to ensure full hardness, while avoiding brittleness, was recognized from the earliest days of the use of steel and, before temperature measurement by pyrometer was introduced about the year 1900, control was by the eye of the skilled smith.

The very great hardness increment is the result of a re-arrangement of the atoms to produce a structure known as martensite. The characteristic 'acicular' (needle-like) structure of martensite is revealed by optical microscopy (*Figure 6.1a*). Martensite is hard because the layers of iron atoms are restrained from slipping over one another by the dispersion among them of the smaller carbon atoms, in a formation which forces the iron atoms out of their normal cubic space lattice and locks them into a highly rigid but unstable structure.

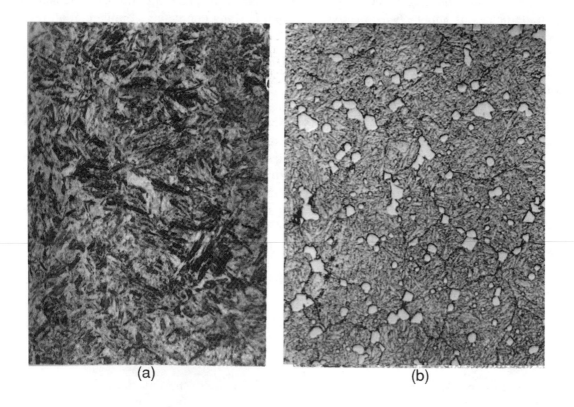

(a)   (b)

FIGURE 6.1   Microstructures of hardened tool steel (a) carbon steel (b) high speed steel

If re-heated (tempered) at a temperature above 200°C, the carbon atoms start to move from their unstable positions and the steel passes through a gradual transformation, losing hardness but increasing in ductility as the tempering time is prolonged or the temperature increased. In most cases hardened tool steels are tempered at temperatures between 200 and 350°C before use to render them less sensitive to accidental damage.

Figure 6.2 shows a tempering curve for a carbon tool steel - the hardness at room temperature after re-heating for 30 min at temperatures up to 600°C. The tool relaxes to a more stable condition and the high hardness can be restored only by again quenching from above 730°C.

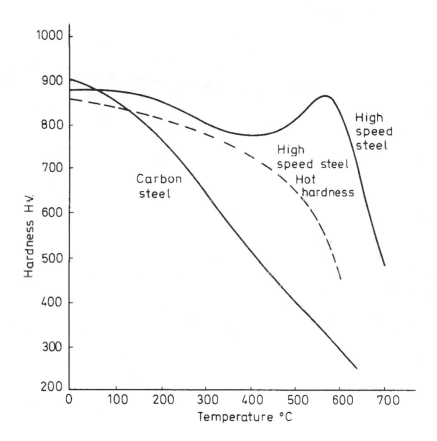

FIGURE 6.2   Tempering curves for carbon and high speed steel

The main mechanical test applied to tool materials is the diamond indentation hardness test, introduced in the 1920s. In use, most tools are stressed mainly in compression but, remarkably, very little data has been published on the behavior of tool materials under compressive stress.

Figure 6.3 shows the stress vs plastic strain curves for a hardened 1%C steel at room temperature and at temperatures up to 400°C. Cylindrical test pieces were step-loaded at increasing stress at each temperature, the permanent deformation being measured after each stress increment. At room temperature the 0.2% proof stress was about 2300 MPa (150 tonf/in^2) and there was considerable strain hardening over the first few per cent of plastic strain to values over 3000 MPa (200 tonf/in^2). There was a reduction in strength at 200°C and a greater reduction at higher temperatures so that at 400°C the 0.2% proof stress was below 800 MPa (50 tonf/in^2). Although strain hardening occurred at 400°C, after 4% strain the yield stress was still below 1300 MPa (85 tonf/in^2).

The stress near the tool edge when cutting steel has been estimated to be of the order of 1100 to 1600 MPa (75 to 100 tonf/in^2) or higher. The yield stress of carbon steel tools is therefore exceeded if the temperature rises above about 350°C and deformation of the tool edge is to be expected.

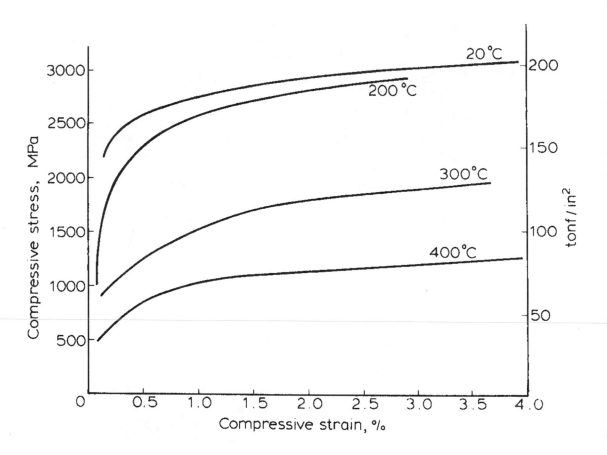

FIGURE 6.3   Stress vs plastic strain in compression tests on 1% C steel, hardened

Figure 6.4a shows a section through the cutting edge of a carbon steel tool used to cut wrought iron at a speed of only 3.3 m min^{-1}(10 ft/min) at a feed of 0.5 mm/rev (0.020 in/rev) for 5 min. *Figure 6.4b* shows the temperature contours in this tool estimated from microhardness measurements. The cutting edge had reached approx 350°C and had been deformed plastically, the first step towards tool failure.

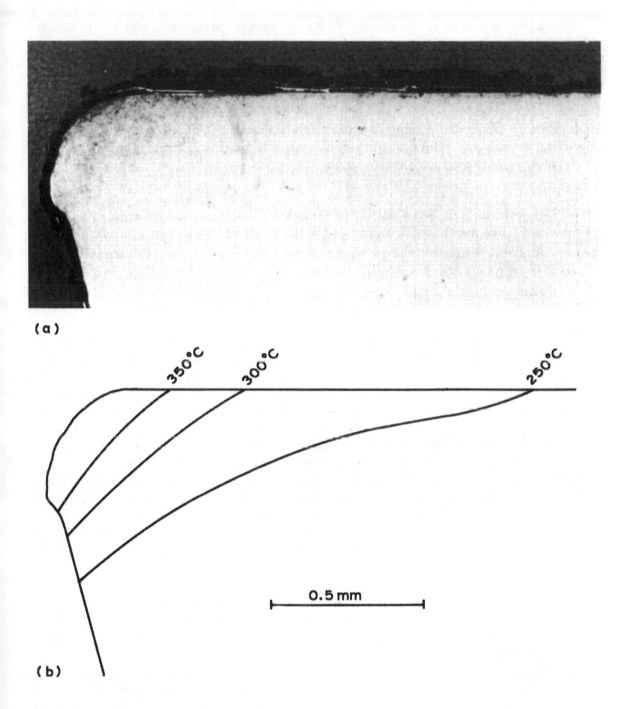

(a)

(b)

350°C 300°C 250°C

0.5 mm

FIGURE 6.4 Section through edge of carbon steel tool used to cut wrought iron at 3.3 m min^{-1} (10 ft/min) (a) photomicrograph showing deformation of edge; (b) temperature distribution

Carbon steel tools were used successfully for cutting copper at speeds as high as 110 m min^{-1} (350 ft/min), but for cutting iron and steel, speeds were normally kept to about 5 or 7 m min^{-1} (16 to 22 ft/min) to ensure a reasonable tool life. At the end of the last century, when the amount of machining was escalating with industrial development, the very high costs resulting from the extremely low productivity of machine tools operating at such very low speeds provided a major incentive to develop improved tool materials. Steel had become the most important of materials in engineering and the criterion of an improved tool material was its ability to cut steel at high rate of metal removal. This remains largely true up to the present and the majority of the references at the end of this chapter are concerned with improving the structure of high speed steels or using them in improved ways to cut steels and other "difficult-to-machine" alloys.[6- 20]

It should be emphasized that high-speed steels are still, today, the main tool materials used for drills, end-milling-cutters, and broaches.[11,19,20] *At the end of Chapter 8, it will be emphasized that by no means have carbide or harder tools taken the place of high speed steel for such day-to-day operations in a typical machine shop.*

6.3   HIGH SPEED STEELS

The earliest commercially successful attempt to improve cutting tools by use of alloying elements was the 'self-hardening' tool steel of Robert Mushet, first made public in 1868.[3] This contained about 6 to 10% tungsten and 1.2 to 2% manganese and, later, 0.5% chromium, with carbon contents of 1.2 to 2.5%. The most outstanding property was that it could be hardened by cooling in air from the hardening temperature and did not need to be quenched in water. Water quenching often brought trouble through cracking, particularly with tools of large size and difficult shape. The 'self-hardening' quality was mainly the result of the high manganese content and the chromium, both of which greatly retard the rate of transformation during cooling.

The tools were also found to give a modest improvement in the speed at which steel could be cut compared with carbon steel tools - for example, from 7 m min^{-1} to 10 m min^{-1} (22 to 33 ft/min). This can be attributed mainly to the tungsten which, after normal heat treatment, results in a significant increase in the yield stress at elevated temperature. *R. Mushet's Special Steel*, being more expensive, was usually reserved by craftspeople for cutting hard materials or for difficult operations.

The hardening procedures had been developed by centuries of experience of heat treatment of carbon tool steels. Generations of blacksmiths, some very competent and highly skilled, had established that tools were made brittle by hardening from temperatures above "cherry-red heat", and this constraint was, usually, applied to the heat treatment of self-hardening alloy steels for more than 20 years after their introduction.

The story of how revolutionary changes in the properties and performance of alloy tool steels were accomplished by modifying the conventional heat treatment is told by F. W. Taylor in his presidential address in 1906 to the A.S.M.E., 'On the Art of Cutting Metals'.[4] This outstanding paper deserves to be read as an historical record of the steps by which an engineer, Taylor, and a metallurgist, Maunsel White, developed high speed steels as part of what was probably the most thorough and systematic program of tests in the history of machining technology.

They first laid the basis of a sound technological method of tool testing. Rough turning of a standard steel was selected as the basic machining operation, and a standard method was adopted of determining the cutting speed which would give a 20 minute tool life. All the variables of the cutting process were then systematically investigated to establish the optimum feed, depth of cut, tool geometry and use of coolants. When using the carbon and the self-hardening steels discussed above, they achieved consistently the efficiency of the best craftspeople, but no great increment in metal removal rate above this.

Their investigation thus turned from the optimization of cutting conditions to the importance of heat treatment. Putting on one side conventional craft wisdom and the advice of academic metallurgy, Taylor and White conducted a series of tests in which tools were quenched from successively higher temperatures up to their melting points and then tempered over a range of temperatures. This work was made possible by use of the thermocouple which had not long been in use in industrial conditions. After each treatment, cutting tests were carried out on each tool steel to determine the cutting speed for a 20 minute tool life. Certain tungsten/chromium tool steels gave the best results. By 1906 the optimum composition was:

C	-	0.67%
W	-	18.91%
Cr	-	5.47%
Mn	-	0.11%
V	-	0.29%
Fe	-	Balance

The optimum heat treatment consisted in heating to just below the solidus (the temperature where liquid first appears in the structure, about 1250-1290°C), cooling in a bath of molten lead to 620°C, and then to room temperature. This was followed by a tempering treatment just below 600°C. Unlike carbon tool steel the tools were not embrittled by heating above cherry red heat. The tools treated in this way were capable of machining steel at 30 m min^{-1} (99 ft/min) under Taylor's standard test conditions. This was nearly four times as fast as when using the self-hardening steels and six times the cutting speed for carbon steel tools. This was a remarkable breakthrough.

At the Paris exhibition in 1906 the Taylor-White tools made a dramatic impact by cutting steel while 'the point of the tool was visibly red hot'. High speed steel tools revolutionized metal cutting practice, vastly increasing the productivity of machine shops and requiring a complete revision of all aspects of machine tool construction. It was estimated that in the first few years, engineering production in the USA had been increased by $8,000 million through the use of $20 million worth of high speed steel.

High speed steels of basically the same chemical composition and heat treated in basically the same way as described by Taylor in 1906 are still, today, one of the two main types of tool material used for metal cutting. Many modifications to the basic composition have been introduced by commercial producers. The steel making and hot working procedures have been refined and the heat treatment has been made much more precise. The numerous high speed steels commercially available have been classified into a small number of standard types or grades according to chemical composition. The *International Metallic Materials Cross Reference* lists typical analysis and hardness for over a dozen of types of high speed steel *(Table 6.1)*.

6.4   STRUCTURE AND COMPOSITION

As can be seen from *Table 6.1*,[5] the room temperature hardness of high speed steels is of the order of 850 HV, rather lower than that of many carbon tool steels. *Figure 6.1b* is a photomicrograph of the structure of one of the most commonly used types, M2. The bulk of the structure *(the matrix)* consists of martensite. The alloying elements, tungsten, molybdenum and vanadium, tend to combine with carbon to form very strongly bonded carbides with the compositions $Fe_3(W,Mo)_3C$ and V_4C_3 and the former can be seen in the structure of *Figure 6.1b* as small, oblong, white areas a few micrometers (microns) across. These micrometer-sized carbide particles play an important part in the heat treatment.

As the temperature is raised, the carbide particles tend to be dissolved, the tungsten, molybdenum, vanadium and carbon going into solution in the iron. The higher the temperature the more of these elements go into solution but even up to the melting point some particles remain intact, and their presence prevents the grains of steel from growing. It is for this reason that high speed steel can be heated to temperatures as high as 1290°C, without becoming coarse grained and brittle. These carbide particles are harder than the martensitic matrix in which they are held; typical figures are

$$Fe_3W_3C = 1,150 \text{ HV}$$

$$V_4C_3 = 2,000 \text{ HV}$$

However, they constitute only about 10 to 15% by volume of the structure and, surprisingly, have only a minor influence on the properties and performance of the tools. *The vital role in producing the outstanding behavior of high speed steel is played by other carbide particles formed after precipitation hardening, during the tempering operation.* This is described below. These particles are much too small to be observed by optical microscopy, being only about one hundredth of the size of those visible in *Figure 6.1b*.

Figure 6.2 shows a typical tempering curve for a high speed steel. At first, as with carbon steel, the hardness begins to drop, but over 400°C it begins to rise again and, after tempering between 500°C and 600°C, hardness is often higher than before tempering. With further increase in tempering temperature the hardness falls off rapidly.

The *secondary hardening,* after tempering at about 560°C, is caused by the formation within the martensite of the extremely small particles of carbides. Much of the tungsten, molybdenum and vanadium taken into solution in the iron during the high temperature treatment is retained in solution during cooling to room temperature. On re-heating to 400-600°C they come out of solution and precipitate throughout the structure forming extremely numerous carbide particles.

This is the process known as precipitation-hardening. High speed steels were probably the first commercial precipitation-hardened alloys, preceding precipitation-hardened aluminum alloys by more than ten years. They resulted from a technological investigation to solve an

TABLE 6.1   Typical compositions of high speeds steels*

Designation	C	Cr	Chemical Composition (Wt%) Mo	W	V	Co	Hardness (HV) Min
T1	0.75	4	-	18	1	-	823
T2	0.8	4	-	18	2	-	823
T4	0.75	4	-	18	1	5	849
T5	0.8	4	-	18	2	9.5	869
T6	0.8	4.5	-	20	1.5	12	969
T15	1.5	4	-	12	5	5	890
M1	0.8	4	8	1.5	1	0.8	823
M2	0.85	4	5	6	2	0.85	836
M4	1.3	4	4.5	5.5	4	-	849
M15	1.5	4	3.5	6.5	5	5	869
M30	0.8	4	8	2	1.25	5	869
M42	1.10	3.75	9.5	1.5	1.15	8	897

* Reproduced from *International Metallic Materials Cross Reference, 2nd ed.* General Electric: 1983.

urgent engineering problem. Taylor and White intuitively understood that what had been achieved was a new sort of hardening, which they called 'red hardness', but the hardening mechanism in high speed steel was not understood for 50 years. In the 1950s electron microscope techniques and physical metallurgy theory had advanced to the stage where the structures of these very complex alloys could be demonstrated.

A study by El-Rakayby and Mills,[17] using analytical electron microscopy, identified the secondary hardening precipitate in M42 high speed steel as the type M_2C with a face-centered cubic structure. Micrographs (*Figure 6.5*) show these particles as being smaller than 0.05 μm. The main metal atom is Mo, with smaller percentages of V and Cr. These micrographs show specimens tempered at 540° and 550°C. Up to 560°C, the particles remain stable for many hours and harden the steel by blocking the dislocations which facilitate slip between the layers of iron atoms. At higher temperatures particularly above 650°C, the particles coarsen rapidly and lose their capacity for hardening the steel matrix. The hardness can then be restored only by repeating the whole heat treatment cycle.

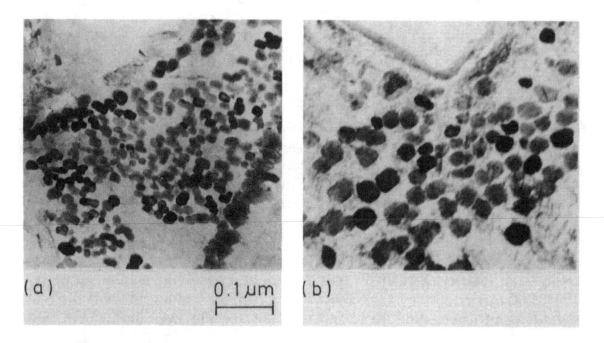

FIGURE 6.5 Carbon replica electron micrographs showing secondary hardening carbide precipitation at peak hardness after tempering for 2 x 2 hours (a) at 540°C (b) at 550°C (Courtesy of El Rakayby and B. Mills)

The useful properties of all grades of high speed steel depend on the development within them of the precipitation-strengthened martensitic structure as a result of high temperature hardening, followed by tempering in the region of 520-570°C. All of them are softened by prolonged heating to higher temperatures, and no development up to the present has greatly raised the temperature range within which the hardness is retained. A few of these grades, notably M2 and TI, are produced in large quantities as general purpose tools, while the others answer the requirements of particular cutting applications. A summary of the role of each of the alloying elements gives a guide to the type of applications for which the different grades are suited.

6.4.1 Tungsten and molybdenum

In the first high speed steels, tungsten was the essential metallic element upon which the secondary hardening was based, but molybdenum performs the same functions and can be substituted for it. Equal numbers of atoms of the two elements are required to produce the same properties, and since the atomic weight of molybdenum is approximately half that of tungsten, the percentage of molybdenum in an M steel is usually about half that of tungsten in the equivalent T-type steel. The heat treatment of molybdenum-containing alloys presents rather more difficulties, but the problems have been overcome, and the molybdenum steels are now the most commonly used of the high speed steels. There is evidence that they are tougher, and their cost is usually considerably lower.

6.4.2 Carbon

Sufficient carbon must be present to satisfy the bonding of the strongly carbide-forming elements (vanadium, tungsten and molybdenum). An additional percentage of carbon is required, which goes into solution at high temperature and is essential for the martensitic hardening of the matrix. Precise control of the carbon content is very important. The highest carbon contents are in those alloys containing large percentages of vanadium.

6.4.3 Chromium

The alloys all contain 4-5% chromium, the main function of which is to provide *hardenability* so that even those tools of a large cross-section may be cooled relatively slowly and still form a hard, martensitic structure throughout.

6.4.4 Vanadium

All the grades contain some vanadium. In amounts up to 1% its main function is to reinforce the secondary hardening, and possibly to help to control grain growth. A small volume of hard particles of V_4C_3 of microscopic size is formed, and these are the hardest constituents of the alloy. When the steels contain as high as 5% V, there are many more of these hard particles, occupying as much as 8% by volume of the structure, and these play a significant role in resisting wear, when cutting abrasive materials.

6.4.5 Cobalt

Cobalt is present in a number of the alloys in amounts between 5 and 12%. The cobalt raises the temperature at which the hardness of the fully hardened steel starts to fall, and, although the increase in useful temperature is relatively small, the performance of tools under particular conditions may be greatly improved. The cobalt appears to act by restricting the growth of the precipitated carbide particles, while not itself forming a carbide.

6.5   PROPERTIES OF HIGH SPEED STEELS

Hardness at room temperature is the most commonly measured property of tool materials. The Vickers diamond pyramid indentation hardness test (HV) is used, and can conveniently be carried out on pieces of many shapes and sizes. It is an effective quality control test and useful as a first indication of the properties of tool steel.

The tempering curves for M2 and carbon steel (*Figure 6.2*) and the hardness data in *Table 6.1* are the results of tests at room temperature. *Hot hardness tests* carried out at elevated temperatures, could be of more direct significance for cutting tool performance, but the test procedure is more difficult and the results are less reliable. The dashed line in *Figure 6.2* shows the results of one set of hardness tests on fully heat treated M2 made at the temperature indicated on the graph. There is a continuous fall in hardness with increasing temperature, with no peak at 550°C, and although the hardness is dropping rapidly, it is still nearly 600 HV at this temperature.

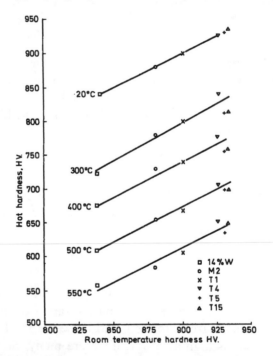

FIGURE 6.6   Hot hardness *vs* cold hardness for six high speed steels (Kirk *et al*[7])

Figure 6.6[7] shows the hot hardness of six high speed steels with different room temperature hardness, at temperatures up to 550°C. The results show that, for a range of high speed steels, room temperature hardness provides an indication of hot hardness also. This relationship is true only for materials within the range of the high speed steels.

When tested in compression, an elastic limit is observed (*Figure 6.7*).[8] Young's modulus is only slightly higher than that of most engineering steels and varies little with heat treatment. Plastic deformation is accompanied by strain hardening, as shown in *Figure 6.7*.

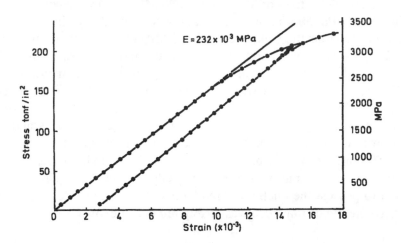

FIGURE 6.7 Compression test on type T1 high speed steel, fully heat treated[8]

The compressive strength of fully heat treated high speed steels is of direct relevance to their performance as cutting tools. *Figure 6.8* shows the compressive stress vs plastic strain curves for fully heat-treated M2 at room temperature and up to 600°C. These curves were obtained by step-loading cylindrical specimens and can be compared with these for a 1% C steel (*Figure 6.3*). At room temperature both the 0.2% yield stress of nearly 3000 MPa (195 tonf/in²) and the rate of strain hardening are greater with high speed steel. At 2% strain the yield stress is over 4000 MPa (260 tonf/in²). It is at elevated temperatures that the advantage of high speed steel over carbon steel is critical.

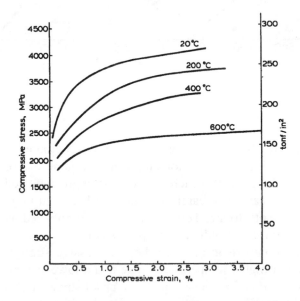

FIGURE 6.8 Compressive strain *vs* plastic stress for fully heat-treated M2 at various temperatures

At 400°C the 0.2% yield stress of M2 is over 2000 MPa (130 tonf/in^2), while with 1% C steel it was less than 800 MPa (50 tonf/in ^{2}). Even at 600°C the 0.2% yield stress is 1800 MPa (115 tonf/in^2) and after 3% strain the yield stress is 2500 MPa (160 tonf/in^2). Thus, even at 600°C the strength of M2 high speed steel is high enough to withstand the stress of the order of 1500 MPa (100 tonf/in^2) which has been demonstrated at the edge of tools when cutting steel. A temperature of 600°C at the cutting edge during the machining of steel has been observed at relatively low cutting speeds (*Figure 5.11*) and at the tool edge this temperature rises only slowly as the cutting speed increases (*Figure 5.13 and 5.18*). The characteristic stress and temperature, and their distribution in tools used to cut steel is such that cutting speeds on the order of 30-50 m min^{-1} (100 to 150 ft/min) can be employed when machining with high speed steels tools, without failure through plastic deformation of the cutting edge. In the early days of high speed steels these were said to possess the quality of "red hardness" by which was meant the ability to cut steel at speeds where the cutting edge was visibly red hot - i.e. over 600°.

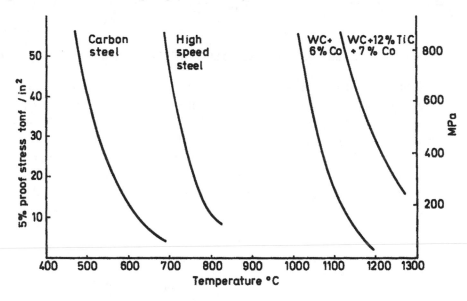

FIGURE 6.9 Hot compression tests: 5% proof stress of tool steels and cemented carbide[9]

Since that time the efforts to strengthen further these precipitation-hardened martensitic tool steels, by modification of the composition and heat treatment, has resulted in some increase in the yield stress at 600°C. The further increase in hot strength achieved by alloying is real but relatively modest after 85 years of research. It seems unlikely that major improvements in respect of this property, fundamental to metal cutting, can be accomplished without a basic change in the structures which control the strength of high speed steels.

Figure 6.9 shows a comparison between the compressive strengths of high speed steel and other tool materials at a higher temperature range and a lower stress level.[9] The property measured is the proof stress after 5% reduction. A 5% proof stress of 770 MPa (50 tonf/in ^{2}) is supported up to 740°C by high speed steel, compared with 480°C for carbon tool steel and 1000°C for a cemented carbide.

Other mechanical properties of a number of high speed steels are given in *Table 6.2*. These are for steels in the fully heat treated condition as recommended by manufacturers for general use as cutting tools. The properties are very dependent on heat treatment and *Table 6.3* shows how the properties of M2 high speed steel vary with the hardening temperature and with the tempering temperature. Tensile tests are not normally carried out on tool steels and the transverse rupture test is commonly used, the reported value of strength being the maximum tensile stress at the bar surface before fracture. The strength is high, typical values being 4-5 GPa (260-320 tonf/in^2).

Another important quality of a tool material for which a measure is being sought is usually called *toughness*. In the context of cutting tool materials toughness means the ability of the tool to continue cutting under difficult conditions for long periods without fracture. When cutting is interrupted - for example, when turning a slotted bar or an irregular shaped casting or forging, or when milling - the tool edge is subjected to impact stresses as well as mechanical and thermal fatigue stresses which may result in local fracture and chipping of the edge. No single test has been generally accepted as an overall measure of toughness in tool materials. The breaking stress in the bend test is proposed as one such measure. Impact tests (Charpy or Izod, using test pieces either unnotched or with a C-notch) place the grades and heat treatments in the same order as transverse rupture tests. In these tests the tool steel specimens show plastic deformation before fracture. The amount of plastic deformation is an indication of the ability to absorb energy before fracture and may, therefore, be related to the probability of tool failure, at least in some conditions of service. In the transverse rupture test the strength value is influenced by the inherent strength of the alloy and by the presence in the test piece of flaws which may greatly reduce the measured strength compared with a flaw-free specimen. There is much scatter in individual test results for this reason. The fracture toughness test is designed to measure the inherent toughness by determining the energy required to propagate sharp crack. The K_{IC} value is given in units of MPa m $^{1/2}$ Fracture toughness tests carried out by a number of research organizations are in reasonable agreement on the values of the K_{IC} parameter for high speed steels. There is some scatter in test results but, in general, the fracture toughness values place the different grades of steel and the different heat treatments in the same order as does the bend test. [10,16,18]

TABLE 6.2 Typical values of mechanical properties of high speed steels heat-treated for general use as cutting tool

Grade	Transverse rupture strength		Fracture toughness, K_{IC}	Izod impact strength (un-notched)	Hardness
	(GPa)	*(ton f/in^2)*	*(MPa m$^{-1/2}$)*	*(ft lbs)*	*(HV)*
T1	4.6	300	18	15	835
M1	4.8	310	18	24	835
M2	4.8	310	17	25	850
T6	3.0	195	16	9	880
M15	4.0	260	15	13	880
M42	3.4	220	10	13	910

TABLE 6.3   Influence of heat treatment on mechanical properties of M2 high speed steel (typical values)

Heat treatment temperature (°C)		Transverse rupture strength		Fracture toughness K_{IC}	Izod impact strength (un-notched)	Hardness
Hardening	Tempering	(GPa)	(ton f/in²)	(MPa m⁻¹ᐟ²)	(ft lbs)	(HV)
1050	560	4.6	300	19.2	42	660
1150	560	4.7	305	18.6	30	785
1200	560	4.5	290	16.2	27	846
1220	560	4.0	260	16.6	25	850
1250	560	3.8	250	15.1	20	870
1220	450	3.3	210	18.0	28	800
1220	500	3.8	250	18.0	26	820
1220	560	4.6	300	16.6	25	850
1220	600	4.7	305	17.4	30	770
1220	650	3.8	250	18.7	40	600

Apart from chemical composition and heat treatment, a further factor with major influence on toughness as experienced in the machine shop is the homogeneity and isotropy of the steel. In high speed steels produced by conventional melting, ingot casting and hot working, the carbide particles are never uniformly dispersed. After casting they are arranged in clusters, which are drawn out into lines of closely spaced carbide particles during hot working. In rolled products the lines are parallel to the rolling direction. The greater the reduction of section in hot working the more homogeneous is the product. Thus small drills, for example, are reasonably uniform, but the inhomogeneity is very pronounced in tools of large section. In this respect the quality varies with the steel making and processing practice of different producers and can be assessed by comparing the metallographic structures of longitudinal sections with standard micrographs (e.g. in American National Standard Institute B212.12-91[5]).

In bend tests, the strength in the transverse direction (i.e. when fracture takes place in a direction normal to the lines of carbide) is higher than in the longitudinal direction and the difference is greater the more the segregation.[10,11] This is in agreement with practical experience. In milling cutters, for example, edge fracture is more frequent when the cutting edge is parallel to the direction of the lines of carbide. The fracture toughness test, however, shows little difference between the transverse and longitudinal directions, even in severely segregated steel.[10] Fracture toughness values are a more fundamental quality than strength determined by bend or impact tests, since they measure a 'material property' and are independent of the size and shape of the test specimen. For measuring the quality of 'toughness' in tool materials and predicting tool performance in relation to toughness, however, bend or impact tests seem superior to fracture toughness tests. For further treatment of this subject see Hoyle.[18] More research is urgently required to establish a test or tests which can be internationally recognized.

One of the main areas in which the experience of the specialist is valuable is in selection of the optimum grade and heat treatment to secure maximum tool life and metal removal rate for each particular application. Inadequate toughness will lead to fracture of the tool, giving a short and erratic tool life. The other factors controlling tool life with high speed steel are now considered.

6.6 TOOL LIFE AND PERFORMANCE OF HIGH SPEED STEEL TOOLS

For satisfactory performance, the shape of the cutting tool edge must be accurately controlled and is much more critical in some applications than in others. (While the following discussions seem to be directed at the grinding and honing needed for high speed steels, they are also relevant for the precise sintering operations and edge preparations needed for carbide and ceramic tools.) Much skill is required to evolve and specify the optimum tool geometry for many operations, to grind the tools to the necessary accuracy, and to inspect the tools before use. This is not merely a question of measuring angles and profiles on a macro-scale, but also of inspecting and controlling the shape of the edge on a very fine scale, within a few tenths of a millimeter of the edge, involving such features as burrs, chips or rounding of the edge.

In almost all industrial machining operations, the action of cutting gradually changes the shape of the tool edge so that in time the tool ceases to cut efficiently, or fails completely. The criterion for the end of tool life is very varied - the tool may be reground or replaced when it fails and ceases to cut; when the temperature begins to rise and fumes are generated; when the operation becomes excessively noisy or vibration becomes severe; when dimensions or surface finish of the workpiece change or when the tool shape has changed by some specified amount. Often the skill of the operator is required to detect symptoms of the end of tool life, to avoid the damage to the part being machined, caused by total tool failure (*see* diagrams at end of Chapter 3).

The change of shape of the tool edge is very small and can rarely be observed adequately with the naked eye. A binocular microscope with a magnification of at least $\times 30$ is needed even for preliminary diagnosis of the character of tool wear in most cases. The worn surfaces of tools are usually covered by layers of the work material which partially or completely conceal them. To study the wear of high speed tools it has been necessary to prepare metallographic sections through the worn surfaces, usually either normal to the cutting edge or parallel with the rake face. The details which reveal the character of the wear process are at the worn surface or the interface between tool and adhering work material, and the essential features are obscured by rounding of the edge of the polished section unless special metallographic methods are adopted. Sections shown here were mostly prepared by:

(1) Mounting the tool in a cold setting resin in vacuum.
(2) Carefully grinding the tool to the required section, using much coolant.
(3) Lapping on metal plates with diamond dust.
(4) Polishing on a vibratory polishing machine using a nylon cloth with one micron diamond paste.

Observations have shown that the shape of the tool edge may be changed by *plastic deformation* as well as by *wear*. The distinction is that a wear process always involves some loss of material from the tool surface, though it may also include plastic deformation locally, so that there is no sharp line separating the two. The use of the term *wear* is, in the minds of many people, synonymous with *abrasion* - the removal of small fragments when a hard body slides over a softer surface, typified by the action of an abrasive grit in a bearing. There are, however, other processes of wear between metallic surfaces, and a mechanism of *metal-transfer* is frequently described in the literature. This term is used for an action in which very small amounts of metal, often only a relatively small number of atoms, are transferred from one surface to the other, when the surfaces are in sliding contact, the transfer taking place at very small areas of actual metallic bonding.

Both of these mechanisms may cause wear on cutting tools, but they are essentially processes taking place at sliding surfaces. There are parts of the tool surface and conditions of cutting where the work material slides over the tool as in the classical friction model, but, as argued in Chapter 3, it is characteristic of most industrial metal cutting operations that the two surfaces are seized together over a large part of the contact area, and under conditions of seizure there is no sliding at the interface. Wear under conditions of seizure has not been studied extensively, and investigations of cutting tool wear are in an uncharted area of tribology. The wear and deformation processes which have been observed to change the shape of high speed steel tools when cutting steel and other high melting point metals are now considered.[12]

6.6.1   Superficial plastic deformation by shear at high temperature

When cutting steel and other high melting-point materials at high rates of speed and feed, a characteristic form of wear is the formation of a *crater*, a hollow in the rake face some distance behind the cutting edge, shown diagrammatically in *Figure 6.10*. The crater is located at the hottest part of the rake surface, as demonstrated by the method of temperature estimation described in Chapter 5 and illustrated in *Figure 5.11*. In *Figure 5.11a* the beginnings of the formation of a crater can be observed. The small ridge just beyond the hottest part on the rake face consists of tool material sheared from the hottest region and piled up behind. This is shown at higher magnification in *Figure 6.11*. In this case the work material was a very low carbon steel, the cutting speed was 183 m min^{-1} (600 ft/min), and the cutting time was 30 s. The maximum temperature at the hottest position was approximately 950°C.

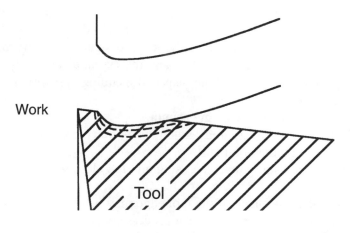

Work

Tool

FIGURE 6.10 Cratering wear in relation to temperature contours

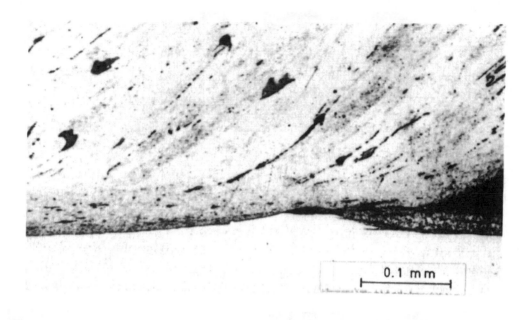

FIGURE 6.11 Section through rake face of tool. As *Figure 5.11*, back of crater in high speed steel tool after cutting iron[12]

The steel chip was strongly bonded to the rake face, and this metallographic evidence demonstrates that the stress required to shear the low carbon steel in the flow-zone, at a strain rate of at least 10^4 s^{-1}, was high enough to shear the high speed steel where the temperature was about 950°C. Shear tests on the tool steel used in this investigation gave a shear strength of 100 MPa (9 tonf/in^2) at 950°C at a strain rate of 0.16 s^{-1}. It is unexpected to find that nearly pure iron can exert a stress high enough to shear high speed steel.

This is possible because:

(1) The yield stress of high speed steel is greatly lowered at high temperature.

(2) The rate of strain of the low carbon steel in the flow-zone is very high, while the rate of strain in the high speed steel, although it cannot be estimated, can be lower by several orders of magnitude.

(3) In both materials the yield stress increases with the strain rate.

This wear mechanism has been observed on tools used to cut many steels. *Figure 6.12* shows cratering of high speed steel tools used to cut austenitic stainless steel.The shearing away of successive layers of tool steel is particularly well seen in *Figure 6.13* at the rear end of the crater. It occurs also when cutting titanium and its alloys as well as nickel and its alloys. When cutting the stronger alloys it occurs at much lower cutting speeds than when cutting the commercially pure metals. This is because the shear stress is higher and the tool is therefore sheared at lower interface temperatures.

However, it always occurs at the regions of highest temperature at the tool work interface, and when cutting nickel alloys, it has been observed at the cutting edge (*Figure 6.14*). It also occurs on a tool flank when the tool is severely worn with accompanying high temperatures.

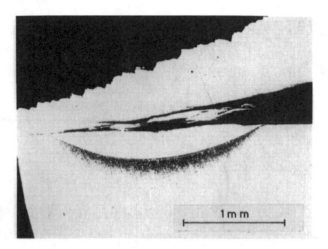

FIGURE 6.12 Crater in high speed steel tool used to cut austenitic stainless steel. Etched to show heat-affected region below crater filled with work material. (After Wright and Trent[12])

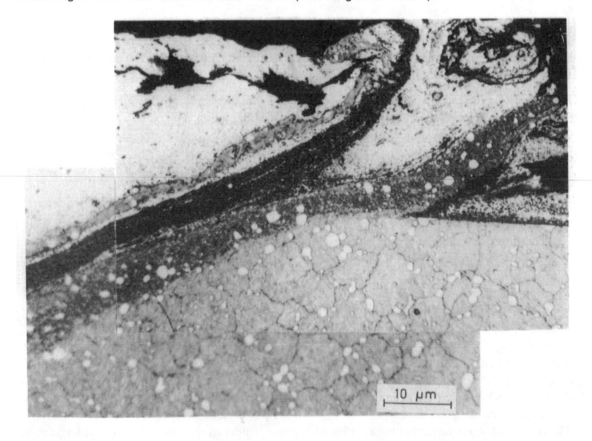

FIGURE 6.13 End of crater after cutting austenitic stainless steel (After Wright and Trent[12])

Shearing is a rapid-acting wear mechanism, forming deep craters which weaken the cutting edge so that the tool may be fractured. This wear mechanism may not be observed very often under industrial cutting conditions, but it is a form of wear which sets a limit on the speed and feed which can be used when cutting the higher melting point metals with high speed steel tools.

It is unlikely that this wear mechanism will be observed when cutting copper-based or aluminum-based alloys with high speed steel tools, since, with these lower melting point materials, both the temperatures generated and the shear yield stresses are very much lower. It has, however, been seen on the severely worn flank of a carbon steel tool used to cut 70/30 brass at 240 m min⁻¹ (800 ft/min), where a temperature over 730°C was generated.

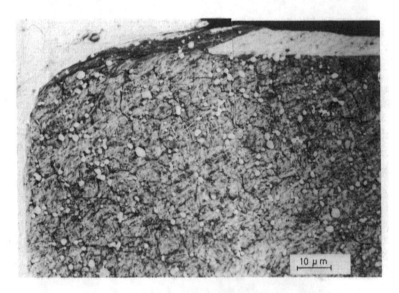

FIGURE 6.14 Shearing of high speed steel tool at cutting edge after cutting nickel-based alloy (After Wright and Trent[12])

6.6.2   Plastic deformation under compressive stress

Deformation of the tool edge has been discussed in dealing with the properties of tool steels. This usually takes a form such as that shown in *Figure 6.15*. In itself, deformation is not a wear process since no material is removed from the tool, but forces and temperature may be increased locally and so the flow pattern in the work material is modified. These more severe conditions bring into play or accelerate wear processes which reduce tool life. Deformation is not usually uniform along the tool edge. Plastic deformation often starts at the nose of a tool with a sharp nose or a small nose radius. Once started, a chain reaction of increased local stress and temperature may result in very sudden failure of high-speed steel tools.

Sudden failure initiated by plastic deformation may be difficult to distinguish from brittle failure resulting from lack of toughness in the tool material. It may be necessary to examine tools after cutting for a very short time to observe the initial stages of plastic deformation. Where failure is initiated in this way, a larger nose radius, where this is possible, often prolongs tool life or permits higher cutting speed.

FIGURE 6.15 Deformation of cutting edge of tool used to cut cast iron (After Wright and Trent[12])

At the higher cutting speeds employed with high speed steel tools, and the resulting higher tool edge temperatures, catastrophic failure occurs after a smaller amount of plastic deformation than with carbon steel tools (*Figure 6.4*). Strain is often not uniform within the strained region but is concentrated in localized shear bands (*Figure 6.16*) and distinct blocks of tool material may be sheared away.

Deformation of the tool edge, together with the previously discussed shearing away of the tool surface, are two mechanisms which often set a limit to the speed and feed which can be used.

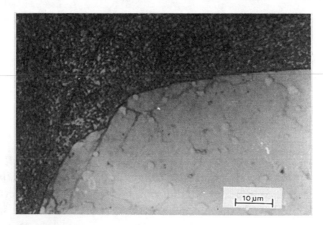

FIGURE 6.16 Deformed edge of high speed steel tool showing localized shear bands

Tools are more likely to be damaged by deformation when the hardness of the work material is high and it is this mechanism which limits the maximum workpiece hardness which can be machined with high speed steel tools, even at very low speed where temperature rise is not important. An upper limit of 350 HV of the steel work material being cut is often considered as the highest hardness at which practical machining operations using high speed steel tools can be carried out on steels; though steel as hard as 450 HV may be cut at very low speed.

6.6.3   Diffusion wear

Metallographic evidence has been given to show that conditions exist during cutting where diffusion across the tool/work interface is probable. Such wear by diffusion has been observed in tools that have been used for longer cutting times.[12] There is metal to metal contact and temperatures of 700°C to 900°C are high enough for appreciable diffusion to take place. Thus tools may be worn by metal and carbon atoms from the tool diffusing into, and being carried away by, the stream of work material flowing over its surface, and by atoms of the work material diffusing into or reacting with the surface layers of the tool to alter and weaken the surface.

Rates of diffusion increase rapidly with temperature, the rate typically doubling for an increment of the order of 20°C. There is strong evidence that wear by diffusion and interaction does, in fact, occur in the high temperature regions of the seized interface when cutting steel and other high melting point alloys at high speed.

The rapid form of cratering caused by superficial plastic deformation has been described and there is clear evidence in that process of deformation of grain boundaries and other features in the direction of chip flow. At somewhat lower cutting speeds, craters form more slowly and there is no evidence of plastic deformation of the tool.

Figure 6.17 is a section through such a worn surface with a thin layer of the work material (steel) seized to the tool. The wear is of a very smooth type and no plastic deformation is observed, the grain boundaries of the tool steel being undeformed up to the surface. The carbide particles were not worn at all, or worn much more slowly, undermined and eventually carried away.

Diffusion wear is a sort of chemical attack on the tool surface, like etching, and is dependent on the solubility of the different phases of the tool material in the metal flowing over the surface, rather than on the hardness of these phases. The carbide particles are more resistant because of their lower solubility in the steel work material, whereas there are no solubility barriers to the diffusion of iron atoms from the tool steel into a steel work material.

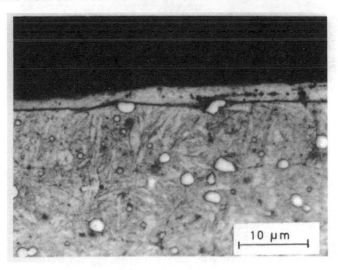

10 μm

FIGURE 6.17 Section through rake face of high speed steel tool after cutting steel. Interface characteristic of diffusion wear (After Wright and Trent[12])

Diffusion wear is well illustrated in sections through tools used to cut austenitic stainless steel. *Figure 6.18* is a section through the rake face with adhering stainless steel after cutting at 23 m min^{-1} (75 ft/min).

At this position, near where the chip left the tool, the temperature was approximately 750°C during cutting. The grain boundaries are undeformed up to the interface. Diffusion of alloying elements from the tool into the flow-zone had caused structural modification, which shows up as a stream of dark etching particles, too small to be identified by optical microscopy. These are being carried away with the chip flowing from left to right.

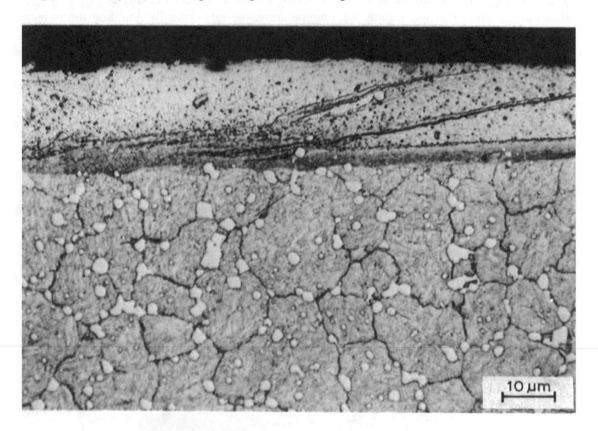

FIGURE 6.18 Section through rake face of high speed steel tool after cutting austenitic stainless steel

That diffusion may also alter the composition of the tool at the interface is shown in *Figure 6.19*. The work material adhering to the rake surface in this case was a low alloy engineering steel. The cutting speed was 18 m min^{-1} (60 ft/min) and the cutting time was 38 minutes. The temperature was estimated as 750° C at this position, near the end of the crater on the rake face. Between the tool and the chip is a layer about 2 μm thick with an unresolved structure. There was no deformation of the grain boundaries below the layer, but the layer itself was flowing in the direction of chip movement, carrying away tool material. In this case an interaction between tool and work material resulted in structural change in the matrix of the tool steel to form a layer which was more easily deformed at 750° C than the tool itself.

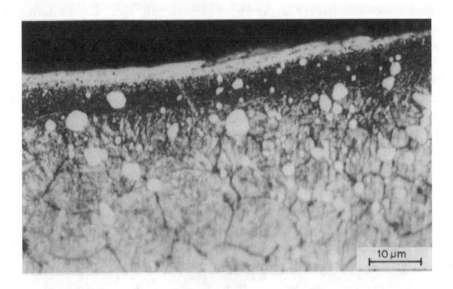

FIGURE 6.19 Section through rake face of high speed steel tool after cutting low alloy engineering steel

The diffusion wear process is one which has not been seriously considered as a cause of wear, except in relation to metal cutting. The rate of diffusion wear is very dependent on the metallurgical relationship between tool and work material and this is important when cutting different metals such as titanium or copper. It is of more significance for cemented carbide tools than for high speed steel and is considered again in relation to them.

With high speed steel tools used in the usual cutting speed range, rates of wear by diffusion are relatively slow because the interface temperatures are relatively low. At higher speeds and higher temperatures, diffusion is accelerated, but diffusion wear is masked by the plastic deformation which is a much more rapid wear mechanism. Diffusion and interaction account for the formation of craters at speeds below those at which plastic deformation begins, and this is probably the most important wear process responsible for flank wear in the higher speed range.

Wear by diffusion depends both on high temperatures, and also on a rapid flow rate in the work material very close to the seized surface, to carry away the tool metal atoms. On the rake face, very close to the cutting edge, there is usually a dead metal region, or one where the work material close to the tool surface flows slowly (*Figures 4.26* and *4.27*), and this, together with the lower temperature, accounts for the lack of wear at this position. The rate of flow past the flank surface is, however, very high when cutting at high speeds, and wear by diffusion can take place on the flank, although the temperature is very much the same as on the adjacent unworn rake face.

6.6.4   Attrition wear

At relatively low cutting speed, temperatures are low, and wear based on plastic shear or diffusion does not occur. The flow of metal past the cutting edge is more irregular, less stream-lined or laminar. Also, a built-up edge may be formed and contact with the tool may be less continuous. Under these conditions larger fragments, of microscopic size, may be torn intermittently from the tool surface, and this mechanism is called attrition.

Figures 6.20 and *6.21* show a section through the cutting edge of a high speed steel tool after cutting a medium carbon steel for 30 minutes at 30 m min^{-1} (100 ft/min). A built-up edge remained when the tool was disengaged. The tool edge had been 'nibbled' away over a considerable period of time. Fragments of grains have been pulled away, with some tendency to fracture along the grain boundaries, *Figure 6.21*, leaving a very uneven worn surface.

The tool and work materials are strongly bonded together over the whole of the torn surface. Although relatively large fragments were removed, this must have happened infrequently once a stable configuration had been reached, because the tool had been cutting for a long time.

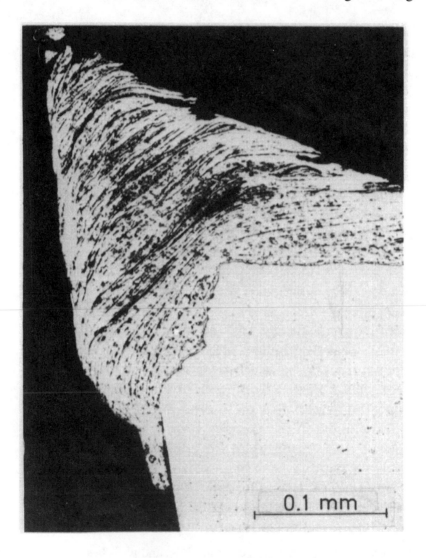

FIGURE 6.20 Section through cutting edge of high speed steel tool after cutting steel at relatively low speed, showing attrition wear (After Wright and Trent[12]).

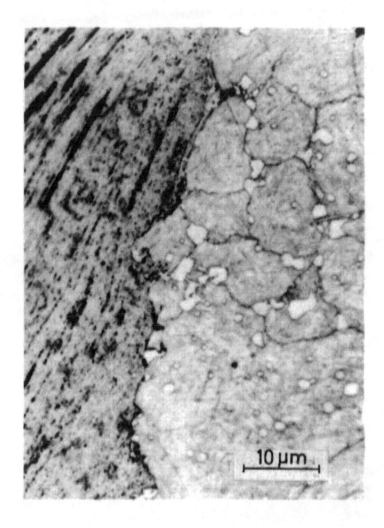

FIGURE 6.21 Detail of *Figure 6.20* (After Wright and Trent[12])

In continuous cutting operations using high speed steel tools, attrition is usually a slow form of wear, but more rapid destruction of the tool edge occurs in operations involving interruptions of cut, or where vibration is severe due to lack of rigidity in the machine tool or very uneven work surfaces. Attrition is not accelerated by high temperatures, and tends to disappear at high cutting speed as the flow becomes laminar. This is a form of wear which can be detected and studied only in metallographic sections. Adhering metal often completely conceals the worn surface and, under these conditions, visual measurements of wear on the untreated tool may be misleading.

6.6.5 Abrasive wear

Abrasive wear of high speed steel tools requires the presence in the work material of particles harder than the martensitic matrix of the tool. Hard carbides, oxides and nitrides are present in many steels, in cast iron and in nickel-based alloys, but there is little direct experimental evidence to indicate whether abrasion by these particles does play an important role in the wear of tools. Some evidence of abrasion of rake and flank surfaces by Ti (C,N) particles is seen in sections through tools used to cut austenitic stainless steel stabilized with titanium.

Figure 6.22 shows a Ti(C,N) particle which had ploughed a groove in the rake face of the tool as it moved from left to right, eventually remaining partially embedded in the tool surface. In this experiment, a corresponding steel without the hard particles showed only slightly less wear when cut under the same conditions. It seems doubtful whether, under conditions of seizure, small, isolated hard particles in the work material make an important contribution to wear. *Figure 6.22* suggests that they may be quickly stopped, and partially embedded in the tool surface where they would act like microscopic carbide particles in the tool structure. It is possible, however, that the reverse process takes place - that carbide particles in the tool steel, undermined by diffusion wear, can be detached from the tool and dragged along its surface, ploughing groove.

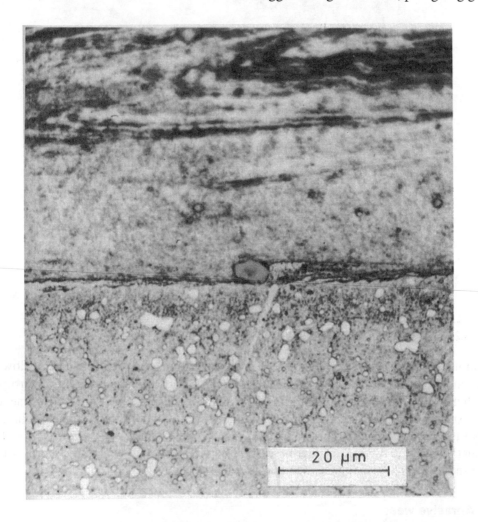

FIGURE 6.22 Section through rake face of high speed steel tool after cutting austenitic stainless steel containing titanium; abrasive action by Ti(C,N) particle (After Wright and Trent[12])

Abrasion is intuitively considered as a major cause of wear and the literature on the subject often describes tool wear in general as abrasive, but this is an area that requires further investigation for normal conditions of cutting. Where the work material contains greater concentrations of hard particles, such as pockets of sand on the surface of castings, rapid wear by abrasion undoubtedly occurs. In such concentrations the action is like that of a grinding wheel, and the surfaces of castings are treated to remove abrasive material in order to improve tool life. Under conditions of sliding at the interface, or where seizure is intermittent, abrasion may play a much more important role than under complete seizure.

6.6.6   Wear under sliding conditions

At those parts of the interface where sliding occurs, either continuously or intermittently, other wear mechanisms can come into play and, under suitable conditions, can cause accelerated wear in these regions. The parts of the surface particularly affected are those shown as areas of intermittent contact in *Figure 3.17*. The most frequently affected are those marked *E, H* and *F* in *Figure 3.17*, on the rake and clearance faces. Greatly accelerated wear on the rake face at the position *EH*, and down the flank from *E* where the original work surface crossed the cutting edge, is shown in *Figure 6.23*. The wear mechanisms operating in these sliding regions are probably those which occur under more normal engineering conditions at sliding surfaces, involving both abrasion and metal transfer, and greatly influenced by chemical interactions with the surrounding atmosphere. This is further discussed in relation to the action of cutting lubricants in Chapter 10.

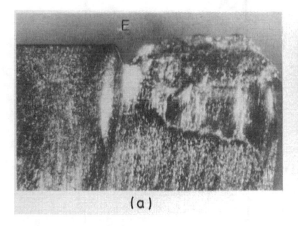 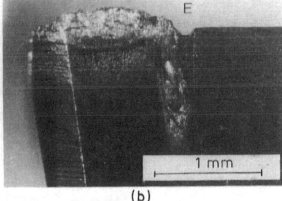

FIGURE 6.23 (a) Rake surface and (b) flank of high speed steel tool used to cut steel, showing built-up edge and sliding wear at position E

6.6.7   Summary

Figure 6.24 concludes the discussion of the wear and deformation processes which have been shown to change the shape of the tool and affect tool life when cutting steel, cast iron, and other high temperature metals and alloys, with high speed steel tools.

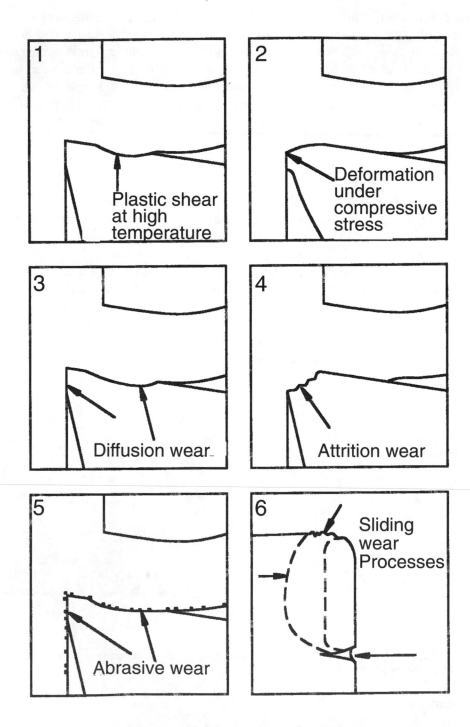

FIGURE 6.24 Wear mechanisms on high speed steel tools: 1) Plastic shear at high temperature, 2) Deformation of the edge under the applied normal stress, 3) Diffusion and solution 4) Attrition or "plucking" wear, 5) Abrasion, 6) Notch wear at the perimeter of contact

The relative importance of these processes depends on many factors - the work material, the machining operation, cutting conditions, tool geometry, and use of lubricants. In general, the first three processes are important at high rates of metal removal where temperatures are high, and their action is accelerated as cutting speed increases. It is these processes which set the upper limit to the rate of metal removal. At lower speeds, tool life is more often terminated by one of the last three - abrasion, attrition or a sliding wear process - or by fracture. Under unfavorable conditions the action of any one of these processes can lead to rapid destruction of the tool edge, and it is important to understand the wear or deformation process involved in order to take remedial action.

6.7   TOOL-LIFE TESTING

Most data from tool-life testing has been compiled by carrying out simple lathe turning tests in continuous cutting, using tools with a standard geometry, and measuring the width of the flank wear land and sometimes the dimensions of any crater formed on the rake face (*Figure 6.25*).

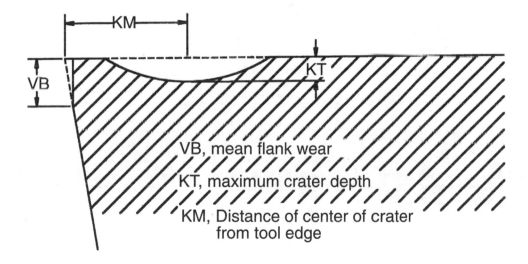

FIGURE 6.25   Conventional tool wear measurements

Steel and cast iron have been the work materials in the majority of reported tests. The results of one such test program, cutting steel with high speed tools over a wide range of speed and three feed rates is given by Opitz and König[13] and one set of results is summarized in *Figure 6.26*. For crater wear the results are simple, the wear rate being very low up to a critical speed, above which cratering increased rapidly. This critical speed is lowered as the feed is increased.

For flank wear the wear rate increases rapidly at about the same speed and feed as for cratering. It is in this region that the temperature-dependent wear mechanisms control tool life. Below this critical speed range, flank wear rate does not continue low, but often increases to a high value as attrition and other wear mechanisms not dependent on temperature become dominant.

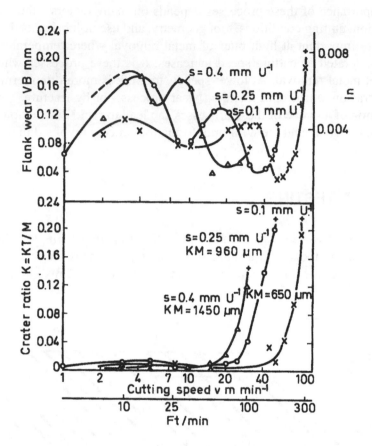

FIGURE 6.26 Influence of cutting speed and feed on flank and crater wear of high speed steel tools after cutting steel: Work material: $Ck_{55}N$ (AISI C 1055); tool material: S 12-1-4-5; depth of cut: a = 2 mm (0.08 in); tool geometry: a = 8°, γ = 10°, λ = 4°, x = 90°, $\mathcal{E}$ = 60°, r = 1 mm; cutting time: T = 30 min (After Opitz and Konig)[13]

Very high standards of systematic tool testing were set by F.W. Taylor in the work which culminated in the development of high speed steel. The variables of cutting speed, feed, depth of cut, tool geometry and lubricants, as well as tool material and heat treatment were studied and the results presented as mathematical relationships for tool life as a function of all these parameters. These tests were all carried out by lathe turning of very large steel billets using single point tools.

Such elaborate tests have been too expensive in time and manpower to repeat frequently, and it has become customary to use standardized conditions, with cutting speed and feed as the only variables. The results are presented using what is called *Taylor's equation*, which is Taylor's original relationship reduced to its simplest form:

$$VT^n = C \tag{6.1}$$

or

$$\log V = \log C - n\log T \tag{6.2}$$

where V = cutting speed, T = cutting time to produce a standard amount of flank wear (e.g. 0.75 mm (0.030 in)) and C and n are constants for the material or conditions used.

Figure 6.27 shows another set of results in which the time to produce a standard amount of flank wear is plotted on a linear scale and on a logarithmic scale against the cutting speed.[14] In spite of considerable scatter in individual test measurements, the results often fall reasonably well on a straight line on the log/log graphs. From these curves the cutting speed can be read off for a tool life of, for example, 60 minutes (V 60) or 30 minutes (V 30) and work materials are sometimes assessed by these numbers. Pronounced and significant differences are demonstrated in *Figure 6.27* between the different steels machined and such graphs are often presented as an evaluation of 'machinability'. They should be considered rather as showing one aspect of machinability or of tool performance - i.e. the tool life for a given tool and work material and tool geometry *when cutting in the high speed range*.

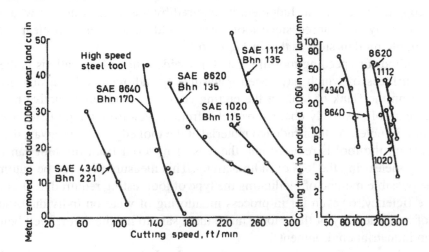

FIGURE 6.27 'Taylor' tool-life curves. Metal removed and cutting time to produce a 0.060 in (1.5 mm) wear land on a high speed steel tool (18-4-1) *vs* cutting time for five annealed steels. Tool angles in degrees: back rake, 0; side rake, 15; side cutting edge angle, 0; end cutting edge angle, 5; relief, 5. Nose radius: 0.005 in (0. 13 mm); feed: 0.009 in/rev (0.23 mm rev^{-1}); depth of cut: 0.062 in (1.57 mm).[14] (By permission from *Metals Handbook*, lst Supplement, Copyright American Society for Metals, 1954)

The *Taylor curves* are valid for those conditions in the high speed range where tool life is controlled by the temperature-dependent wear processes involving deformation and diffusion. It could be implied from *Figure 6.27* that, if cutting speed were reduced to still lower values, the tool life would become effectively infinite - the tools would never wear out. Such extrapolations to lower cutting speeds are not valid. As shown in *Figure 6.26*, the rate of flank wear may increase again at lower speeds because other wear mechanisms come into play.

In many practical operations it is not possible to use high cutting speeds - for example along the cutting edge of a drill, when turning or forming small diameters, or when broaching, planing or shaping. For these operations the 'Taylor curves' are not suitable for predicting tool life. It is relatively easy to predict from laboratory tool tests the upper limit to the speed and feed which can be employed with a given tool and work material. Below this upper limit it is much more difficult to predict the life of tools or the optimum speed and feed, but this information is often urgently required in industry for process planning. Even for continuous turning operations with single point tools the number of machine shop parameters is very high and tool life may be very greatly influenced by tool rake and clearance angles, approach angle, nose radius, cutting lubricant, etc. When machining operations require tools of more complex shapes as in milling, drilling, tapping, or parting off, prediction of the desired tool life/cutting speed and feed relationships is much more difficult; preliminary modeling for such predictions is introduced in Chapter 12.

Adequate recording of industrial experience is essential. Books of data and data banks are available to suggest to users starting points for determining optimum cutting conditions for maximum metal removal efficiency for particular types of operation.[19,20] These must often be supplemented by tool life tests under the user's actual conditions of machining. Such tests are apt to be expensive in material and manpower, not least because of the large scatter in individual test results. This work involves very careful measurement of the very small amounts of wear, the use of a microscope being essential. Judgment is required by the investigator on what is significant and what can safely be ignored since tool wear is seldom as even and clearly defined as is implied by simple models such as that in *Figure 6.25*.

A recent method for speeding tool life tests for a wide range of operations is the use of tools the wearing surfaces of which have been made lightly radio-active.[15] In one investigation, a small volume on the flank of tools at the cutting edge was made radio-active by exposing the surface to a beam of positively charged particles from a cyclotron, the particles are usually protons or helium-3 nuclei. As the irradiated material at the tool edge is worn away, the total amount of radiation from the tool is reduced and the loss of material from the tool can be accurately determined by measuring the loss of radio-activity. This measurement can be automated, so this presents one possible method of facilitating the type of tool testing required by users for optimizing cutting efficiency, or even for in-process monitoring of wear on individual tools. The very low levels of radio-activity of these tools make them safe to handle, being well below the limits set for use in industrial environment.

Quantitative determination of the amount of wear should be accompanied by intelligent investigation of the causes of wear. Worn tools are metallurgical failures and should be treated as such.

6.8 CONDITIONS OF USE

An extremely wide range of sizes and shapes of high speed steel tools is in use today. Many of these are in the form of solid tools, consisting entirely of high speed steel, which are shaped by machining close to the final size when the steel is in the annealed condition. After hardening, the final dimensions are achieved by grinding and this operation requires much skill and experience. There is a considerable difference in the difficulty of grinding between one grade and another, the high vanadium grades causing most problems, with very high rates of grinding wheel wear.

In some cases the high speed steel is in the form of an insert which is clamped, brazed or welded to the main body of the tool, which is made of a cheaper carbon or low alloy steel. This is a tendency which is increasing, and recent developments include drills where the high speed steel cutting end is friction welded to a carbon steel shank, and band saws in which a narrow band of high speed steel, with teeth, is electron-beam welded to a low alloy steel band. Most steel tools are designed to be reground when worn, so that each tool edge may be prepared for use many times. The 'throw-away' tool tip, commonly used in carbide tooling, has only recently become a serious commercial tooling type in high speed steel. For reasons connected with recent developments in steel production and tool manufacture, the high speed steel indexable insert, mechanically clamped into a tool holder, is now being more widely used, a trend which seems likely to develop in the future.

6.9   FURTHER DEVELOPMENT

Among recent innovations in high speed steel technology, four that seem likely to find a permanent place are:
- Tool steel manufacture by powder metallurgy processing.
- Spray deposition of molten steel.
- Powder metal combinations of high speed steel and carbide
- Wear resistant coatings on tools produced by chemical vapor deposition (CVD) and physical vapor deposition (PVD).

6.9.1   Powder metal high speed steel

A number of different routes have now been explored and developed commercially for the production of high speed steel by powder metallurgy. All of these start with powder produced by atomization, a stream of molten steel being broken into droplets by jets of gas or water. The particles, usually about 50 to 500 μm across, solidify within a fraction of a second or a few seconds and therefore have a very fine-grained structure. The carbide particles are much smaller and more evenly dispersed than in ingots, where rates of cooling are thousands of times slower.

The powder can be consolidated by a number of different processes. The highest quality is produced by sealing the powder, which has been atomized in a nitrogen atmosphere, into evacuated steel canisters which are heated to a temperature of the order of 1150°C and subjected to isostatic pressure up to 1500 bar (10 tonf/in^2). This process is carried out in highly engineered and expensive hot isostatic pressing units (HIP). Billets up to 0.7 m diameter and 1.5 m long are produced in this way with only very slight residual porosity.

Hot rolling or forging to the required size eliminates any remaining pores and the product has a very uniform distribution of carbide particles. In a variant of this process high speed steel powder, sealed into canisters, can be consolidated by hot extrusion. Tools are made from the billets by conventional methods.

An alternative powder metal route starts with water-atomized powder. The carbon content can be adjusted by blending carbon with the steel powder. The blended powder is consolidated in dies under high pressure to produce small tools or blanks from which tools can be made. The cold-pressed compacts, which are very weak and of high porosity, are consolidated by sintering. This operation is carried out *in vacuo* or in a protective atmosphere at a temperature very close to the solidus temperature (where a liquid phase first appears in the structure).

The sintering temperature must be controlled with extreme accuracy and the heating cycle is critical if a uniform, fine-grained structure is to be produced, with well-dispersed carbide particles and almost free from porosity. It is doubtful whether the cold pressed and sintered tools can achieve the freedom from porosity of the isostatically hot pressed material, but performance in many industrial applications is satisfactory.

Sintered tools of a variety of shapes can be made to good dimensional accuracy, eliminating many shaping operations required to produce the same tools from bars. For the production of large numbers of identical tips, this process has considerable economic advantage and it is ideal for production of the small indexable inserts.

Powder metal high speed steels have certain major advantages which should ensure their continued development on a commercial scale. The superior structures, free from segregation, ensure good and nearly uniform mechanical properties in all directions. This is particularly important for tools of large size which are likely to be of inferior quality when made by conventional processing. The advantage is likely to be a lower incidence of premature failure rather than a reduction in rate of wear. The economic potential of the cold-pressed and sintered tools may be fully realized only as the techniques of using clamped-tip high-speed steel tools are more fully developed.

6.9.2   Spray deposition

An alternative method which promises to become of increasing importance starts, as in the gas atomization process for producing powder, with a stream of molten steel which is broken into a spray of droplets by jets of nitrogen gas. Instead of the droplets being allowed to solidify as separate particles of a powder, they are sprayed onto a solid substrate while they are still molten or only partially solidified. The particles coalesce and are very rapidly solidified, heat being conducted into the substrate.

The product is of very low porosity with a fine grained, uniform and isotropic structure. Tubes, bars, flat products and a variety of shapes can be produced which can be further processed by rolling, or forging. Up to ten tons per hour can be deposited by this method, one of the best known commercial processes being the 'Osprey' process. [18,21]

6.9.3   Powder metal combinations of high speed steel and carbide

Sintered tools of cemented carbides such as titanium carbide (TiC) are the subject of the next chapter. Generally speaking, all such tools are harder yet less tough than high speed steel. Not surprisingly, research has investigated the possibility of creating tool materials that are "half way between high speed steel and carbide." Often, the goal is to create a material that can be used in interrupted milling and drilling operations, where a combination of wear resistance and impact resistance is desirable.

Various matrix mixtures of 40-55% TiC and 45-60% steel {with possible additions of chromium (in the range of 3-17.5%), molybdenum (0.5-4%), nickel (0.5-12%), cobalt (5-5.7%), titanium (0.5-0.7%), and carbon (0.4-0.85%)} have been developed.[22] Compacted TiC powder is sintered at high temperature and then infiltrated with molten steel under vacuum: this creates rounded carbide grains in a spheroidite steel matrix. After annealing, the tool material can be machined and ground into desirable shapes. After subsequent quenching and tempering the steel part of the matrix becomes a fine martensite structure, similar to *Figure 6.1b*.

Titanium nitride (TiN) may also be used, instead of TiC, as the complementary carbide in these "mixed matrix" powdered metal tools.[23] 30-60% of fine grain size (~0.1 µm) TiN is generally used. Again, this material can be heat-treated and ground in the annealed state, prior to quenching and tempering. The outer surfaces can also be coated with TiCN or TiN by physical vapor deposition (PVD). This further adds to the wear resistance, especially for drills.

6.9.4   Coated steel tools

The concept of improving performance by forming hard layers on the working surfaces of tools to reduce friction or wear is not a new one. Many treatments to form such layers have been patented and used in industrial applications. Two examples are the formation of blue oxide film on tool surfaces by heating in a steam atmosphere, and treatment of tools in salt-baths to introduce high levels of carbon, nitrogen and/or sulfur into the wearing surfaces.

Many users have found advantage in these surface treatments to prolong tool life or to reduce pick-up of work material on the tool surfaces. They have been used more frequently for forming tools and drills rather than turning tools and there have been few studies of the mechanism by which advantages are achieved.[18]

Recently, development of the processes of chemical vapor deposition (CVD) and physical vapor deposition (PVD) have resulted in the commercial availability of high speed steel cutting tools coated with thin layers of refractory metal carbide or nitride.

Although first proposed for the coating of steel, large scale commercial development of the CVD process was for coatings on carbide tools. CVD layers on steel tools are usually less than 10 µm in thickness. Specialized equipment is required in which deposition takes place at temperatures in the range 850-1050°C in a sealed chamber in a hydrogen atmosphere.

The metal (e.g. titanium) and non-metal (e.g. carbon or nitrogen) atoms are introduced into the atmosphere as gaseous compounds. Very fine grained, solid coatings of metallic carbide or nitride are deposited and adhere strongly to the tool surfaces. An example is shown in *Figure 6.28*, which is a section normal to the coated surface.

FIGURE 6.28 Section through CVD coating on high speed steel tool

Because of the high temperatures required for the CVD process, tools must be hardened and tempered after coating and special precautions taken to preserve the very thin coating intact. A vacuum heat treatment system is usually employed for the high temperature hardening. This may alter the precise shape of tools, particularly ones of large size and complex shape, and correction cannot be made without removing the coating.

To avoid this type of problem a PVD process is usually employed for coating of high speed steel. This also is carried out in sealed vessels at reduced gas pressure, but at temperatures in the range 400-600°C. Lower temperatures are possible because the coating atoms are ionized and attracted to the tool surfaces which are at a negative potential, these surfaces having first been cleaned by bombardment (sputtering) with ions of a neutral gas. Because of the lower temperatures of PVD coating, complete heat treatment of high speed steel tools can be carried out by normal methods before coating. PVD seems likely to be the main process used for coating high speed steel (*Figures 6.29* and *6.30*).

Titanium carbide (TiC) and nitride (TiN), hafnium nitride (HfN) and alumina have been proposed as coatings and, of these, TiN has probably the most to commend it. TiN is a cubic compound, isomorphous with the better known TiC. It is not as hard as TiC, measured by indentation hardness test, but is its equal or superior in terms of wear resistance in many cutting operations. The bright gold color of TiN has the advantage of allowing coated tools to be easily identified. TiC-coated tools cannot readily be distinguished from uncoated tools.

Both laboratory tests and machine shop experience demonstrate considerable advantages for TiN-coated high speed steel tools when machining cast iron and most types of steel. Cutting forces when using coated tools are lower than with uncoated tools on the same operation.[24,25] This is related to a corresponding reduction of contact area on the rake face of the tool. This may be an advantage, for example, in reducing the incidence of fracture in twist drills and facilitating swift removal. The rate of wear may be reduced by several orders of magnitude. Improvements in tool life have been reported of from 2 to 100 times in different operations.[24,25]

When using coated tools the design of the cutting edge may be modified for further improvement in tool life or rate of metal removal. At present the most commonly coated steel tools are twist drills. Efficient use of coated drills requires reshaping without great loss in tool life. In regrinding drills the coating is removed from the clearance faces but retained in the flutes. Experience indicates that even after regrinding most of the advantage of the coating is retained.

There is evidence that, when cutting steel at relatively low speed, the built-up edge is either absent or is very much smaller with TiN coated tools. Another advantage claimed for coated tools is that the surface finish may be greatly improved by elimination of the built-up edge. Laboratory tests have demonstrated that, during high speed cutting of steel, seizure occurs even between coated tools and the work material. The flow-zone at the interface is still the heat source and there is usually only a small reduction of the maximum temperature at the interface. When the high speed steel beneath the coating is heated to high temperature, its strength is greatly reduced, it is plastically deformed and the coating is broken up. *Figure 6.30* illustrates the break-up of a coating where the substrate had been deformed - the adhesion of lengths of the coating in spite of very severe deformation of the steel demonstrates the very strong adhesion of a CVD coating. The coating can resist wear at very high cutting speeds but the use of high speed steel as a substrate always limits the use of these tools when cutting steel. A speed increase of 25-50% over that of uncoated tools is typical.[25]

For many types of coated tooling re-grinding is not possible and the throw-away indexable tool tip with several cutting edges is of great advantage. With these, increased efficiency comes not only from the longer tool life and increased metal removal rates, but from reduced tool-changing time. PVD coating of high speed steel is an important addition to the range of tool materials available and more development can be expected.

Thus, in summary, it perhaps important to emphasize that CVD and PVD coatings on high speed steel give the most benefit where a) speeds are not increased too much but b) where the user is seeking longer tool life or smoother cutting at more or less the existing speed.

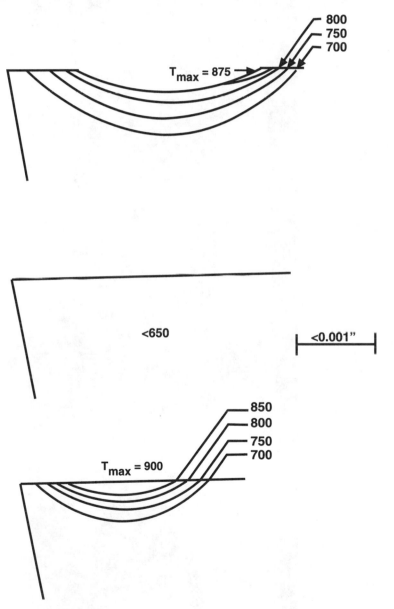

FIGURE 6.29 Temperatures in a) uncoated M34 high speed steel at 100 m min^{-1} b) TiN coated tool at 100 m min^{-1} c) TiN coated tool at 150 m min^{-1}. Cutting low carbon iron (AISI 1004 steel)

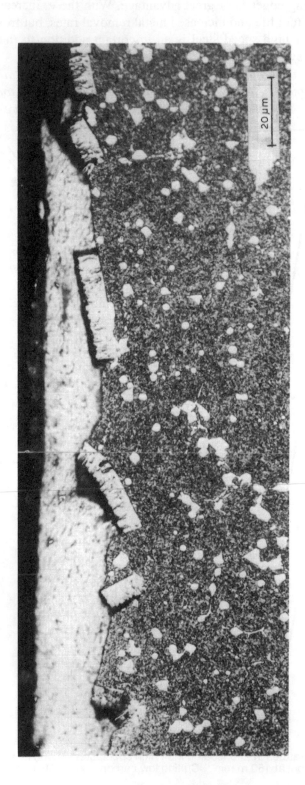

FIGURE 6.30 Section through rake face of CVD-coated high speed tool after cutting steel at high speed. Coating broke when substrate deformed.

6.10 CONCLUSION

High speed steel tools, first demonstrated in 1900, were the product of an intense technological research program to determine the most efficient tool steel composition and heat treatment for the rough machining of steel. They made possible the cutting of steel at about four times the rate of metal removal achieved with carbon steel tools. This advance was made possible by retention of compressive strength of the order of 1500 MPa to temperatures in the range 550-650° C. These properties are the result of precipitation hardening within the martensitic structure of the chromium, tungsten, molybdenum, and vanadium tool steels after a very high temperature heat treatment. The critical strength is retained at the same temperatures generated at the cutting edge of tools when used to cut a wide range of engineering steels over a range of cutting speeds and feeds useful in industrial machining.

The most important development has been the coating of tools with thin layers ($\sim$ 10 µm) of TiN by a PVD process. Such coatings prolong tool life in most situations and in others give a smoother surface finish on the machined part. *Figures 3.25* to *3.30* are worth revisiting at this point. The evidence indicates that the TiN coatings have several beneficial effects. They keep the all-important cutting edge smooth and blemish free; they reduce friction on the rake face and allow a smoother chip flow process - consequently both forces and temperatures are lower; they reduce any heat due to rubbing on the clearance face. All these effects reduce the amount of sub-surface strain going into the machined surface.

Correctly heat-treated high speed steel has adequate toughness to resist fracture in machine shop conditions, and it is the combination of high-temperature strength and toughness that has made this class of tool material a successful survivor for 100 years in the evolution of cutting tool materials. It is significant that in those 100 years no new class of tool steel has superseded it. The loss of strength and permanent changes in the structure of high speed steel when heated above about 650° C limit the rate of metal removal when cutting high melting-point metals and alloys.

6.11 REFERENCES

1. 'Tool Steels', *Iron & Steel*, Special Issue, (1968)
2. Brookes, K.J.A., *World Directory and Handbook of Hard Metals*, Engineers Digest (1987)
3. Osborn, F.M., *The Story of the Mushets*, Nelson (1952)
4. Taylor, F. W., Trans. A.S. M. E., **28**, 31 (1907)
5. American National Standards Institute B212.4-86, B212.12-91
6. Mukherjee, T., I.S.I. *Publication*, **126**, 80 (1970)
7. Kirk, F.A., Child, H.C., Lowe, E.M. and Williams, T.J., *I.S.I. Publication*, **126**, 67(1970)
8. Weaver, C., Unpublished work
9. Trent, E.M., *Proc. Int. Conf M.T.D.R., Manchester,* 1967, 629 (1968)
10. Hellman, P. and Wisell, H., *Bulletin du Cercle d'Etudes des Metaux*, p. 483 (Nov. 1975)
11. Ekelund, S., *Fagersta High Speed Steel Symposium*, p. 3 (1981)
12. Wright, P.K. and Trent, E.M., *Metals Technology*, **1**, 13 (1974)

13. Opitz, H. and Konig, W., I.S.I. *Publication*, **126**, 6 (1970)
14. *A.S.M. Handbook* (1948), also *Metal Progress*, 15 July, **141**, (1954)
15. Amini, E. and Winterton, R.H.S., *Proc. Inst. Mech. Eng.*, **195** (21), 241 (1981)
16. Horton, S.A. and Child, H.C., *Metals Technol.*, **10**, 245 (1983)
17. El-Rakayby, A.M. and Mills, B., *Mat. Sci. & Tech.*, **2**, 175 (1986)
18. Hoyle, G., *High Speed Steels*, Butterworths, London (1988)
19. *Machining Data Handbook,* Vol. 1 Machinability Data Center, Cincinnati (1990)
20. *ASM Handbook*, 9th Edition, Vol. 16 (1989)
21. Leatham, A., Ogilvy, A., Chesney, P. and Wood, J. V., *Metals & Materials*, 140 (March 1989)
22. Tarkan, S.E., and Mal, M.K., *Metal Progress*, **105**, 99 (1974)
23. Sandvik Coromant Company, *Modern Metal Cutting- A Practical Handbook*, page III-41, (1994)
24. Barrell, R. and Rickerby, D.S., *Metals & Materials*, (August 1989)
25. Chow, J.G., *Ph.D. Dissertation*, Carnegie Mellon University, Pittsburgh PA., (1985)

CHAPTER 7

CUTTING TOOL MATERIALS II: CEMENTED CARBIDES

7.1 CEMENTED CARBIDES: AN INTRODUCTION

Many substances harder than quenched tool steel have been known from ancient times. Diamond, corundum and quartzite, among many others, were natural materials used to grind metals. These could be used in the form of loose abrasive or as grinding wheels, but were unsuitable as metal-cutting tools because of inadequate toughness. The introduction of the electric furnace in the last century, led to the production of new hard substances at the very high temperatures made available. The American chemist Acheson produced silicon carbide in 1891 in an electric arc between carbon electrodes. This can be used loose as an abrasive and, when bonded with porcelain, is very important as a grinding wheel material, but is not tough enough for cutting tools.

Many scientists, engineers and inventors in this period explored the use of the electric furnace with the aim of producing synthetic diamonds. In this they were not successful, but Henri Moissan at the Sorbonne made many new carbides, borides and silicides - all very hard materials with high melting points. Among these was tungsten carbide which was found to be exceptionally hard and had many metallic characteristics. It did not attract much attention at the time, but there were some attempts to prepare it in a form suitable for use as a cutting tool or drawing die.

There are, in fact, two carbides of tungsten - WC which decomposes at 2,600°C, and W_2C which melts at 2,750°C. Both are very hard and there is a eutectic alloy at an intermediate composition and a lower melting point (2,525°C). This can be melted and cast with difficulty, and ground to shape using diamond grinding wheels, but the castings are coarse in structure, with many flaws. They fracture easily and proved to be unsatisfactory for cutting tools and dies.

In the early 1900s the work of Coolidge led to the manufacture of lamp filaments from tungsten, starting with tungsten powder with a grain size of a few micrometers (microns).

The use of the powder route of manufacture eventually solved the problem of how to make use of the hardness and wear resistant qualities of tungsten carbide. In the early 1920s Schröter, working in the laboratories of Osram in Germany, heated tungsten powder with carbon to produce the carbide WC in powder form, with a grain size of a few microns.[1,2] This was thoroughly mixed with a small percentage of a metal of the iron group - iron, nickel or cobalt - also in the form of a fine powder.

Today, the raw materials used in carbide cutting tools are made by melting and subsequent particle-size reduction by ball-milling (comminution) to a fine powder. The comminution process limits the final size of the powder to about a size of one micron. Newer, chemical routes, such as the sol-gel technique can produce ultrafine materials in nanocrystalline size. These materials are also relatively freer from defects. Cost permitting, they will find a niche for high-speed milling.

Cobalt was found to be the most efficient metal for bonding WC. The mixed powders were pressed into compacts which were sintered by heating in hydrogen to above 1,300°C. Unlike many powder-metal products, tungsten carbide/cobalt mixtures can be sintered to full density - free from porosity - in a single heating cycle.

This is because a liquid phase solution of WC in cobalt is formed at about 1,300°C. This wets and pulls together all the remaining WC particles. On cooling to room temperature, the liquid phase solidifies and the product is fully dense. The liquid phase sintering process, introduced by Schröter, became the technological basis for the *cemented carbide* industry and is used universally, not only for alloys of WC and cobalt, but for many other combinations of metal carbides and nitrides with cobalt, nickel and iron.

The cemented carbides have a unique combination of properties which has led to their development into the second major "genus" of cutting tool materials in use today.

7.2　STRUCTURES AND PROPERTIES

Tungsten carbide is one of a group of compounds including the carbides, nitrides, borides and silicides of transition elements of Groups IV, V and VI of the Periodic Table.[1-4] Of these, the carbides are important as tool materials, and the dominant role has been played by the mono-carbide of tungsten, WC.

Table 7.1 gives the melting point and room temperature hardness of some of the carbides. All the values are very high compared with those of steel.

The carbides of tungsten and molybdenum have hexagonal structures, while the others of major importance are cubic. These rigid and strongly bonded compounds undergo no major structural changes up to their melting points, and their properties are therefore stable and unaltered by heat treatment, unlike steels which can be softened by annealing and hardened by rapid cooling.

These carbides are strongly metallic in character, having good electrical and thermal conductivities and a metallic appearance. Although they have only slight ability to deform plastically without fracture at room temperature, the electron microscope has shown that they are deformed by the same mechanism as are metals by movement of dislocations.

TABLE 7.1   Properties of carbides (There are large differences in reported values for all of the carbides. The values given here are representative.)

Carbides	Melting point	Diamond indentation hardness
	(°C)	*(HV)*
TiC	3,200	3,200
V_4C_3	2,800	2,500
TbC	3,500	2,400
TaC	3,900	1,800
WC	2,750	2,100

They are sometimes included in the category of ceramics, and the cemented carbides have been referred to as 'cermets', implying a combination of ceramic and metal, but this term seems quite inappropriate, since the carbides are much closer in character to metals than to ceramics.

Figure 7.1 shows the hardness of four of the more important carbides measured at temperatures from 15°C to over 1,000°C.[5,6] All are much harder than steel, though not as hard as diamond (6,000-8,000 HV in this temperature range).

The hardness of the carbides drops rapidly with increasing temperatures, but they remain much harder than steel under almost all conditions. This hardness, and the stability of the properties under a wide range of thermal treatments, encourage the use of carbide cutting tools.

In cemented carbide alloys, the carbide particles constitute about 55-92% by volume of the structure, and those alloys used for metal cutting normally contain at least 80% carbide by volume. *Figure 7.2* shows the structure of one alloy, the angular gray particles being WC and the white areas cobalt metal.

In high speed steels the hard carbide particles of microscopic size constitute only 10-15% by volume of the heat-treated steel (*Figure 6.1b*), and play a minor role in the performance of these alloys as cutting tools, but in the cemented carbides they are the decisive constituents. The powder metal production process makes it possible to control accurately both the composition of the alloy and the size of the carbide grains.

7.3   TUNGSTEN CARBIDE-COBALT ALLOYS (WC-Co)

This group, technologically the most important, is considered first. These are available commercially with cobalt contents between 4% and 30% by weight - those with cobalt contents between 4% and 12% being commonly used for metal cutting - and with carbide grains varying in size between 0.5 μm and 10 μm across. The performance of carbide cutting tools is very dependent upon the composition and grain size: *Tables 7.1* to *7.3* include the range of the properties and usage of WC-Co grades.

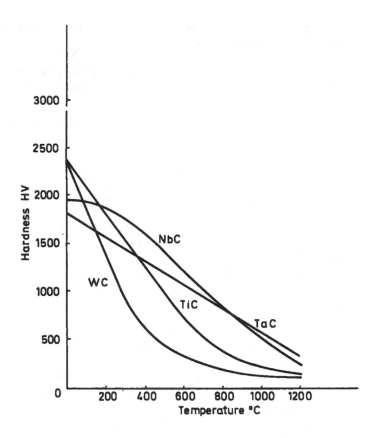

FIGURE 7.1 Hot hardness tests on mono carbides of four transition elements (After Atkins and Tabor[5]; Miyoshi and Hara[6])

FIGURE 7.2 Structure of a cemented carbide (WC-Co) of coarse grain size

The structure of tungsten carbide cobalt alloys should show two phases only - the carbide WC and cobalt metal (*Figure 7.2*). The carbon content must be controlled within very narrow limits. The presence of either free carbon (too high a carbon content) or 'eta phase' (a carbide with the composition Co_3W_3C, the presence of which denotes too low a carbon content) results in reduction of strength and performance as a cutting tool. The structures should be very sound showing very few holes or non-metallic inclusions.

Table 7.2 gives properties of a range of WC-Co alloys in relation to their composition and grain size, and *Figure 7.3* shows graphically the influence of cobalt content on some of the properties. Both hardness and compressive strength are highest with alloys of low cobalt content and decrease continuously as the cobalt content is raised.[7] For any composition the hardness is higher the finer the grain size and over the whole range of compositions used for cutting, the cemented carbides are much harder than the hardest steel.

As with high speed steels, the tensile strength is seldom measured, and the breaking strength in a bend test or 'transverse rupture strength' is often used as a measure of the ability to resist fracture in service. *Figure 7.3* shows that the transverse rupture strength varies inversely with the hardness, and is highest for the high cobalt alloys with coarse grain size.

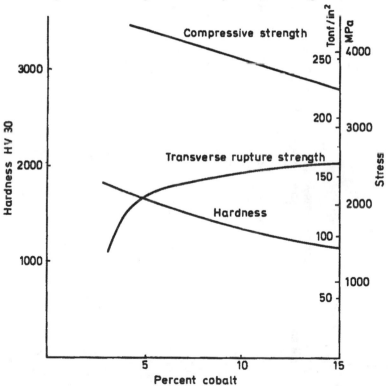

FIGURE 7.3   Mechanical properties of WC-Co alloys of medium-fine grain size

In tensile or bend tests, fracture occurs with no measurable plastic deformation, and cemented carbides are, therefore, often characterized as 'brittle' materials in the same category

TABLE 7.2   Properties of WC-Co alloys

Co %	Mean WC grain size	Hardness	Transverse rupture strength		Compressive Strength		Young's modulus		Fracture toughness K_{IC}	Specific gravity
	(μm)	(HV 30)	(MPa)	(tonf/in²)	(MPa)	(tonf/in²)	(GPa)	(tonf/in² x 10³)	(MPa m$^{-1/2}$)	
3	0.7	2,020	1,000	65	-	-	-	-	-	-
	1.4	1,820	-	-	-	-	-	-	8	-
6	0.7	1,800	1,750	113	4,550	295	-	-	-	-
	1.4	1,575	2,300	148	4,250	275	630	40.7	10	14.95
	0.7	1,670	2,300	148	-	-	-	-	-	-
9	1.4	1,420	2,400	156	4,000	260	588	38.0	13	14.75
	4.0	1,210	2,770	179	4,000	260	-	-	-	-
15	0.7	1,400	2,770	179	-	-	538	34.8	-	-
	1.4	1,160	2,600	168	3,500	225	-	-	18	14.00

as glass and ceramics. This is not justified because, under conditions where the stress is largely compressive, cemented carbides are capable of considerable plastic deformation before failure.

Figure 7.4 shows stress-strain curves for two cemented carbide alloys in compression, in comparison with high speed steel. Both cemented carbides have much higher Young's modulus (*E*), and the 6% Co alloy has higher yield stress than the steel. Above the yield stress, the curve departs gradually from linearity, and plastic deformation occurs accompanied by strain hardening raising still further the yield stress. The amount of plastic deformation before fracture increases with the cobalt content.[8,9] The ability to yield plastically before failure can also be demonstrated by making indentations in a polished surface.

Indentations in a carbide surface, made with a carbide ball, are shown in *Figure 7.5*, the contours of the impressions being demonstrated by optical interferometry. Indentations to a measurable depth can be made before a peripheral crack appears, and the higher the cobalt content the deeper the indentation before cracking. It is this combination of properties - the high hardness and strength, together with the ability to deform plastically before failure under compressive stress - which makes cemented carbides based on WC so well adapted for use as tool materials in the engineering industry (*Table 7.3*).

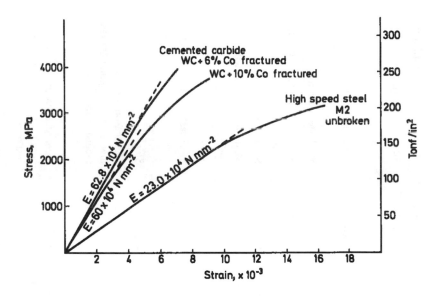

FIGURE 7.4   Compression tests on two WC-Co alloys compared with high speed steel

There is no generally accepted test for toughness of cemented carbides and the terms *toughness* and *brittleness* have to be used in a qualitative way. This is unfortunate because it is on the basis of toughness that the selection of the optimum tool material must often be made.

Fracture toughness tests on cemented carbide have now been reported from a number of laboratories and there is reasonably good agreement on the values of K_{IC}. *Table 7.2* shows that K_{IC} values for WC-Co alloys increase with cobalt content, varying from 10 MPa m$^{-1/2}$ for a 6% Co

TABLE 7.3   Classification of carbides according to use[10]

Symbol	Broad categories of material to be machined	Designation	Material to be machined	Use and Working Conditions
P	Ferrous metals with long chips	P 01	Steel, steel castings	Finish turning and boring; high cutting speeds, small chip section, accuracy of dimensions and fine finish, vibration-free operation.
		P 10	Steel, steel castings	Turning, copying, threading and milling, high cutting speeds, small or medium chip sections.
		P 20	Steel, steel castings; Malleable cast iron with long chips	Turning, copying, milling, medium cutting speeds and chip sections, planing with small chip sections.
		P 30	Steel, steel castings; Malleable cast iron with long chips	Turning, milling, planing, medium or low cutting speeds, medium or large chip sections, and machining in unfavorable conditions
		P 40	Steel; Steel castings with sand inclusion and cavities	Turning, planing, slotting, low cutting speeds, large chip sections with the possibility of large cutting angles for machining in unfavorable conditions and work on automatic machines
		P 50	Steel; Steel castings of medium or low tensile strength, with sand inclusion and cavities	For operations demanding very tough carbide: turning, planing, slotting, low cutting speeds, large chip sections, with the possibility of large cutting angles for machining in unfavorable conditions and work on automatic machines.
M	Ferrous metals with long or short chips and non-ferrous metals	M 10	Steel, steel castings, manganese steel, gray cast iron, alloy cast iron	Turning, medium or high cutting speeds. Small or medium chip sections
		M 20	Steel, steel castings, austenitic or manganese steel, gray cast iron	Turning, milling. Medium cutting speeds and chip sections

Symbol	Broad categories of material to be machined	Designation	Material to be machined	Use and Working Conditions
		M 30	Steel, steel castings, austenitic steel, gray cast iron, high temperature resistant alloys	Turning, milling, planing. Medium cutting speeds, medium or large chip sections
		M 40	Mild free cutting steel, low tensile steel	Turning, parting off, particularly on automatic machines
			Non-ferrous metals and light alloys	
K	Ferrous metals with short chips, non-ferrous metals and non-metallic materials	K 01	Very hard gray cast iron, chilled castings of over 85 Shore, high silicon aluminium alloys, hardened steel, highly abrasive plastics, hard cardboard, ceramics	Turning, finish turning, boring, milling, scraping
		K 10	Gray cast iron over 220 Brinell, malleable cast iron with short chips, hardened steel, silicon aluminium alloys, copper alloys, plastics, glass, hard rubber, hard cardboard, porcelain, stone	Turning, milling, drilling, boring, broaching, scraping
		K 20	Gray cast iron up to 220 Brinell, non-ferrous metals: copper, brass, aluminium	Turning, milling, planing, boring, broaching, demanding very tough carbide
		K 30	Low hardness gray cast iron, low tensile steel, compressed wood	Turning, milling, planing, slotting, for machining in unfavorable conditions and with the possibility of large cutting angles
		K 40	Soft wood or hard wood, Non-ferrous metals	Turning, milling, planing, slotting, for machining in unfavorable conditions and with the possibility of large cutting angles

alloy to 18 MPa m$^{-1/2}$ for a 15% Co alloy. Considering the scatter in individual test results, this is a relatively small difference, while in practice, there is a very large difference between the performances of 15% Co and 6% Co alloys subjected to applications with severe impact. This may be because the test measures the energy required to propagate a crack under tensile stress, while in many practical applications performance depends on the energy required to initiate a crack under localized compressive stress.

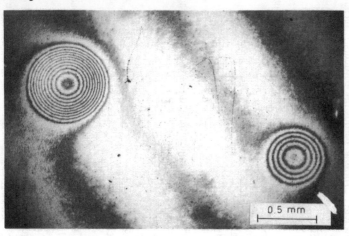

FIGURE 7.5   Ball indentations in surface of cemented carbide, revealed by optical interferometry

More research should be directed towards understanding the quality of toughness of cemented carbides as observed in practice and towards developing adequate tests for its measurement. Frequently the best tool material for a particular application is the hardest combined with adequate toughness to resist fracture. In practice the solution to this problem has been achieved by the 'natural selection' process of industrial experience.

Just as for high speed steels, where a general purpose grade M2, has become established for general machining work, so with the WC-Co alloys, a grade containing 6% Co and a grain size near 2 μm (*Figure 7.6*) has become acknowledged as the grade for the majority of applications for which this range of alloys is suitable, and this grade is made in much larger quantities than the others. If this grade proves too prone to fracture in a particular application, a grade with a higher cobalt content is tried, while, if increased wear resistance is needed, one with a finer grain size or lower cobalt content is recommended.

Trial by industrial experience is a very prolonged and expensive way of establishing differences in toughness. It is difficult for the user to assess the claims of competing suppliers regarding the qualities of their carbide grades. The user must rely on classifications such as that of the International Organization for Standardization (ISO).[10] *Table 7.3* shows the performance of cutting tools as related to the *relative* toughness and hardness of the different WC-Co alloys, arranged in a series from K01 to K40. It is the responsibility of the individual manufacturer to decide in which category their products shall be placed for guiding the end-user.

Both hardness and compressive strength of cemented carbides decrease as the temperature is raised (*Figures 7.7* and *6.9*). The comparison of compressive strength at elevated temperatures with that of high speed steel[11] (*Figure 6.9*) shows that a WC-Co alloy with 6% Co withstands a stress of 750 MPa (50 tonf/in^2) at 1000°C, while the corresponding temperature for high speed steel is 750°C.

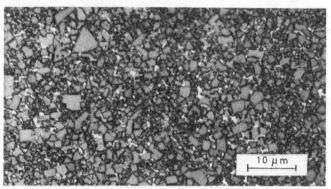

FIGURE 7.6   Structure of medium fine grained alloy of WC with 6% Co

With cemented carbides, the temperature at which this stress can be supported drops if the cobalt content is raised or the carbide grain size is increased.[12] The coefficient of thermal expansion is low - about half that of most steels. Thermal conductivity is relatively high: a 6% Co-94% WC alloy has a thermal conductivity, of 100 W/mK compared with 31 W/mK for high speed steel. Oxidation resistance at elevated temperatures is poor, oxidation in air becomes rapid over 600°C and at 900°C is very rapid indeed. Fortunately this rarely becomes a serious problem with cutting tools because the surfaces at high temperature are usually protected from oxidation (see Chapter 10).

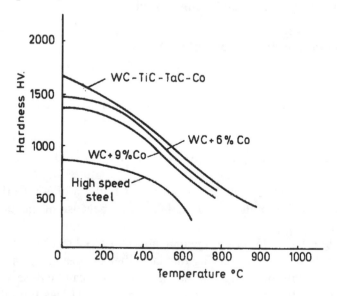

FIGURE 7.7   Hot hardness of cemented carbides

7.4   TOOL LIFE AND PERFORMANCE OF TUNGSTEN CARBIDE-COBALT TOOLS

Cemented tungsten carbide tools were introduced in the early 1930s and inaugurated a revolution in productive capacity of machine tools as great at that which followed the introduction of high speed steels. Over a period of years the necessary adaptations have been made to tool geometry, machine tool construction, and operation methodology, enabling these tools to be used efficiently. Having learned how to avoid premature fracturing, machinists could use carbide tools to cut many metals and alloys at much higher speeds and with much longer tool life than had previously been possible. There are limits to the rate of metal removal at which one can maintain a reasonable tool life.

As with high speed steel tools, the shape of the edge gradually changes with continued use until the tool no longer cuts efficiently. The mechanisms and processes which change the shape of the edge of tungsten carbide-cobalt tools when cutting cast iron, steel and other high melting-point alloys are now considered, looking first at those which set the limits to the rate of metal removal.[13]

7.4.1   Superficial plastic deformation by shear at high temperature

Similar to the situation that was examined with high speed steel tools, with cemented tungsten carbide-cobalt tools the work material is seized to the tool over much of the worn rake and flank surfaces when cutting cast iron and most steels at medium and high cutting speeds. A flow-zone or built-up edge is formed at the interface, of the same character as described for steel tools. Nevertheless, the significantly higher hot-strength of carbides seems to protect them from the wear mechanism due to superficial plastic deformation by shear at high temperature (compare *Figure 6.13* and the top left of *Figure 6.24* for high speed steel.)

7.4.2   Plastic deformation under compressive stress

Even though there is yet no experimental evidence for superficial plastic deformation by shear, carbides are nonetheless prone to deformation by compression under the influence of high temperatures and compressive stress. Thus, the rate of metal removal as cutting speed or feed are raised is often limited by deformation of the tool under compressive stress on the rake face. Carbide tools can withstand only limited deformation, even at elevated temperature, and cracks form which lead to sudden fracture. *Figure 7.8* shows such a crack in the rake face of a tool, this surface being stressed in tension as the edge is depressed.[14] Failure due to deformation is more probable at high feed rates, and when cutting materials of high hardness. Carbide grades with low cobalt content can be used to higher speed and feed rate because of increased resistance to deformation.

Deformation can be detected at an early stage in a laboratory tool test by lapping the clearance face optically flat before the test, and observing this face after the test using optical interferometry. Any deformation in the form of a bulge on the clearance face can be observed and measured as a contour map formed by the interference pattern. *Figure 7.9* shows the interference fringe pattern on the clearance face of a carbide tool used to cut steel at high speed and feed. The maximum deformation is seen to be at the nose, a common feature when cutting steel and iron. A tool with a sharp nose, or a very small nose radius, starts to deform at the nose and thus fails at much lower speed than tools with a large nose radius. In many cutting operations the precise form of the nose radius is critical for the performance of the tool. The ability of tungsten carbide-cobalt tools to resist deformation at high temperatures is the most important property permitting higher rates of metal removal compared with steel tools.

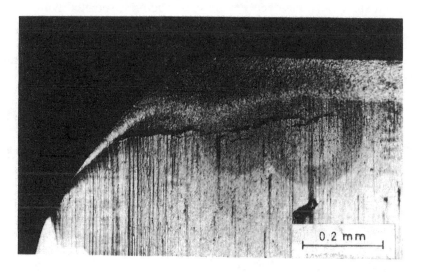

FIGURE 7.8 Cracks across the nose of cemented carbide tool deformed during cutting[14]

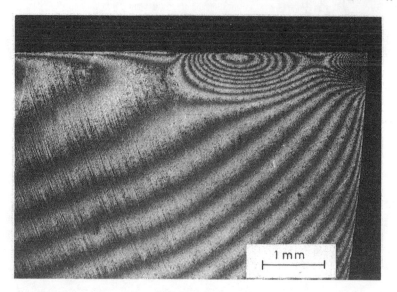

FIGURE 7.9 Deformation on clearance face of cemented carbide tool, after cutting at high speed and feed, revealed by optical interferometry[13]

7.4.3   Diffusion wear

When cutting steel at high speed and feed, a crater is formed on the rake face of tungsten carbide-cobalt tools, with an unworn flat at the tool edge (*Figure 7.10*). Dearnley's[15] measurement of temperature in carbide tools (using iron-bonded carbide) shows the same pattern of temperature as with steel tools (*Figure 7.11*). This was predictable since the heat source is of the same character: a thin flow-zone in metallic contact with the rake face of the tool. The crater is in the same position as on steel tools, the deeply worn crater being associated with the high temperature regions, and the unworn flat with the low temperatures near the edge. Sections through the crater in WC-Co tools show no evidence of plastic deformation by shear in the tool material, in contrast to the deformation observed in high speed steel tools (*Figures 6.11* and *6.13*). The grains are smoothly worn (*Figure 7.12*) with no evidence that particles are broken away from the surface.

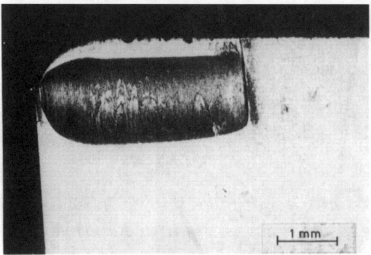

FIGURE 7.10   Crater on rake face of WC-Co alloy tool after cutting steel at high speed and feed

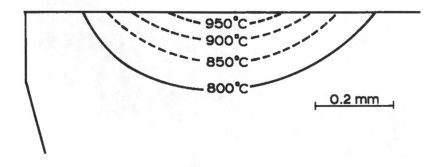

FIGURE 7.11   Temperature distribution in cemented carbide tool used to cut 0.4% C steel at a speed of 183 m min⁻¹ (600 ft/min); estimated from structural change in iron bonded carbide tool [15]

FIGURE 7.12 Section through crater surface in WC-Co tool and adhering steel, showing interface characteristics of diffusion wear

The wear process in the cratering of WC-Co tools is one of diffusion wear. *Metal and carbon atoms of the tool material, diffuse into the work material that is seized to the rake face surface. These atoms are then carried away in the body chip.* An extreme example of this sort of process was observed when studying sections through WC-Co bullet cores embedded in steel plates. These showed a layer in which WC and Co were dissolved in the steel which had been melted at the interface.[16]

Figure 7.13 shows such a section with steel at the top, the cemented carbide at the bottom, and the white, fused layer in between containing partially dissolved, rounded WC grains. The melting point of the eutectic between WC and cobalt or WC and iron is about 1300°C. Evidently, this temperature had been reached at the thin layer at the interface. In this case, the speed on impact exceeded 1000 ms^{-1} and a very large amount of energy had been converted into heat in a very short interval of time.

In a paper published in 1952,[17] Trent suggested that the cratering of WC-Co tools when cutting steel was also the result of fusion at the interface at a temperature of 1300°C. With cutting tools, however, the speeds involved are more than 100 times lower than in the case of the bullet core.

Evidence of temperature deduced from observations of worn high speed steel (*Figure 5.12*) and iron-bonded carbide tools (*Figure 7.11*) suggests that the maximum sustained temperatures at the interface, for conditions where cratering of WC-Co carbide tools occurs, are in the range 850° to 1200 °C. While these temperatures are too low for fusion, they are high enough to allow considerable diffusion to take place in the solid state, and the characteristics of the observed surfaces are consistent with a wear process based on solid phase diffusion.

FIGURE 7.13 Section through interface of WC-Co bullet core embedded in steel plate[16]

Worn carbide tools can be treated in HCl or other mineral acids to remove adhering steel or iron so that the worn areas of the tool surface can be studied in detail. While the acids dissolve the cobalt binder to some extent they do not attack the carbides, and wear on carbide tools can thus be examined by a method not possible for high speed steel tools.

Figures 7.14 and *7.15* show the worn crater surfaces of WC-Co alloy tools after cutting steel. The carbide grains are mostly smoothly worn through, but sometimes have an etched appearance, as in the large grain in *Figure 7.15* where steps etched in the surface are parallel to one of the main crystallographic directions. Smooth ridges are often seen on the surface, originating from large WC grains and running in the direction of chip flow, as in *Figure 7.14*. The smooth polished surfaces are characteristic of areas where the work material flows rapidly immediately adjacent to the surface, while the etched appearance is seen where the temperature is high enough, but the flow is less rapid or there is a stagnant layer at the interface.

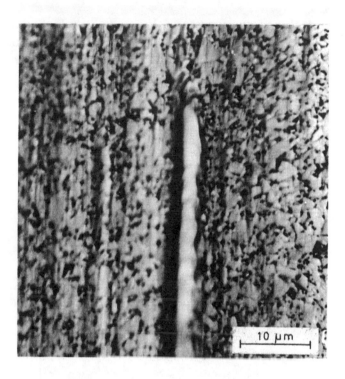

FIGURE 7.14 Worn crater surface in WC-Co tool after cutting steel at high speed. Very smooth surface with ridge characteristic of diffusion wear

FIGURE 7.15 Worn crater surface on WC-Co tool, showing large WC grain 'etched' by action of the hot steel at the interface[13]

Cratering wear showing these features occurs only at relatively high rates of metal removal, where a flow-zone exists at the interface. This and other features of wear on cemented carbide tools can be conveniently summarized by 'machining charts' such as *Figure 7.16*.[14]

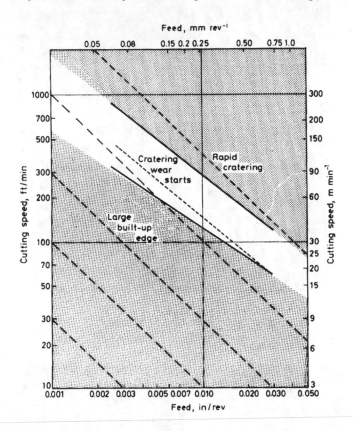

FIGURE 7.16   'Machining chart' for WC + 6% Co tool cutting 0.4% steel with hardness 200 HV[14]

The coordinates on the chart are the cutting speed and feed plotted on a logarithmic scale, and the diagonal dashed lines are lines of equal rate of metal removal. On each chart are plotted the occurrence of the main wear features for one combination of tool and work material using a standard tool geometry. Two lines show the conditions under which cratering wear was first detected, and the conditions under which it became so severe as to cause tool failure within a few minutes of cutting. *Figure 7.16* is for a WC-6% Co tool cutting a medium carbon steel.

The occurrence of crater wear by diffusion is a function of both cutting speed and feed and an approximately straight line relationship is found to exist over a wide range of cutting conditions. For example, the limit to cutting speed caused by rapid cratering was 90 m min^{-1} (300 ft/min) at a feed of 0.25 mm rev^{-1} (0.01 in/rev), while the critical speed is reduced to 35 m min^{-1} (120ft/min) at a feed of 0.75 mm rev^{-1} (0.030 in/rev). These charts and the figures for maximum cutting speeds, illustrate the importance of the cratering-diffusion wear in limiting the rates of metal removal when using WC-Co alloys to cut steel. The maximum rates of metal removal are not much higher than those when using high speed steel tools, and for this reason *WC-Co alloys are not often used for cutting carbon and low alloy steel*. They are recommended mainly as the K01 to K40 grades, for cutting cast iron and non-ferrous metals, although they can be used to advantage to give longer tool life when cutting steel at relatively low speeds.

Since this mechanism of wear is of such importance with carbide cutting tools, it is worth considering in more detail.[13] It is referred to here as 'diffusion wear' because the observed features of the worn tool surface and the interface are consistent with a diffusion wear hypothesis. It is a hypothesis because there is little direct evidence that loss of material from the tool surface to form the crater takes place by migration of individual atoms into the work material flowing over the surface. There is some electron-analytical evidence of increased concentration of tungsten and cobalt atoms in the flow zone within one or two micrometers of the interface,[18] but this evidence is too slight to form the main support for diffusion wear theory. Convincing evidence of this type is unlikely to be found because concentrations of metallic elements from the tool in the flow-zone as close as 2 µm from the interface are likely to be very low because they are so rapidly swept away (*see Figure 5.5* and discussion on flow-zone, Chapter 5).

With high speed steel tools, structural changes observed in layers several micrometers in thickness at the interface are definite evidence for wear involving diffusion and interaction on an atomic scale (*Figures 6.17* and *6.18*). With cemented carbide tools there is no direct evidence of structural change. Transmission electron microscopy of the seized interface between WC-Co tools and steel bonded to the interface has shown only WC grains in contact with ferrite (*Figure 3.12*), at magnifications such that modified layers thicker than 5 nanometers would have been detected.

It has been suggested that it is the rate of solution - i.e. the rate at which atoms leave the tool surface - rather than the rate at which they diffuse through the work material, which determines the wear rate,[19] and that this should be styled a 'solution wear' mechanism. This term might over-simplify a complex process. It might suggest that the rate of wear is dependent only on the composition of tool and work materials and could be calculated from thermo-chemical consideration of the bonding of the carbides in the tool material. The evidence presented[19] demonstrates that the bonding energy is a major factor, but atoms leaving the tool surface must have to migrate to distances of 1µm or greater before they are swept away because of unevenness in the tool surface and in the work-material flow very close to the interface. Such migration would be controlled by diffusion across thousands of atom spacings. Diffusion wear still seems the more appropriate term even though solution is involved.

While the diffusion wear hypothesis can be asserted with confidence, it is important not to over-simplify a process of great complexity. Rates of diffusion determined from static diffusion couples cannot be used by themselves to predict rates of tool wear because conditions at the tool/work interface are very different from the static conditions in diffusion tests. The flow of work material carries away tool atoms taken into solution, greatly restricting the build-up of a concentration gradient. More important, the work material in the flow-zone has a very high concentration of dislocations and is undergoing dynamic recovery and recrystallization. In these thermoplastic shear bands, new grain boundaries are constantly being generated, and grain boundary diffusion is more rapid than diffusion through a lattice, even under static conditions.

The evidence of formation of martensite in thermoplastic shear bands (*Figure 5.6*) indicates the rapidity of diffusion of carbon in steel over distances on the order of 1 µm, in time periods on the order of 1 ms, at temperatures of ~800°C. Rates of diffusion and solution must increase greatly in such structures, as Loladze[20] has pointed out, but no experimental data exist for these conditions. There is also evidence of a much higher concentration of dislocations in carbide grains at the interface than in the body of the tool.[19] These could accelerate the loss of metal and carbon atoms from surfaces subjected to high shear stress. Mechanical removal of discrete particles of tool material, too small to be observed by electron microscopy with sizes less than about 5 nanometers - may also occur in the complex wear process which is called here 'diffusion wear'.

There must be some solubility for diffusion to take place at all. It has been shown that 7% of WC can be dissolved in iron at 1250°C. The rate of diffusion wear depends on what is sometimes called the "compatibility" of the materials; large differences in diffusion wear rate occur with different tool and work materials. The rate of wear is more dependent on the chemical properties than on the mechanical strength or hardness of the tool, provided the tool is strong enough to withstand the imposed stresses. It is for this reason that the higher hardness of cemented carbides with fine grain size is not reflected in improved resistance to diffusion wear. In fact coarse-grained alloys are rather more resistant than fine-grained ones of the same composition, but the difference is small.

The rate of diffusion wear depends on the rate at which atoms from the tool dissolve and diffuse into the work material and consideration is now given to the question - "Which atoms from the tool material are most impregnating"? Recall that in the case of high speed steels, the iron atoms from the matrix diffused into the work until the isolated alloy carbide particles, which remain practically intact, were undermined and carried away bodily (*Figures 6.17, 6.18* and *6.19*). With cemented carbide tools also, the most rapid diffusion is by the cobalt atoms of the tool, and the iron atoms of the work material. The carbide grains, however, are not undermined and carried away for two reasons. First, because the carbide particles are not isolated, but constitute most of the volume of the cemented carbide, supporting each other in a rigid framework. Second, when cobalt atoms diffuse out of the tool, iron atoms diffuse in, and iron is almost as efficient in "cementing" the carbide as is cobalt. Carbon atoms are small and diffuse rapidly through iron, but those in the tool are strongly bonded to the tungsten and are not free to move away by themselves.

TEM observations (*Figure 3.12*) show no structural changes in carbide grains within distances of 0.01μm of the interface. Changes would be observed if carbon atoms were lost from the carbide without loss of tungsten atoms. *It is the rate of diffusion of tungsten and carbon atoms together into the work material which controls the rate of diffusion wear.* This depends not only on the temperature, but also on the rate at which they are swept away - i.e. on the rate of flow of the work material very close to the tool surface - at distances of 0.001-1 μm. Just as the rate of evaporation of water is very slow in stagnant air, so the rate of diffusion wear from the tool is low where the work material is stationary at, and close to, the tool surface. At the flank of the tool, the rate of flow of the work material close to the tool surface is very high (*Figure 5.2a*), and diffusion may be responsible for a high rate of flank wear even when the nearby rake face surface is practically unworn. In *Figure 3.7* the carbide grains on the flank can be seen to be smoothly worn through. Under conditions of seizure, the smooth wearing through of carbide grains can be regarded as a good indication that a diffusion wear process is involved.

When cutting at relatively high speeds where the flank wear is based on diffusion, the wear rate increases rapidly as the cutting speed is increased. *Figure 7.17* shows a typical family of curves of flank wear against cutting time for cutting steel with carbide tools in the higher range of cutting speeds. In the WC-Co alloys, the percentage of cobalt influences the rate of wear by diffusion, the flank wear rate rising with increasing cobalt content, but, within the range of grades commonly used for cutting, the differences in wear rate are not very great provided the tools do not become deformed.

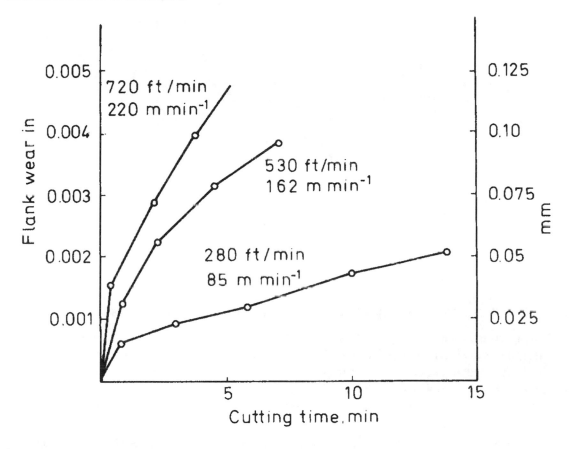

FIGURE 7.17   Flank wear *vs* time for increasing cutting speeds when cutting steel with WC-Co tools where wear is mainly by diffusion[13]

7.4.4   Attrition wear

When cutting at relatively low speeds, attrition takes over as the dominant wear process. The condition for this is a less laminar and more intermittent flow of the work material past the cutting edge, of which the most obvious indication is the formation of a built-up edge. *Figure 7.16* is typical of the charts for many steels, in that it shows the presence of a built-up edge at relatively low rates of metal removal. Below the built-up edge line on the chart, wear is largely controlled by attrition.

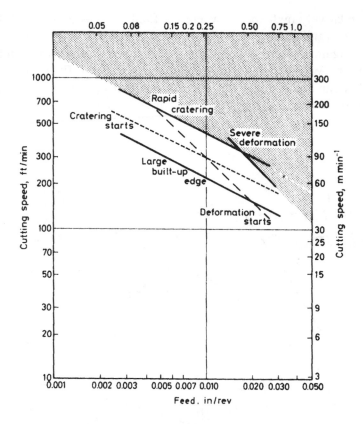

FIGURE 7.18 'Machining chart' for WC+6% Co tool cutting pearlitic flake graphite cast iron[14]

During cutting, the built-up edge is continually changing, work material being built on to it and fragments sheared away (*Figure 3.21*). If only the outer layers are sheared, while the part of the built-up edge adjacent to the tool remains adherent and unchanged, the tool continues to cut for long periods of time without wear. For example, under some conditions when cutting cast iron, the built-up edge persists on WC-Co tools to relatively high cutting speeds and feed, as shown on the chart, *Figure 7.18*.

With gray cast iron the built-up edge is infrequently broken away and tool life may be very long. Thus, WC-Co alloy tools are commonly used for cutting cast iron, and the recommended speeds are those where a built-up edge is formed. The wear rate is low but wear is of the attrition type, whole grains or fragments of carbide grains being broken away leaving the sort of worn surface shown in the sections *Figures 3.9* and *3.10*.

When cutting steel, however, under conditions where a built-up edge is formed, the edge of a WC-Co alloy tool may be rapidly destroyed by attrition. As with high-speed steel tools, fragments of the tool material of microscopic size are torn from the tool edge, but whereas this is a slow wear mechanism with steel tools, it may cause rapid wear on carbide tools. If the built-up edge is firmly bonded to the tool and is broken away as a whole, as frequently happens where cutting is interrupted, relatively large fragments of the tool edge may be torn away as shown in *Figure 7.19*.

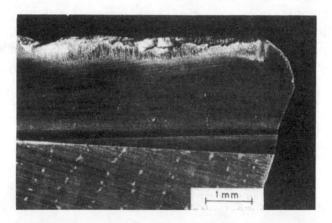

FIGURE 7.19   Edge chipping of carbide tool after cutting steel at low speed, with a built-up edge

Where the machine tool lacks rigidity, or the work piece is slender and chatter and vibration occur, the metal flow past the tool may be very uneven and smaller fragments of the tool are removed. *Figure 7.20* shows WC grains being broken up and carried away in the stream of steel flowing over the worn flank of a WC-Co tool. Fragments are broken away because localized tensile stresses are imposed by the unevenly flowing metal. Steel tools are stronger in tension, with greater ductility and toughness, and for this reason have greater resistance to attrition wear. *This is one of the main reasons why high-speed steel tools are employed.*

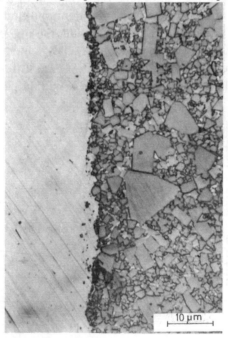

FIGURE 7.20   WC-Co tool used for cutting steel at low speed, showing attrition wear[13]

Worn surfaces produced by attrition are very rough compared with the almost polished surfaces resulting from diffusion wear. There is, however, no sharp dividing line between the two forms of wear, both operating simultaneously, so that worn surfaces, when adhering metal has been dissolved, often show some grains smoothly worn and others torn away, *Figure 7.21*.

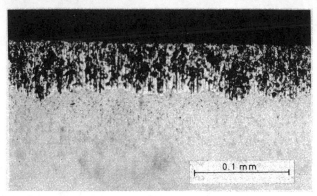

FIGURE 7.21 Worn flank of cemented carbide tool, adhering steel removed in acid. Surface shows evidence of both diffusion and attrition wear

The rate of wear by attrition is not directly related to the hardness of the tool. With WC-Co tools, the most important factor is the grain size, fine-grained alloys being much more resistant than coarse-grained ones. *Figure 7.22* shows the rates of wear of a series of tools all containing 6% cobalt when cutting cast iron in laboratory tests under conditions of attrition wear. The hardness figures are a measure of the grain size of the carbide, the highest hardness representing the finest grain size of less than 1μm. By comparison, the cobalt content has a relatively minor influence on the rate of attrition wear. *Figure 7.23* shows the small difference in rates of wear of carbide tools with 5.5-20% cobalt, all of the same grain size with large differences in hardness.

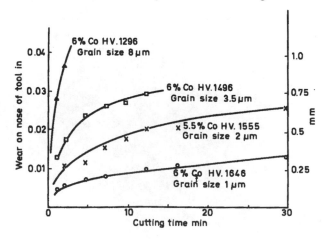

FIGURE 7.22 Flank wear on WC-Co tools showing influence of carbide grain size when cutting cast iron under conditions of attrition wear

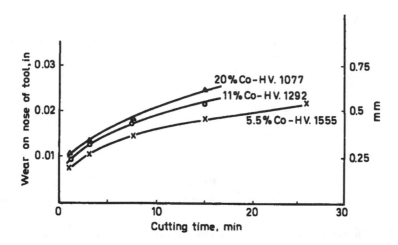

FIGURE 7.23 Flank wear on WC-Co tools showing influence of cobalt content when cutting cast iron under conditions of attrition wear

Consistent performance under these conditions depends on the ability of the manufacturer to produce fine-grained alloys with close control of the grain size. Some manufacturers now produce ultra-fine grained grades (e.g. 0.6 μm) for resistance to attrition wear.

Since the metal flow around the tool edge tends to become more laminar as the cutting speed is increased, the rate of wear by attrition is quite likely to increase if the cutting speed is reduced. *Figure 7.24* shows a family of curves for the flank wear rate when cutting cast iron under mainly attrition wear conditions and should be compared with *Figure 7.17* for diffusion wear. To improve tool life where attrition is dominant, attention should be paid to reducing vibration, increasing rigidity and providing adequate clearance angles on the tools.

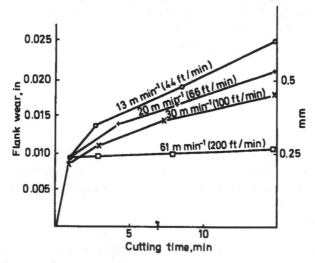

FIGURE 7.24 Flank wear *vs* time for increasing cutting speeds when cutting with WC-Co tools where wear is by attrition - compare *Figure 7.17*[13]

7.4.5   Abrasive wear

Because of the high hardness of tungsten carbide, abrasive wear is much less likely to be a significant wear process with cemented carbides than with high speed steel. There is little positive evidence of abrasion except under conditions where very large amounts of abrasive material are present, as with sand on the surface of castings. The wear of tools used to cut chilled iron rolls, where much cementite and other carbides are present, may be by abrasion, but most of the carbides, even in alloy cast iron, are less hard than WC and detailed studies of the wear mechanism in this case have not been reported.

It seems very unlikely that isolated small particles of hard carbide or of alumina in the work material can be effective in eroding the cobalt from between the carbide grains under conditions of seizure. Where sliding conditions exist at the interface there is a greater probability of significant abrasive wear.

Worn surfaces sometimes show sharp grooves which suggest abrasive action. The abrasion could result from fragments of carbide grains or whole grains, broken from the tool surface, being dragged across it, ploughing grooves and removing tool material. To resist abrasive wear, a low percentage of cobalt in the cemented carbide is the most essential feature, and fine grain size also is beneficial.

7.4.6   Fracture

Erratic tool life is often caused by fracture before the tool is much worn. The importance of toughness in grade selection, and the improvement in this property which results from increased cobalt content or grain size have already been discussed. Considerable care is necessary in diagnosing the cause of fracture to decide on the correct remedial action.

It is rare for fracture on a part of the tool edge to occur while it is engaged in continuous cutting. More frequently the tool fractures on starting the cut, particularly if the tool edge comes up against a shoulder so that the full feed is engaged suddenly. Interrupted cutting and operations such as milling are particularly severe and may involve fracture due to mechanical fatigue.

A frequent cause of fracture on the part of the edge not engaged in cutting is impact by the chip curling back onto the edge or entangling the tool. This is particularly damaging if the depth of cut is uneven, as when turning large forgings or castings.

Fracture may be initiated also by deformation of the tool, followed by crack formation (*Figure 7.8*), the mechanical fracture being only the final step in tool failure. This case emphasizes the importance of correct diagnosis, since the action to prevent failure in this case would include the use of a carbide with higher hardness, to prevent the initial plastic deformation, but less toughness. Prevention of fracture is rarely a problem which can be solved by changes in the carbide grade alone, and more often involves also the tool geometry and the cutting conditions.

7.4.7   Thermal fatigue

Where cutting is interrupted very frequently, as in milling, numerous short cracks are often observed in the tool, running at right angles to the cutting edge, *Figure 7.25*. These cracks are caused by the alternating expansion and contraction of the surface layers of the tool as they are heated during cutting, and cooled by conduction into the body of the tool during the intervals between cuts. The cracks are usually initiated at the hottest position on the rake face, some distance from the edge, then spread across the edge and down the flank.

Carbide milling cutter teeth frequently show many such cracks after use, but they seem to make relatively little difference to the life of the tool in most cases. If cracks become very numerous, they may join and cause small fragments of the tool edge to break away. Also they may act as stress-raisers through which fracture can be initiated from other causes. Many carbide manufacturers have therefore selected the compositions and structures least sensitive to thermal fatigue as the basis for grades recommended for milling.

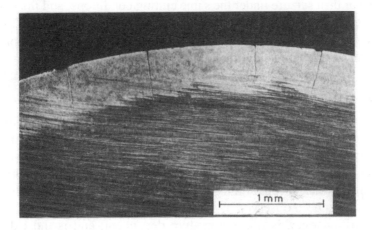

FIGURE 7.25   Thermal fatigue cracks in cemented carbide tool after interrupted cutting of steel

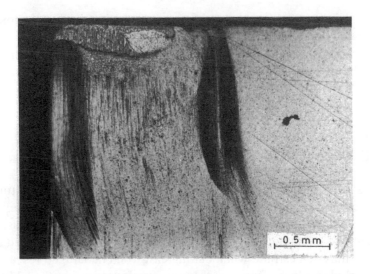

FIGURE 7.26   Sliding wear under edges of chip on rake face of carbide tool used for cutting steel

7.4.8   Wear under sliding conditions

Accelerated wear often occurs at those positions at the tool work interface where sliding occurs, as with high speed steel tools (*Figures 6.23* and *6.24*). A pronounced example in the case of a WC-Co tool used to cut steel in air is in *Figure 7.26*. The deep grooves are at the positions where the edges of the chip slid over the rake face of the tool. While abrasion may account for some wear under sliding conditions, the main wear mechanism in the sliding regions at the periphery of the contact area involves reactions with the atmosphere, and is discussed further in Chapter 10 in relation to cutting lubricants. It is of interest in showing that sliding may cause much more rapid wear than seizure under the same cutting conditions, and the elimination of seizure is, therefore, not a desirable objective in many cutting tool operations. The rate of wear in the sliding areas is mainly controlled by a chemical interaction and depends more on the composition of the tool material than on its hardness or other mechanical properties.

7.4.9   Summary

These mechanisms and processes appear to be the main ways in which *carbide tools* are worn or change shape so that they no longer cut efficiently. Many of the mechanisms are the same as with high speed steel tools and, in summing up, *Figure 6.24* can be referred to again.

Mechanisms *2* and *3*, based on deformation and diffusion, are temperature dependent and come into play at high cutting speeds, limiting the rate of metal removal which can be achieved. (Mechanism *1* has not yet been observed in micrographs on carbide tools).

Mechanism *4*, attrition wear, is not temperature dependent, and is most destructive of the tools in the low cutting speed range, where high speed steels often give equal or superior performance.

Mechanism *5*, abrasion, is probably a minor cause of wear, of less significance than on steel tools.

Mechanisms *6* and *7*, fracture and thermal fatigue, are quite similar phenomena, involving stress- and heat-induced cracks in the tool.

Sliding wear processes, Mechanism *8*, may be as important as with steel tools and occur at the same positions on the tool. In addition, carbide cutting edges are more sensitive to failure by fracture and thermal fatigue.

7.5   TUNGSTEN-TITANIUM-TANTALUM CARBIDE BONDED WITH COBALT ({WC+ TiC + TaC} -Co)

The *WC-Co tools* developed in the late 1920s proved very successful for cutting cast iron and non-ferrous metals, at much higher speeds than is possible with high speed steel tools. However, they *are less successful for cutting steel*. This is because of the cratering-diffusion type of wear which causes the tools to fail rapidly at speeds not much higher than those used with high speed steel. The success of tungsten carbide when cutting cast iron encouraged research and development in cemented carbides. Much effort was then put into investigating the addition of 5 to 40% of other carbides of the other transition elements to the basic WC-Co structure (*Table 7.4*).

TABLE 7.4   Properties of steel-cutting grades of cemented carbide ({WC+TiC+TaC} - Co)

Composition			Mean grain size	Hardness	Transverse rupture strength	Young's modulus	Specific gravity
Co%	TiC%	TaC%	(μm)	(HV 30)	(MPa)	(GPa)	
9	5	-	2.6	1,475	1,890	566	13.35
9	9	12	3.0	1,450	1,930	510	12.15
10	19	15	3.0	1,525	1,410	455	10.30
5	16	-	2.5	1,700	1,230	537	11.40

The most successful additions have been titanium carbide (TiC), tantalum carbide (TaC), and niobium carbide (NbC). All of these can be bonded by nickel or cobalt, using a powder metallurgy liquid-phase sintering technology similar to that developed for the WC-Co alloys. In spite of all the effort put into their development, none of these has so far been as successful as the WC-Co alloys in combining the properties required for a wide range of engineering applications. In particular, these other cemented carbides lack the toughness which enables tools based on WC to be used, not only for heavy machining operations, but also for rock drills, coal picks and a wide range of wear resistant applications. Two of these carbides, however, have come to play a very important role as constituents in commercial cutting tool alloys.

In the early 1930s tools based on TaC and bonded with nickel were marketed in USA under the name of 'Ramet' and were used for cutting steel because they were more resistant to cratering wear than the WC-Co compositions. Alloys based on TaC never became a major class of tool material, but they did indicate the direction for development of carbide tool materials for cutting steel, i.e. incorporation into WC-Co alloys of proportions of one or more of the cubic carbides - TiC, TaC or NbC. In the early years of their development, tools made from these alloys were often more porous than those of WC-Co, but as a result of the research and development work by Schwarzkopf[2] and many others, the technology of production was fully mastered, and their quality is very high. They are generally known as "*the steel-cutting grades of carbide*", and all major manufacturers of cemented carbides produce two main classes of cutting tool materials:

- WC-Co alloys (often called *the straight grades*) for cutting cast iron and non-ferrous metals
- WC-TiC-TaC-Co alloys (often called the *mixed-crystal grades* or usually *the steel-cutting grades*) for cutting steels

There are, worldwide, a very large number of manufacturers of cemented carbide tools. Some individual producers put on the market as many as 40 or 50 grades of different composition and grain size. *The World Directory and Handbook of Hard Metals* (Brookes[23]) gives details of composition, structure and properties for each grade where available, and is updated every few years.

7.5.1　Structure and properties of these "steel-cutting grades"

The compositions and some properties of representative steel-cutting grades of carbide are also given in *Table 7.4*. Alloys containing from 4 to 60% TiC and up to 20% TaC by weight are commercially available, but those with more than about 20% TiC are made and sold in very small quantities for specialized applications.

The proportions of cobalt and the grain size of the carbides are in the same range as for WC-Co alloys. Micro-examination shows that, in the structures of alloys containing up to about 25% TiC, two carbide phases are present instead of one, *Figure 7.27*. Both angular blue-gray grains of WC and rounded grains of the cubic carbides, which appear yellow-brown by comparison, are present.

FIGURE 7.27　Structure of steel cutting grade of carbide containing WC, TiC, TaC and Co

In alloys with more than 25% by weight of TiC, no WC grains can be detected, and, at lower TiC contents, it is obvious that the proportion of the cubic carbides is much greater than could be accounted for by the relatively small percentage by weight of TiC present. Two factors explain this structural effect. The first relates to density. Tungsten has a high atomic weight (184), while that of titanium is low (48). Correspondingly, the specific gravities of WC and TiC are 15.68 and 4.94. A given weight of TiC occupies about three times the volume of the same weight of WC, and the microstructure reflects the percentage by volume rather than by weight. *Table 7.5* compares the percentages by volume with the weight percentages of the constituents in four grades of carbide.[24] The composition by volume is more directly related to the microstructure and also to the performance as cutting tool materials.

TABLE 7.5   Comparison of volume percent with weight percent of 4 cemented carbide grades[24]

ISO class	Weight	%			Volume	%		
	WC	TiC	Ta, NbC	Co	WC	TiC	Ta,NbC	Co
K20	92	-	2	6	87	-	3	10
M40	77	4	8	11	64	10.5	9	16.5
P20	68.5	12	10	9.5	49.5	27.5	11	12
P10	55.5	19	16	9.5	36	39	14	11

The second factor which reduces the amount of free WC in the structure, relates to the composition of the two phases. WC does not contain any titanium, but TiC takes into solid solution more than 50% of WC by weight. In alloys containing TaC, all of this carbide is also in solution in the TiC.[4] The rounded grains in the structure of these alloys are crystals based on TiC, with a face-centered cubic structure in which some atoms of titanium are replaced at random by atoms of tungsten and tantalum. A fourth carbide NbC is usually present together with TaC. This is because Ta and Nb occur in the same ores and are difficult to separate. Like TaC, NbC is completely soluble in TiC. The cubic phase in these alloys was first shown to be a solid solution in Germany, where the word *mischkristalle* is the term for a *solid solution*, and those engaged in cemented carbide technology refer to it as the mixed *crystal phase*.

Table 7.4 also shows that the hardness of the steel-cutting grades is in the same range as that of the WC-Co alloys of the same cobalt content and grain size (*Table 7.2*). Increasing the (Ti, Ta) content reduces the transverse rupture strength, and practical experience in industry confirms that the toughness is reduced. For this reason the most popular alloys for cutting tools contain relatively small amounts of TiC/TaC. TiC is much cheaper than TaC, titanium being plentiful and tantalum being a rare metal, but TaC is considered to cause less reduction in toughness and to increase the high temperature strength. Most present day steel-cutting grades contain some TaC. The addition of the cubic carbides lowers the thermal conductivity, which is high for WC-Co alloys. The conductivity of an alloy with 15% TiC is approximately the same as that of high speed steel, and less than half of that for the corresponding WC-Co alloy.

7.6   PERFORMANCE OF (WC+TiC+TaC) -Co TOOLS

The steel-cutting grades of carbide have two major advantages. The hardness and compressive strength at high temperature are higher than those of WC-Co alloys. *Figure 6.9* shows the compressive strength as a function of temperature for a steel-cutting grade of carbide containing 12% TiC. The 5% proof stress was 800 MPa (52 tonf/in^2) at a temperature of 1100°C, compared with 1000°C for comparable WC-Co alloy and 700°C for high speed steel.

There are few published figures for the high temperature compressive strength, but *Figure 7.28* shows measured values of 0.2% proof stress in compression for an alloy containing 19%

TiC, 16% TaC and 9.5% Co, over a range of temperatures.[24] This alloy would support a stress of 1500 MPa (100 tonf/in^2) at 870°C with 0.2% plastic deformation. A stress of 1500 MPa is typical of the level near the cutting edge of tools when machining steel (Chapter 4). Much more and accurate testing of the stress/strain relationships of these alloys at high temperature is required.

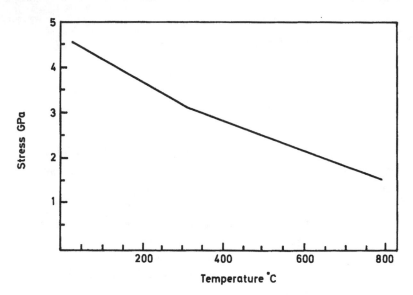

FIGURE 7.28   0.2% compressive proof stress as a function of temperature. Cemented carbide: 19% TiC, 16% TaC, 9.5% Co, 5.5% WC. (Data from Sandvik Coromant Research [24])

The second advantage of the steel-cutting grades, which is more important than their high temperature strength, is that they are much more resistant to the diffusion wear which causes cratering and rapid flank wear on WC-Co alloys when machining steel at high speed. The steel-cutting grades can be used at speeds often three times as high as the tungsten carbide 'straight grades', as is shown by comparing the machining chart in *Figure 7.29* for a tool containing 15% TiC with that in *Figure 7.16* for a WC-Co tool cutting the same steel. Rapid cratering occurred at 90 m min^{-1} (300 ft/min) at a feed of 0.25 mm (0.010 in) per rev on the WC-Co tool and at 270 m min^{-1} (900 ft/min) on the tool with 15% TiC.

The higher the proportion of cubic carbides in the structure, the higher is the permissible speed when cutting steel, but even 5% of TiC has a very large effect. Not only cratering, but also flank wear is reduced when cutting steel in the higher cutting speed range. Metal removal rate when cutting steel with these alloys is limited not by the ability of the tool to resist cratering but by its ability to resist deformation or by too rapid flank wear. The choice of grade for any application depends on achieving the correct balance in the tool material between toughness and resistance to flank wear and deformation.

With WC-Co alloys there are two major variables - cobalt content and grain size. The number of parameters which influence the properties and performance of the steel cutting carbide grades is much larger. Not only are there two other constituents, TiC and TaC, but the grain size of the two carbide phases can be varied independently and the quality and performance of the product is more greatly influenced by variables in the production of the carbides, the sintering and other operations. A very large number of commercial steel-cutting grades is available, each claiming distinctive virtues for a range of applications, and no system of standardization completely satisfactory to the user, has been evolved.

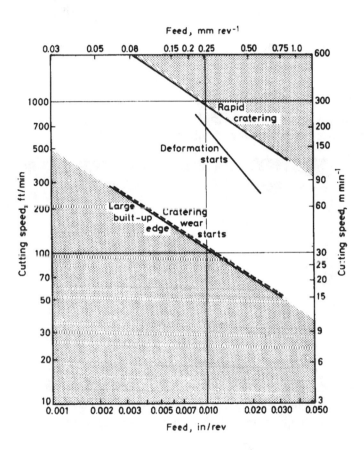

FIGURE 7.29 'Machining chart' for steel cutting grade of carbide cutting 0.4% steel with hardness 200 HV[14]

The ISO system (*Table 7.3*) is a classification which arranges the grades according to application. In each of three main series they are placed in order of the severity of machining:

- the "P" series are specifically steel-cutting grades
- the "K" series are for cutting cast iron and non-ferrous metals
- the "M" series may be used for a wide range of applications in both areas

Within each series the number increases in order of decreasing hardness and increasing toughness. The description of the material and the working conditions guide the user to an initial choice of grade for a particular application.

For example, for turning steel castings at medium cutting speed a P20 carbide might be selected. If tool failures due to fracture were frequent a P30 grade would be tried, or, if tool life was short because of rapid wear, a P10 grade should give longer life. It is the responsibility of the manufacturers to decide what ISO rating should be given to each of their grades and there are considerable differences in composition, structure and properties between grades offered by different manufacturers in any category. It is, therefore, always worthwhile for the user to compare grades supplied by different manufacturers. The steel-cutting grades are used regularly for machining steel at speeds in the range 60 to 170 m min^{-1} (200-600 ft/min) at feeds up to 1 mm (0.040 in) per rev. The process by which the cubic carbides function to reduce wear rate can be demonstrated by metallography. *Figure 7.30a* is the crater on the rake face of a WC-Co alloy tool after cutting a medium carbon steel at 61 m min^{-1} (200 ft/min) and 0.5 mm (0.020 in) per rev feed for 5 minutes, while *Figure 7.30b* is the rake surface of a tool with 15% TiC used for cutting under the same conditions for 25 minutes.

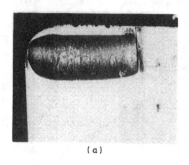 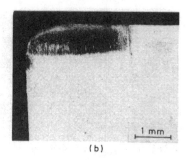

(a)   (b)

FIGURE 7.30   (a) WC-Co tool; (b) smaller crater on rake face of WC+TiC-Co tool

Micro examination of the worn surface shows that the cubic 'mixed crystal' grains are worn at a very much slower rate than the WC grains, and are left protruding from the surface to a height as great as 4 micrometers. *Figure 7.31* is a scanning electron microscope picture of such a worn surface after cleaning by removal of the adhering steel in acid.

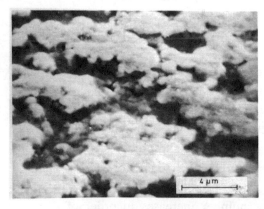

FIGURE 7.31   Scanning electron micrograph of crater surface of WC+TiC+TaC-Co tool

Figure 7.32 is an optical microscope picture of a worn crater surface. The surface of the worn WC grains is characteristic of chemical attack (diffusion) rather than mechanical abrasion. The 'mixed crystal' grains also are worn by diffusion but at a very much slower rate at the same temperature. If cutting is continued, the mixed crystal grains may be undermined and carried away bodily in the under surface of the chips. Cratering is therefore observed on tools of the steel cutting grades of carbide after cutting steel at high speed, but at a slower rate than on WC-Co tools.

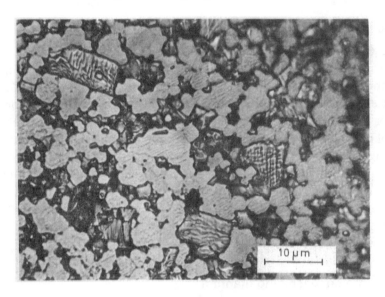

FIGURE 7.32   Crater surface of WC ᵢ TiC-Co tool showing 'etching', diffusion on WC grains[13]

7.7   PERSPECTIVE: "STRAIGHT" WC-Co GRADES versus THE "STEEL-CUTTING" GRADES

The introduction of the steel-cutting grades of carbide - ({WC+TiC+TaC} + Co) - completed a phase in the development of machine shop practice which was initiated by the WC-Co alloys. Their importance lies in the fact that such a very large proportion of the total activity of metal cutting is concerned with the machining of steel. The basic development of the steel-cutting carbides had been accomplished by the late 1930s, and they played a large role in production on both sides in the second World War.

Productivity of machine tools was increased often by a factor of several times and a complete revolution in machine tool design was required. The importance of this advance was equal to that which followed the introduction of high speed steel 40 years earlier. Research on the steel-cutting grades of carbide has continued, and tool tips of very high quality and uniform structure, now give very consistent performance.

Development of the steel-cutting grades was the result of intelligent experimentation and empirical cutting tool tests, similar to Taylor and White's experiments on high speed steel. Explanation of their performance, in terms of a deeper understanding of the wear processes during cutting, followed much later and is still continuing. Details of the mechanism of diffusion wear are still a subject of research, but knowledge of the main features of this wear process can form one of the useful guidelines for those involved in tool development and application.

Minor modifications to composition and structure were introduced to give improvements for particular applications such as milling, but until the 1970s, with the introduction of coated tools, no major new line of development achieved outstanding success.[11-22]

7.7.1   A special note on machining at low speeds

The very large improvement in tool performance achieved by the introduction of the cubic carbides results from the low rate at which TiC and TaC diffuse into (dissolve in) the hot steel moving over the tool surface. The improvement is specific to *the cutting of steel at high rates of metal removal.* If either the temperature at the tool-work interface is too low, or work material other than steel or iron are machined, the advantages gained by using the steel-cutting grades disappear and tool performance is often better when using one of the WC-Co grades.

When machining steel at speeds and feeds in the region of the built-up edge line (*Figures 7.16* and *7.29*) or below, the usual wear mechanism is attrition, and the steel-cutting grades are more rapidly worn by attrition than are the WC-Co grades, the cubic carbides being more readily broken up than WC. For this reason, the straight WC-Co alloys are commonly used when machining steel where high speeds are impossible, for example on multi-spindle automatic machines fed with small diameter bars.

7.7.2   A special note on machining titanium

Titanium is a high melting point metal and, in machining titanium at high speed, high temperatures are generated at the tool/work interface. When machining a titanium work material, there is evidence that the tungsten in the tool diffuses less rapidly into the chip underside than does the titanium in the mixed-crystal cubic carbides.[21] As a consequence, tools of the steel-cutting grades of cemented carbide are worn more rapidly when machining titanium than are WC-Co tools. *The WC-Co tools are always used for machining titanium.* This behavior demonstrates that the superiority of the grades containing TiC and TaC in machining steel is not due to superior resistance to abrasion but to their greater resistance to chemical attack at high temperature by the work material (see also Chapter 9).

7.8   PERFORMANCE OF "TiC ONLY" BASED TOOLS

Steel-cutting grades of carbide which contain large percentages of TiC are difficult to braze and were unpopular for this reason when brazed tools were the norm. With throw-away tool tips, alloys with higher proportions of TiC could more readily be used and consideration was given to tools based on TiC instead of WC, because of its resistance to diffusion wear in steel cutting.

Of all the cubic carbides, TiC has the most obvious potential. Titanium is a plentiful element in the earth's crust, the oxide TiO_2 is commercially available in purified form, and TiC can be readily made by heating the oxide with carbon at temperatures about 2000°C. Cemented TiC alloys can be made by a powder metallurgy process differing only in detail from that used for the production of the WC-based alloys. The most useful bonding metal has been nickel and usually the alloys for cutting contain 10-20% Ni. The difficult problems of producing a fine grained alloy of consistently high quality, free from porosity, have largely been overcome.[22] The addition of about 10% molybdenum carbide (Mo_2C) is often made to facilitate sintering to a good quality. Some commercial suppliers now include a TiC-based grade in their cataloge.[23]

The TiC-based carbides have hardness in the same range as that of the conventional cemented carbides. Experience, both in laboratory tests and in many industrial applications, shows that the TiC-based tools have lower rates of wear when cutting steel at high speed, and can be used to higher cutting speeds than the conventional steel-cutting grades of carbide. The advantage in terms of increased metal removal rate, however, in changing from conventional steel-cutting grades to TiC-based carbides is not so great as that which was achieved when changing from WC-Co alloys to the steel-cutting grades.

The TiC-based tools appear to lack the reliability and consistency of performance of the conventional cemented carbides and operators do not have confidence in their ability to apply them to a wide range of applications without trouble. In spite of their clear advantage in terms of lower wear rate, they have not yet accounted for more than a very small percentage of the tools in commercial use. There is probably a lack of toughness in the alloys so far made. This aspect of tool materials needs further study, because, with the relative shortage of tungsten supplies, a substitute for the WC-based alloys will eventually be required, and those based on TiC seem the most likely candidates so far.

7.9   PERFORMANCE OF LAMINATED AND COATED TOOLS

7.9.1   Laminated tools

Taking advantage of the freedom afforded by throw-away tool tips, laminated tools were first introduced in which the rake face was coated with a thin layer, about 0.25 mm (0.010 in) thick, of a steel cutting grade of carbide, while the main body of the tip consisted of a tough grade of WC-Co composition with high thermal conductivity. Production of such composite bodies by powder metallurgy techniques was possible using automatic presses. When using these laminated tips, reduced rates of wear were achieved compared with conventional cemented carbides, and increased cutting speeds were possible when cutting both steel and cast iron.

7.9.2   Coated tools using chemical vapor deposition (CVD): an overview

Further development of laminated tools has, today, been superseded by the coating of conventional carbide tool tips with a very thin layer of a hard substance by CVD.[25] Commercial development of CVD coatings on cemented carbide tools began in the early 1970s. Very thin layers, usually 10 μm thick or less, are strongly bonded to all the surfaces of the tool tips (*Figure 7.33*).

Coatings at present available consist of titanium carbide (TiC), titanium nitride (TiN) titanium carbonitride (Ti(C,N)), hafnium nitride (HfN) or alumina (Al_2O_3). Attempts to create diamond coatings on carbides are ongoing at the time of writing - this topic is revisited in the next chapter. The CVD deposition process is carried out by heating the tools in a sealed chamber in a current of hydrogen gas (at atmospheric or reduced pressure) to which volatile compounds are added to supply the metal and non-metal constituents of the coatings.

For example, to produce coatings of TiC, titanium tetrachloride ($TiCl_4$) vapor is used to provide the Ti atoms, and methane (CH_4) may be used to supply the carbon atoms for the coating. The temperature is in the region of 800-1050°C and the heating cycle lasts several hours. Optimization of the process parameters makes possible consistent deposition of layers of uniform thickness, strongly adherent to the carbide substrate, with uniform structure and wear-resistant properties. The grain size of the coatings is very fine, usually equi-axed grains with a diameter of a few tenths of a micron. TiN coatings are sometimes columnar with the axis normal to the coated surface. Most commercial cemented carbide manufacturers supply coated tools, and, today, these have become more varied and sophisticated as competing companies have endeavored to improve the performance. Early coatings were TiC, which is gray in color. Since very thin coatings do not conceal the grinding or lapping marks, it is difficult or impossible to be certain whether a tool is coated with TiC by visual inspection alone. TiN coatings are golden in color, while various shades are produced by coating with layers consisting of the solid solution Ti(C,N). Thin coatings of Al_2O_3 are less easy to observe than TiC, but are electrically insulating while the carbide and nitride layers are good electrical conductors. Many commercial coatings now consist of several layers, often one or two micrometers in thickness or even less, of nitride, carbide and alumina (*Figure 7.33*). Some products contain as many as 12 layers[26] including a layer or layers of Al_2O_3 with implanted nitrogen ions.

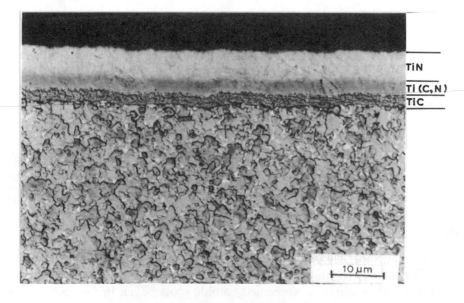

FIGURE 7.33 Multiple layer CVD coating on cemented carbide tool

The substrate cemented carbide on which the coating is deposited is usually a fairly tough steel-cutting grade, and may be varied for use in different applications. Because of their extreme thinness it has not been possible to carry out hardness or other significant mechanical tests on the layers. The coefficient of thermal expansion of the coatings is generally higher than that of the substrate. Cooling from the coating temperature therefore introduces tensile stresses into the layers and polishing of coating surfaces generally reveals a network of very fine cracks. These penetrate no deeper than the coating, but transverse rupture tests show a drop in strength compared with un-coated carbide.[27] There is no evidence that this influences the performance of clamped-tip tools, the main type of tooling for which coated tools are used.

7.9.3   Performance of coated tools

A large proportion of indexable inserts for cutting iron and steel is now coated. The success of coatings is based on their proven ability to extend tool life by a factor of 2 or 3 by reducing the rate of wear in high speed turning of cast iron and steel. Many users consider that more important economies are made by increasing cutting speed by 25 to 50% without reduction in tool life. *Figure 7.34* is a typical set of laboratory test results comparing the rate of flank wear on coated and uncoated carbide tools. Coated tools cannot be universally applied even in turning; operations involving severe interruptions of cut may fracture and flake away the coatings. Improvement in the bonding has made them more suitable for milling operations.[27-30]

Research has indicated that the wear resistant qualities of TiC, TiN and alumina coatings are mainly because of their high resistance to diffusion wear on the parts of the tool surface where seizure occurs.[15] Evidence from quick-stops demonstrates that the coatings do not prevent seizure when cutting steel at high speed[31] (though they may eliminate the built-up edge at low speed). The heat source raising the tool temperature is still a flow-zone and the temperatures appear to be only slightly lower with coated than with uncoated tools. For this reason the rate of metal removal, particularly when cutting high strength materials, is limited by the ability of the substrate to withstand the stress and temperature at the cutting edge. These very thin coatings on cemented carbide substrates may be worn-through locally at the positions of most rapid wear, such as the center of the crater or on the flank just below the edge, but the coating will continue to be effective in reducing the rate of wear as long as it remains intact at the cutting edge.[15,28]

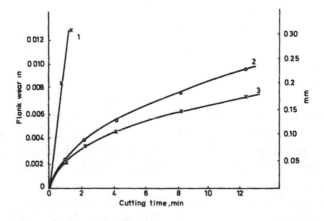

FIGURE 7.34 Tool wear vs time for coated and uncoated cemented carbide tool tips. (1) Uncoated steel cutting grade; (2) coated WC-Co alloy and (3) coated steel cutting grade

7.9.4   Comparisons between different types of coating

There are differences between the different coatings in respect of their resistance to flank and crater wear. Most reported results on steel cutting suggest that TiN is more resistant than TiC to crater wear and that TiC is more resistant to flank wear. The crater and flank wear rates of alumina-coated tools appear to be not greatly different from those of carbide and nitride coated tools. There is evidence that the crater wear of alumina coatings is caused by superficial plastic deformation, rather than by a diffusion mechanism.[15] The greatest difference in wear rate is reported to be that in the notch at the outer edge of the chip (*Figures 6.23* and *6.24*) where sliding rather than seizure occurs. Wear at this position was minimal with alumina coatings. TiN coatings were nearly as good as alumina, but notch wear on TiC coatings was very much greater and not greatly different from that on uncoated tools.

7.9.5   Multiple layer and nanolayer coatings: a design goal

It is clear that there is a place for more than one type of coating, and therefore multi-layer coatings may provide a type of tool which will cope well with a variety of operations. Colloquially speaking, could research work discover a tooling combination that would "give the best of all worlds"? It might consist of a tough carbide core - probably one of the WC-Co, "K-grades" - coated with several layers: one that would bond reliably to the core; another that would buffer thermal-expansion discrepancies; another that would provide insulation; another that would buffer diffusion wear; another on the very outside that would have ultra-low friction characteristics.

Will such an "all-purpose", multi-layer "sandwich" eventually be discovered over the next few years? This is the new challenge to cutting tool developers. Some progress is reviewed in the next two sections.

7.9.6   Multiple layer coatings

For such multiple-layer "sandwiches" the first layer is usually TiN or TiCN created by the PVD process. These can be created on straight WC-Co substrates, or on mixed carbide substrates of varying cobalt content. TiN or TiCN minimize or eliminate the formation of η-phase $(Co_xW_yC_z)$ at the interface with the substrate. It is also common to create the last, or outer, layer of the "sandwich" with TiN using PVD. This induces beneficial residual compressive stresses on the complete structure.

In today's industrial inserts of this type, multiple layers of only TiN may in fact be successful. Multiple layering allows the final thickness of the "sandwich" to be as much as 10-15 μm. This is greater than can be achieved by trying to create the whole layer in one operation. When used in turning and milling operations, the successive layers of very thin coats, rather than one heavy coat, give a more stable structure. In other, triple-layer "sandwiches" the middle layer is often TiCN as shown earlier in *Figure 7.33*.

Beyond this, Komanduri[32] reports that many "sandwich" combinations are being investigated today including the following: {TiN-Al_2O_3-TiC-TiCN}, or {TiN-TiCN-TiC}, or {TiN-Al_2O_3-TiCN}, or {Al_2O_3-TiC}, or {Al_2O_3-TiC-TiCN}, or {TiN-Al_2O_3-TiC}, or {TiN-Al_2O_3-TiN-Al_2O_3-TiN-Al_2O_3-TiCN}. In addition, Komanduri reports that one grade of TiN-Al_2O_3 on a

Si_3N_4 substrate is also available, but that the technology to coat crystalline Al_2O_3 by PVD is not feasible today.

7.9.7   Multiple nanolayer coatings

Nanotechnology also involves sputtering, i.e. deposition, by PVD. It is being used today in a wide variety of industries. These include CMOS semiconductor wafer manufacturing, and the manufacture of the wafers that become the read-write heads for computer disc drives. Magnetron sputtering with many types of targets can be employed and the wafers are rotated with respect to the target. The sputtering rate and the rotational speed of the turntable that holds the wafers, determine the thickness of the coated layer.

Such coating technologies are also being investigated for cutting tools because unusual combinations of layers can be built up. The key concept is to deliberately mix alternating hard and tough layers. Shaw[33] and colleagues report that *a crack that might be initiated in a hard, brittle layer is arrested when it meets the tough layer* thereby increasing the overall fracture toughness of the whole ensemble. Almost any refractory hard material can be used as the hard nanolayer; almost any compatible metal can be used as the tough nanolayer. An example of the {hard carbide + metal} nanolayered system might be {B_4C + W}. Or perhaps the B_4C might be replaced by {HfC, SiC} and the W by {Al, Cr, Ti, Si, Mg, Zr}.Oxidation resistant layers and low-friction layers of molybdenum disulphide (MoS_2) can also be mixed in. When using introducing nanolayers of MoS_2, the adjacent, non-MoS_2 nanolayers interrupt the formation of columnar grains of MoS_2 which are ineffective as lubricants. This interruption produces a uniform equiaxed structure of MoS_2 throughout the coating, which are very effective as lubricants.

Whereas each layer of the multiple-layer "sandwich" may only be a few nanometers thick, the total thickness can be in the range of 2-5 μm. Komanduri[32] reports that while today, nanocoatings for cutting tool applications are not being used commercially, the number of potential combinations of material systems that can be used for nanolayer coatings is virtually unlimited. Thus the technology is expected to grow in the future. [32-34]

7.10   PRACTICAL TECHNIQUES OF USING CEMENTED CARBIDES FOR CUTTING

7.10.1   Early methods using brazed-on carbide tips

High speed steel tools are normally made in one piece, but from their earliest days cemented carbides were made in the form of tool tips: small inserts which were brazed into a seat formed in a steel shank. Except for very small tools, it was not economical to make the whole tool of cemented carbide (which typically costs 20 times as much as high speed steel, weight for weight) and there was normally no advantage in terms of performance. A brazed carbide was used until worn and was then reground. Most grinding of carbide tools required the use of bonded diamond grinding wheels, and more skill was required and the cost was greater than for grinding high speed steel tools.

7.10.2 The transition to throw-away tool tips or indexable inserts

Because cemented carbides are less tough than high speed steel, the tool edge must be more robust. With steel tools the rake angle may be as high as 30°, so that the tool is a wedge with an included angle as small as 50° (*Figure 2.2*). The rake angle of carbide tools is seldom greater than + 10° and is often negative (*Figure 2.2*). A tool with an included angle of 90° set at a negative rake angle of - 5° is particularly useful for conditions of severe interruption of cut, or when machining castings or forgings with rough surfaces.

FIGURE 7.35 Indexable insert tool tips of cemented carbide

This common usage of tools with low positive rake, or negative rake angles led to the concept of *throw-away tool tips*. These are small 'tablets' of carbide, which are usually square or triangular in plan, though other shapes may be used (*Figure 7.36*). Each tip has a number of cutting edges. Positive rake, triangular tips have three cutting edges. Negative rake square tips have eight cutting edges.

In use, the tips are mechanically clamped to the tool shank using one of a number of different methods. When a cutting edge is worn, and the tool no longer cuts efficiently, the tip is unclamped and rotated to an unused corner, and this is repeated until all the cutting edges are worn. Then, the tip is discarded rather than reground. Used in this way, the tips are called *indexable inserts*. Very little of the tool material is worn off when the tool tip is thrown away, but, in spite of this, in many operations very considerable economies are achieved compared with the possible use of a brazed tool which would have to be reground. The high cost of regrinding is eliminated, the time required to change the tool is greatly reduced, and the cost of brazing is avoided.

The powder metallurgy process is well adapted to mass production of carbide tool tips. These are made by pressing on automatic presses, followed by a sintering operation, which produces tips of high dimensional accuracy, so that little grinding is required. The introduction of throw-away tooling has made possible important new lines of development in tool shapes and materials. The tool edges can be given considerable protection against chipping caused by mechanical impact, by slightly rounding the sharpened edge. Many manufacturers supply edge-honed tips, on the cutting edges of which there is a small radius, usually about 25 to 50 μm (0.001-0.002 in). Such tips can give much more consistent performance in many applications, without the rate of wear being increased.

Tips are also made with grooves formed in the rake face close to the edge which curl the chips into shapes which can be readily cleared from around the tool. Clearance of chips from the cutting area is of increasing importance for efficiency of machining operations on automated CNC machine tools. The ability to cut for long periods without stopping to clear chips can have a large effect on machining costs. Carbide tool manufacturers have put great effort into designing chip grooves of optimum shape for operation in wide ranges of speed, feed, depth of cut and work material. A large number of patents cover these edge shapes. The research of Jawahir and colleagues[35] is also of particular interest in this regard.

Figure 7.36 shows chip grooves on indexable inserts marketed by one carbide tool producer to meet requirements for chip control when cutting specific work materials at different speed, feed and depth of cut.

FIGURE 7.36 Chip groove shapes on indexable inserts produced by one carbide tool manufacturer. (Courtesy of Kennametal Inc.)

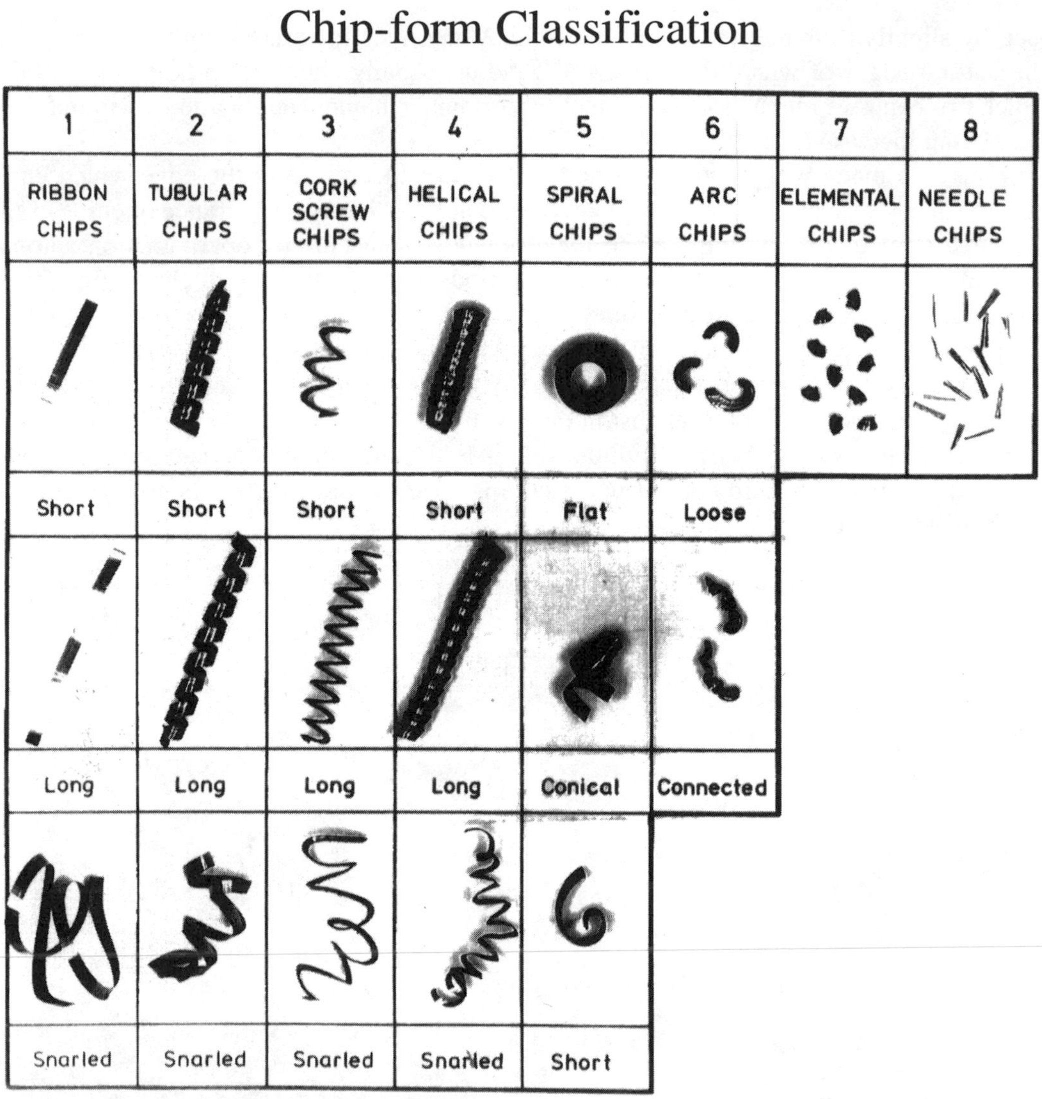

Chip-form Classification

1	2	3	4	5	6	7	8
RIBBON CHIPS	TUBULAR CHIPS	CORK SCREW CHIPS	HELICAL CHIPS	SPIRAL CHIPS	ARC CHIPS	ELEMENTAL CHIPS	NEEDLE CHIPS
Short	Short	Short	Short	Flat	Loose		
Long	Long	Long	Long	Conical	Connected		
Snarled	Snarled	Snarled	Snarled	Short			

FIGURE 7.37 ISO-based chip-form classification (*Figures 7.37* to *7.42* are reproduced with the permission of Professor I.S. Jawahir)

Figure 7.37 shows the most commonly known ISO-based chip-form classification.[36] *Figure 7.38* shows a typical chip chart showing the different regimes of chip formation for various feeds and depths of cut. The representation of chip-forms as a chip-chart enables immediate visual recognition of acceptable and unacceptable regimes of chip formation. The chip chart, thus, not only provides the performance capability of a grooved cutting tool, but also serves as an important tool in selecting cutting parameters (such as feed, depth of cut, cutting speed) for a particular operation.

A Typical Chip Chart

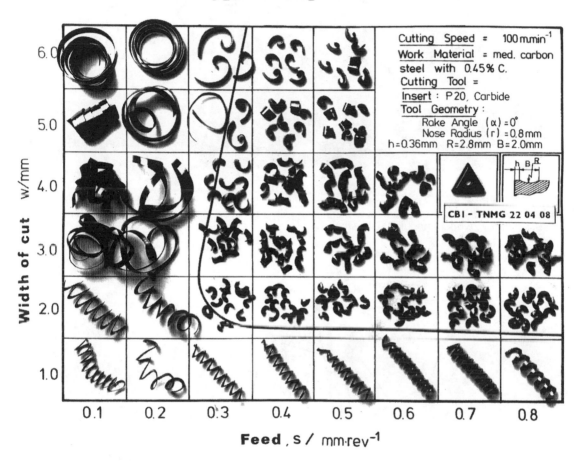

FIGURE 7.38 A representative chip chart showing the regimes of chip formation for varying feeds and depths of cut (Cutting speed = 100 m min^{-1}, rake angle = 0 deg., nose radius = 0.8 mm, work material = 1045 steel)

One of the significant results from the research on chip-forms and chip breakability[37,38] has been the establishment of chip breaking as a cyclic process. Recent work has shown the effect of the cutting tool thermal conductivity on the tool-chip contact length and the chip-form in orthogonal cutting with a grooved tool. Regions of sticking friction, sliding friction and the contact at the groove backwall have been identified (*Figure 7.39*).

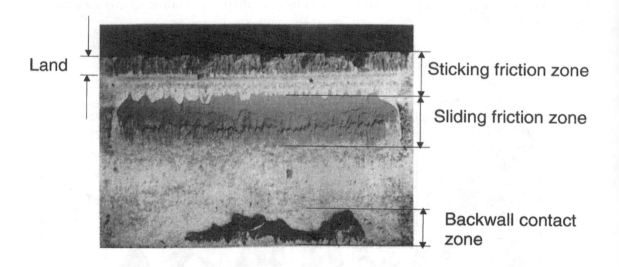

Land

Sticking friction zone

Sliding friction zone

Backwall contact zone

FIGURE 7.39 Tool-chip contact zones in machining with a grooved tool (Cutting speed = 100 m min^{-1}, feed = 0.3 mm/rev., work material = 1045 steel)

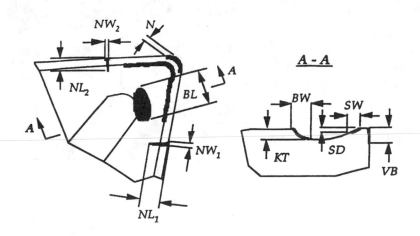

VB	flank wear	N	nose wear
BW	width of groove backwall wear	NL_1	notch wear length on main cutting edge
BL	length of groove backwall wear	NW_1	notch wear width on main cutting edge
KT	depth of groove backwall wear	NL_2	notch wear length on secondary cutting edge
SW	width of secondary face wear	NW_2	notch wear width on secondary cutting edge
SD	depth of secondary face wear		

FIGURE 7.40 Measurable tool-wear parameters in a grooved tool

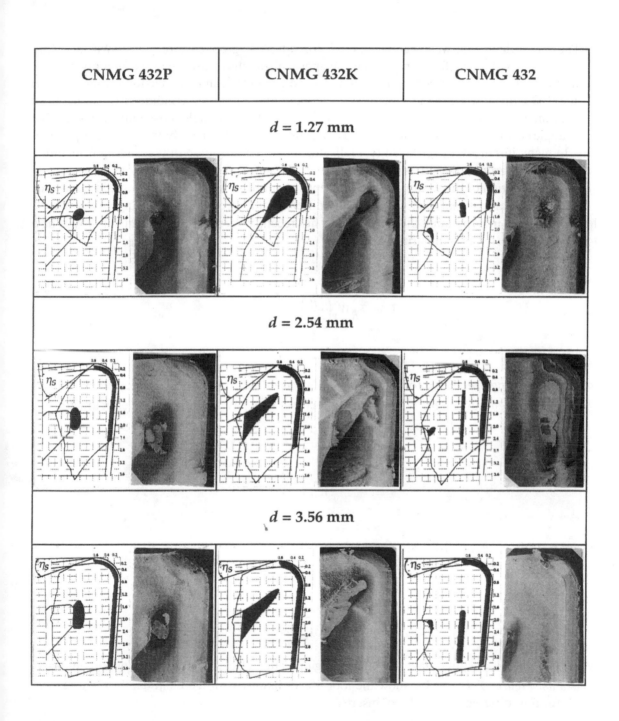

FIGURE 7.41 Variation of tool-wear patterns in three chip-grooves at varying depths of cut (cutting speed = 274 m min^{-1}, feed = 0.25 mm/rev., work material = 1037M steel)

The measurable tool-wear parameters[39,40] for a typical grooved tool are shown in *Figure 7.40*. Section A-A in the figure represents the approximate chip-flow direction for a given set of cutting conditions. The influence of the chip-groove geometry can be understood from its effect on the 3-D chip flow on the tool rakeface. The variations in cutting conditions (feed, depth of cut, cutting speed) and the tool geometry (inclination angle, rake angle, nose radius, side-cutting edge angle) also significantly affect the 3-D chip flow, and consequently the tool-wear parameters. In many instances, the grooved tool fails due to unfavorable chip flow resulting from either improper chip-groove design or inappropriate application of cutting conditions. *Figure 7.41* shows the variation of the wear regions with depth of cut for three different grooved tool inserts.

Tool-life[41] varies significantly when using different chip-grooves and coatings. A new tool-life relationship incorporating these two effects has recently been developed. This improved new tool-life relationship is summarized as follows:

$$T = T_R W_g \left(\frac{V_R}{V} \right)^{W_C \frac{1}{n}} \tag{7.1}$$

where V = cutting speed; T = tool-life; n = Taylor's tool-life exponent; T_R = reference tool-life, V_R = reference cutting speed, W_c = coating effect factor and W_g = chip-groove effect factor.

$$W_C = \frac{n}{n_c} \tag{7.2}$$

where n_c is the actual tool-life slope modified by the coating effect.

$$W_g = \frac{km}{f^{n_1} a^{n_2}} \tag{7.3}$$

where f = feed; d = depth of cut; m = machining operation effect factor; n_1, n_2 and k = empirical constants. *Figures 7.42a* and *b* show the tool-life variations with feed and depth of cut respectively for four different grooved tool inserts.

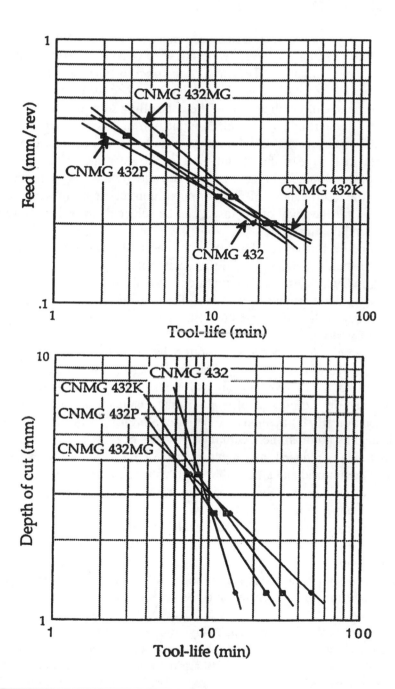

FIGURE 7.42 Variation of tool-life with (a) feed (depth of cut = 2.54 mm), and (b) depth of cut (feed = 0.25 mm/rev.) for four different grooved tool inserts (cutting speed = 274 m/min., work material = 1037M steel)

7.11   CONCLUSION ON CARBIDE TOOLS

Comparison of the performance of the cemented tungsten carbides with and without the addition of the cubic carbides demonstrates very well that wear rate is very dependent on the type of wear process or mechanism involved. Thus *wear resistance is not a unique property of a tool material which can be determined by one simple laboratory test, or correlated with one simple property such as hardness.* Correct diagnosis of the controlling wear mechanism for a particular operation can often be the starting point in selecting the optimum tool material.

The concept of indexable inserts and the ready availability of different types of tools with clamps for the inserts, in a range of shapes and sizes for different operations, offers the potential for many new lines of development in tooling materials and geometries. In particular, the material on the surface of the insert can be selected for maximum resistance to wear for specific conditions, while material for the body of the insert can be selected for resistance to bulk stresses causing deformation and fracture.

There is much scope for further development of coated tools. The existing commercial range of coatings has demonstrated that very hard, brittle materials such as TiC, TiN and Al_2O_3 when in the form of very thin layers - less than 10 μm thick can resist the stresses of cutting operations, when supported on a rigid but tough substrate. Solid tools made from the same materials would fracture if used as tools in the same operations. Diamond coatings are the subject of intense research at the time of writing.

So far coatings have been developed almost entirely by tests on high speed cutting of cast iron and steel. Coatings may be found for cutting materials other than iron and steel and, in parallel with modified tool design, the range of operations for coated tools is likely to be extended.

Further developments in the process of coating may result in improvements. A number of physical vapor deposition (PVD) methods are being explored.[29,30] In these, the surfaces to be coated are first cleaned by bombardment with inert gas ions in a low pressure chamber. Next, the pressure in the chamber is reduced, and metal atoms for the coating are evaporated into the chamber. The atoms are ionized and attracted by high voltage to the tool surface in the presence of a low pressure gas, which contains nitrogen or carbon atoms. Adherent coatings can be formed on the tool at temperatures of 500°C or lower.

The lower deposition temperature of PVD, results in lower residual stress in the coatings compared with CVD. This may be advantageous, although the residual stress and stress-cracking in CVD coatings does not appear to affect significantly their performance in cutting applications. PVD coatings produced so far have been less strongly adherent to the substrate than CVD coatings. This disadvantage may be eliminated and the potential of PVD for development of new coatings will continue.

The greatest potential of the PVD approach may well be the multiple and nanolayer coatings. These create a variety of properties on the surface of the tough carbide core. The nanolayers in particular provide alternate layers of tough and hard materials. The evidence indicates that such "sandwiches" arrest microcracks at the interfaces. There is also the possibility of adding extra layers for oxidation resistance, and solid lubrication for example with nanolayers of molybdenum disulphide.[32,33,34]

These composite nanolayer materials will provide exciting new possibilities in the future. Referring back to the beginning of Chapter 6, perhaps some of the "early magic" of cutting tool development will return in the 21st century, if these nanolayer materials can create an all purpose cutting tool that can be manufactured to "give the best of all worlds" - highly wear resistant at high cutting speeds; but extremely tough in milling and interrupted cutting. Viewed from another perspective, such an "all purpose" tool would be greeted with great enthusiasm by the purchasing department in any organization. Today, the great proliferation of grades and styles causes much overstocking and, in the end, probably much unused, wasted material in the average "tool room".

7.12 REFERENCES

1. Dawihl, W., *Handbook of Hard Metals.* (English translation), H.M.S.O., London (1955)
2. Schwarzkopf, P. and Kieffer, R., *Cemented Carbides*, Macmillan, New York (1960)
3. Goldschmidt, H.J., *Interstitial Alloys*, Butterworths, London (1967)
4. Schwarzkopf, P. and Kieffer, R., *Refractory Hard Metals*, Macmillan, New York (1960)
5. S Atkins, A.G and Tabor, D., *Proc. Roy. Soc.* (A), **292**, 491 (1966)
6. Miyoshi, A. and Hara, A., *J. Japan Soc. Powder and Powder Met.*, **12**, 78 (1965)
7. Exner, H.D, and Gurland, J., *Powder Metal.* **13**, (5), 13 (1970)
8. Featherby, M., *Ph.D. Thesis,* University of Birmingham (1968)
9. Burbach, J., *Sonderdruk aus Technishe Mitteilungen.* Krupp, **26**, S 71, 80 (1968)
10. *Classification of Carbides According to Use*, ISO 513 (1975)
11. Trent, E.M., *Proc. Int. Conf.* M.T.D.R., *Manchester*, 629 (1968)
12. Aschan, I.J., *et al., Proc. 4th Nordic High Temp. Symp.*, 1, 227 (1975)
13. Trent, E.M , *I.S.I. Report No. 94*, pp. 11, 77, 179 (1967)
14. Trent, E.M., *I.P.E.J.*, **38**, 105 (1959)
15. Dearnley, P.A. and Trent, E.M., *Metals Technol.* **9**, (2), 60 (1982)
16. Trent, E.M., *J. Birmingham Met. Soc.*, **40**, 1 (1960)
17. Trent, E.M., *Proc. Inst. Mech. Eng.*, **166**, 64 (1952)
18. Naerheim, Y. and Trent, E.M., *Metals Technol.* **4**, (12) 548 (1977)
19. Kramer, B.M. and Hartung, P.D., *Cutting Tool Materials*, p.57, Am. Soc. for Metals. (1981)
20. Loladze, T.N., *Wear of Cutting Tools,* Moscow (1958)
21. Freeman, R.M., *Ph.D. Thesis*, University of Birmingham (1975)
22. Mayer, J.E., Moskowitz, D. and Humenik, M., *I.S.I. Publication* 126, P. 143 (1970)
23. Brookes, K.J.A., *World Directory and Handbook of Hard Metals*, Engineers' Digest Publication (1987)
24. Sandvik, Coromant AB, Sweden, Private communication
25. Schintlmeister, W., and Pacher, O., *J. Vac. Sci. Technol.*, **12**(4), 743 (1975)
26. Reiter, R.T., *Engineers' Digest Int. Conf New Tool Materials*, paper 6 London (March 1981)
27. Stjernberg, K.G., *Met. Sci*, **14**, (5), 189 (1980)
28. Hale T.E., and Graham, D.E., *Cutting Tool Materials*, p 175, Am. Soc. for Metals (1981)
29. Bunghah, K.F., Proc. *9th Plansee Seminar*, Paper 22 (1974)
30. Yoshihiko, D., et al., *Cutting Tool Materials*, p. 193, Am. Soc. for Metals (1981)
31. Dearnley, P.A., *Surface Engineering* **1**, (1), 43 (1985)

32. Komanduri, R., *Tool Materials*, Encyclopedia of Chemical Technology, Fourth Edition, Volume 24, 390 (1997) John Wiley and Sons Inc.

33. Shaw, M.C., Marshall, D.B., Dadkhah, M.S., and Evans, A.G., *Acta Metall. Mater.* **41**, (11), 3311 (1993)

34. Keem J.E., and Kramer, B.M., U.S. Patent 5,268,216 to Ovonic Synthetic Materials, (Dec. 7, 1993)

35. Fang, X.D., and Jawahir, I.S., Wear, **160**, 243 (1993)

36. Jawahir, I.S., *Annals of the CIRP*, **37**, (1), 121 (1988)

37. Jawahir, I.S., *Annals of the CIRP*, **39**, (1), 47 (1990)

38. Balaji, A.K., Sreeram, G., Jawahir, I.S.,and Lenz, E., *Annals of the CIRP*, **48**, (1), (1999)

39. Jawahir, I.S., Li, P. X., Ghosh R., and Exner, E. L., *Annals of the CIRP*, **44**, (1), 49 (1995)

40. Li, P. X., Jawahir, I.S., Fang X. D., and Exner, E. L., *Trans. NAMRI/SME*, **24**, 33 (1996)

41. Jawahir, I.S., Fang, X.D., Li P.X., and Ghosh, R., *Method of Assessing Tool-Life in Grooved Tools*, United States Patent, # 5,689,062, November 18, 1997

CHAPTER 8

CUTTING TOOL MATERIALS III: CERAMICS, CBN DIAMOND

8.1   INTRODUCTION

In this chapter, the properties and performance of ceramic tools, aluminum-based composites, sialons, cubic boron nitride and diamond will be reviewed. A final survey of the three classes of tool materials from Chapters 6, 7, and 8 is also provided, and many of the cited references cross-compare the abilities of these new materials.[1-25]

8.2   ALUMINA (CERAMIC) TOOLS

Refractory oxides have been among the many substances of high hardness and melting point investigated as potential cutting tool materials. Throw-away tool tips consisting of nearly 100% Al_2O_3 (alumina) have been available commercially for more than 30 years, and have been used in many countries for machining steel and cast iron.[1]

The successful tool materials consist of fine-grained (less than 5 μm) Al_2O_3 of high relative density, i.e. containing less than 2% porosity. Several different methods have been used to make tool tips which combine these two essential structural features, including:

(1) Pressing and sintering of individual tips by a process similar to that used for cemented carbides. Sintering is carried out in air and the tool tips are white.

(2) Hot pressing of large cylinders of alumina in graphite molds, the tool tips being cut from the cylinders with diamond slitting wheels. The tool tips are dark gray.

The possibility of 'cementing' alumina particles together with a metal bond, using a process similar to the bonding of carbide by cobalt, has been explored, but no satisfactory metal bond has been found. (Recall in Section 7.1, the liquid phase solution of WC in cobalt wets and pulls together the WC particles.) However, many additions, e.g. MgO and TiO, have been made to promote densification and retain fine grain size. The basic raw material, alumina, is cheap and plentiful, but the processing is expensive and the tool tips are therefore not cheap compared with cemented carbides. The room temperature hardness of alumina tools is in the same range as that of the cemented carbides, 1550-1700 HV.

The room temperature compressive strength is reported to be approximately 2750 MPa (180 tonf/in^2). Their major advantages are:

(1) Retention of hardness and compressive strength to higher temperatures than with carbides.
(2) Much lower solubility in steel than any carbide - they are practically inert to steel up to its melting point.

To offset these advantages, their toughness and strength in tension are much lower. The transverse rupture strength is an inadequate guide to toughness but values reported range from 390-780 MPa (25-50 tonf/in^2), about one third that of cemented carbides. The reported values of the fracture toughness parameter K_{IC} for alumina tool materials are 1.75 to 4.3 MPa m$^{-1/2}$. This is also much lower than those for cemented carbide (*Table 7.2*).

Alumina is non-metallic in character, with an ionic rather than a metallic bond. Consequently, it is an electrical insulator with low thermal conductivity - thus favored for use in the integrated circuit (IC) packaging industry. Alumina is a true ceramic, the pure form being white, translucent and looking like porcelain to which it is similar. To the new student in the field, it is quite a surprise to find that a material similar to pottery or porcelain can withstand the rapid fluctuations of temperature and stress involved in metal cutting operations. Nevertheless, alumina tools can be used to cut steel at speeds much higher than can be used with conventional cemented carbides or TiC-based alloys.

Negative rake throw-away tool tips are nearly always used, and it is not difficult to demonstrate that cutting speeds of 600-750 m min^{-1} (2000-2500 ft/min) can be sustained, at a feed of 0.25 mm (0.010 in) per rev, for long periods without excessive wear when cutting cast iron and many steels. Tools used at high speed show flank wear and a type of cratering starting close to the cutting edge.

There is evidence to demonstrate that wear, under conditions where the interface temperature is high, is caused by very superficial plastic deformation and flow of a very thin layer on the tool, rather than by diffusion into the work material.[2] Transmission electron micrographs (TEM) show a high concentration of dislocations just below the worn surface, while the body of the tool is practically free from dislocations.[3] Gradual failure involving flaking of thin fragments from the rake or clearance face is sometimes observed. It seems to be an advantage of alumina tools that such fractures do not always lead to sudden and massive failure of the whole tool edge with major damage to the workpiece, such as would occur after fracture of the edge of carbide tools.

The potential speeds using alumina tools, which are three to four times higher than those normally used with carbide tools, represent an increase in metal removal rate as great as that achieved by high speed steel and by cemented carbides at their inception. However, in spite of the efforts of dedicated enthusiasts to introduce these tools on a large scale, the numbers in use are only a very small percentage of the carbide tool applications. Their main continuous usage is in cutting gray cast iron where a very good surface finish is required. On clutch facings and brakes for cars they are being used at speeds up to 600 m min^{-1} (2000 ft/min) to give a surface finish good enough to eliminate a subsequent grinding operation.

8.2.1   Attitudes and Economics on the Factory Floor

It is first difficult to understand why alumina tools have not yet come into more wide spread use. There are several cases where concentrated efforts in the 'Research and Development' laboratories of a large firm have resulted in ceramic tools being introduced into individual factories for cutting both steel and cast iron.

In the long run, however, their continued use has been abandoned except for specific operations, mainly on cast iron. Cast iron, (*see* Chapter 9), machines with a "shower" of discontinuous chips at low cutting forces. It is easy to machine in this regard and once the operation is "up-and-running" not much supervision is needed.

However, for other materials with longer chips and higher cutting forces, the inadequate toughness of ceramics leads to unreliable performance under machine shop conditions. To obtain consistent day-to-day performance (on materials other than cast iron) continuous attention to every precise detail of tool setting and monitoring is required to prevent premature tool failure of ceramics. Such minor, yet frustrating, interruptions to production are bad enough. Worse still, damage to the component being machined, caused by occasional catastrophic failure of the ceramic tool edge, *will more than off-set the economies gained by peak performance.* And, of course, once a ceramic tool gets a "bad reputation" on a particular factory floor, it takes a long time to recover and to be considered for use again.

The enormous economies which resulted from the introduction of high speed steel and cemented carbides naturally whet the appetite for a further round of speed increases on the same scale.

However, in summary, there seems to be less potential advantage to be gained by increasing cutting speeds beyond those achieved by cemented carbides, for two reasons:

- when cutting speeds are high, e.g. 200 m min^{-1} (600 ft/min), the time required to load and unload the work in the machine may be quite a large proportion of the cycle time, and further increases in cutting speed do not achieve a proportional reduction in total machining time.
- at very high speeds, certain difficulties become accentuated - for example the chips coming off at very high speed are difficult to clear and may be a hazard to the operator. The chips are less of a problem when cutting cast iron. The machining of cast iron does not form a continuous ribbon, and this is the main reason for the successful application of ceramic tools to cast iron.

8.3   ALUMINA-BASED COMPOSITES (Al₂O₃ + TiC)

The problems discussed above, with materials that are nearly 100% pure alumina, naturally led to investigations on a mixture of (ceramic + carbide). Ceramic composite tool materials consisting of alumina with 30% or more of a refractory carbide - usually TiC or (Ta,Ti)C - have been commercially available since the 1960s. The mixture is hot pressed to full density and is dark gray in color. The structure consists of fine-grained Al₂O₃ with dispersed carbide grains a few microns in diameter. These tools are used mainly in high speed machining of cast iron and, as might be expected, experience shows that they can be applied in a wider range of applications than pure Al₂O₃ because of the increased toughness that results from the addition of the TiC.

The 1990s was a period in which the producers of tool materials improved the performance of alumina tools by exploring the possibility of 'alloying' with other substances. The main objective was to increase toughness, and development was much influenced by the concepts of fracture mechanics and their application in the area of composite materials - particularly the factors which influence crack propagation and the fracture-toughness value K_{IC}.

Alumina with the addition of up to 10% ZrO_2 is reported to increase the K_{IC} value by about 25% and to improve the capability for machining both cast iron and nickel-based alloys. The crack-retarding action is attributed to compressive stresses brought about by phase changes associated with the ZrO_2 particles (*Table 8.1*).

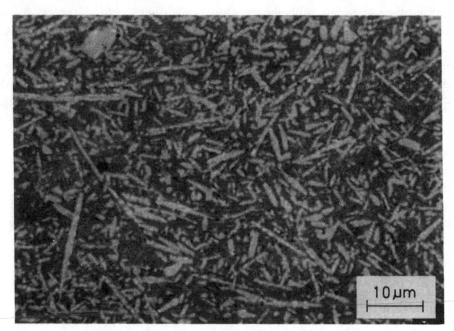

FIGURE 8.1   Structure of Al_2O_3/SiC (30 Vol%) whisker ceramic tool (Courtesy of Sandvik)

8.3.1   SiC whisker reinforced alumina

A more recent innovation is incorporation in alumina of up to 25% of SiC "whiskers", the mixed powders being consolidated by hot pressing. These whiskers are about 1-2 μm in diameter and about 20 μm long. They are very hard and strong, free from structural defects and are randomly distributed in the alumina matrix (*Figure 8.1*).

This material has a higher K_{IC} value and reports indicate considerably improved performance when machining certain nickel-based alloys, hard cast iron and hardened steel.

This is attributed mainly to higher resistance to crack propagation due to deflection of the cracks at interfaces, or to relief of stress at the crack tip when SiC whiskers are pulled from their sockets in the alumina.

In future years, such SiC whisker-reinforced aluminas or carbides may be the basis upon which many new tool materials eventually evolve. Since titanium aerospace work materials are chemically reactive with most tool materials, it will be interesting to continue to search for whisker-like additions to alumina that are inert to titanium.

8.3.2 Tool design of alumina based tools

In parallel with these composition changes, a steady development has taken place in adaptation of the tool edge shape to the machining of different work materials and different cutting conditions. A commonly used edge shape is shown in *Figure 8.2*. To a large extent, the data below and the diagram on the next page are related to all brittle type materials. In production, as much benefit can result from these edge preparations as from the actual choice of material.

A chamfer, C, has a width usually about 0.75 times the feed at an angle β to the rake face (of between 10° and 30° depending on the work material and the severity of the operation).

The edge itself may also have a honed radius, while the tool tip is presented to the work with a negative 5° rake angle. Tools with this type of edge configuration are less susceptible to premature failure, and are now being used in a wider range of operations and for cutting other work materials. As well as gray and chill cast iron, steels with hardness up to 500 HV, martensitic stainless steels, plastics and carbon are being machined.

Recall in Chapter 4 that the normal stress is highest on the cutting edge. Also, in some cutting conditions, there is a tendency for tensile stresses to arise at the rear of the contact length (*Figures 4.21* and *4.22*). Based on intuition, it is clear that the chamfer, C, redistributes the stress at the tool point. Perhaps the normal stress is reduced and the largest stress at the edge now acts inwards towards the tool (on a vector normal to the land measured by "C" in *Figure 8.2*) where the material can bear the stress more easily. Or, perhaps there is now less tendency for the normal stress to "bend the rake face like a cantilever beam" and create the tensile loads in *Figures 4.21* and *4.22*. These intuitive thoughts of the authors deserve further research - perhaps with the birefringent sapphire, ground and polished with different values of chamfer land and chamfer angle.

8.4   SIALON

A group of ceramics known as 'sialons' has been intensively investigated because of their outstanding properties as high strength refractory materials. Since 1976 their use as cutting tools has been explored and they have been successful in several applications.

Sialons (Si-Al-O-N) are silicon nitride-based materials with aluminium and oxygen additions. Silicon nitride (Si_3N_4) itself has useful properties, including high hardness (*ca.* 2000 HV), bend strength better than that of alumina (*ca.* 900 MP) and a low coefficient of thermal expansion (3.2 $\times 10^{-6}$), giving good resistance to thermal shock. It has been used as a cutting tool material but has not been accepted widely, partly because it can be produced in high-density form only by hot pressing, so that the cost of accurately shaped tools is high.

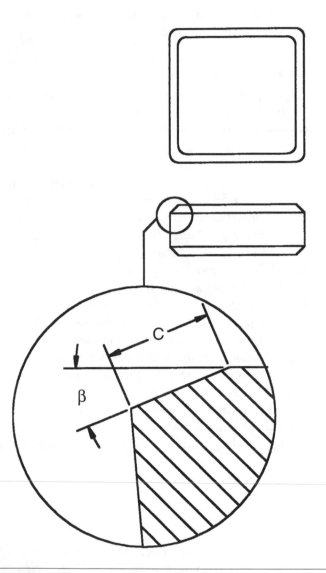

FIGURE 8.2   Edge chamfer commonly used on ceramic tools: c = 0.08-0.5 mm; β = 10 to 30°. Recall - at the end of Chapter 4 - it was shown that the largest normal stress on the rake face is at the cutting edge. Values of 1.5 to 2 times the average stress are likely. Also recall that under some cutting conditions - perhaps those where the feed rate is heavy- there is a tendency for tensile stresses to occur at the rear of the chip-tool contact length (*Figures 4.21* and *4.22*). The evidence suggests that the chamfer on ceramic tools redistributes the stresses and ameliorates both these conditions. The authors suggest that the chamfer causes the peak stress at the edge to act inwards towards the bulk of the tool, thereby causing less pressure on the delicate cutting edge itself. At the same time, there is less of a tendency for the rake face to bend like a cantilever beam reducing the tensile stresses on the tool face near the rear of the contact length. Both these effects allow the inherently brittle ceramics to perform better without premature fracture.To investigate these issues further, a fruitful research area would be to use the birefringent sapphire tools polished to exhibit a variety of chamfer length and chamfer angle styles. The effect of feed rate (undeformed chip thickness) is probably the most important machining parameter to vary in such tests.

Research on sialons has demonstrated that tool inserts can be produced by a process similar to that used for cemented carbide (as opposed to the alumina-type tools described in Section 8.1).

One manufacturer has stated[4] that the starting materials may be Si_3N_4, aluminium nitride (AlN) and alumina together with an 'alloying' addition of several per cent of yttria (Y_2O_3). These are milled together, dried, pressed to shape and sintered at a temperature of the order of 1800°C. The production processes are the subject of patents. In this case the production of a consistent material of optimum properties for use as tools depends on very careful balancing of the composition.

The oxide, yttria, reacts to form a silicate which is liquid at the sintering temperature. The liquid phase formed is the key to the achievement of nearly full density (98%) in a single sintering operation. During cooling after sintering the liquid solidifies as a glassy phase bonding together the silicon nitride-based crystals.

These hexagonal crystals are of fine grain size[4] (*Figure 8.3*), about 1 μm. In this form the high temperature strength is lower than that of hot pressed silicon nitride, but an annealing treatment at about 1400°C precipitates fine crystals of yttrium-aluminum garnet, 'YAG' ($3Y_2O_3.5Al_2O_3$), in the glassy phase. This heat treatment raises the high temperature strength to a value close to that of fully dense silicon nitride.

Table 8.1 gives properties of sialon and other ceramic tool materials. Sialon has higher transverse rupture strength and fracture toughness K_{IC} values than Al_2O_3, but rather lower values than Al_2O_3/SiC whisker ceramics. (It is unfortunate that reported values for K_{IC}, on which much emphasis is placed as indicating toughness of ceramics, are seldom accompanied by statistical evidence of scatter in repeated test results. Since values for different ceramics often differ by as little as 1 MPa $m^{-1/2}$, it is difficult to assess the validity of these data.)

Further advantages of sialon are low coefficient of thermal expansion and high thermal conductivity, which provide increased resistance to thermal shock and thermal fatigue compared with alumina-based ceramics.

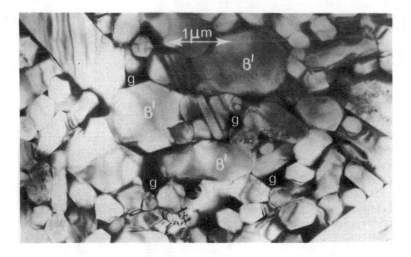

FIGURE 8.3 TEM showing structure of as-sintered sialon with yttria addition: β′ = Si_3-N_4-based phase containing substituted Al and O; g = glassy phase. (Courtesy of Lucas Industries)

TABLE 8.1   Properties of ceramic and ultra-hard tool materials

Tool material	Transverse rupture strength		Compressive Strength		Fracture toughness (K_{IC})	Hardness	Thermal expansion coefficient	Thermal conductivity at 20°C
	(MPa)	(Tonf/in²)	(MPa)	(Tonf/in²)	(MPa m$^{-1/2}$)	(HV)		(W/m°C)
Al$_2$O$_3$	550	35	3000	194	4	1600	8.2×10^{-6}	10.5
Al$_2$O$_3$/TiC	800	52	4500	290	4.5	2200	8.0×10^{-6}	16.7
Al$_2$O$_3$/SiC whiskers	900	58	-	-	7	1925	6.4×10^{-6}	13.0
Al$_2$O$_3$/1% ZrO$_2$	700	45	-	-	5.5	2230	8.5×10^{-6}	10.5
Sialon (sintered)	800	52	3500	230	6.5	1870	3.2×10^{-6}	20-25
WC-Co K10	2000	130	5500	350	9	1500	5.0×10^{-6}	100
Polycrystalline diamond			4740	310	8.8	50 GPa Knoop	3.8×10^{-6}	560
Polycrystalline cubic boron nitride			3800	250	4.5	28 GPa Knoop	4.9×10^{-6}	100

As with alumina ceramics, sialon tools are normally used with negative rake and often with similar chamfers at the edge (*Figure 8.2*). Industrial usage has been largely in machining ferrous materials. Successful use has been reported on roughing cuts and in operations involving rough surfaces and interruptions such as holes. Gray cast iron can be machined at 600 m min^{-1} (2000 ft/min) with a feed of 0.25 m/rev. Milling cutters tipped with sialon inserts are being used for cutting cast iron. Steel hardened to 550 HV has been successfully cut at speeds of up to 60 m min^{-1} (200 ft/min) with a feed rate of 0.12 mm/rev.

8.4.1 A special note on machining aerospace nickel alloys with sialon tools

In the machining of aerospace alloys, nickel-based gas turbine discs are being faced using sialon tips at 180 to 300 m min^{-1} (600-1000 ft/min) at a feed of 0.2 mm/rev, whereas carbide tools can be used at only 60 m min^{-1} (200 ft/min). Use in this application is significant for the confidence placed in the reliability of the tool, since tool failure could result in scrapping of very costly components

Both crater and flank wear are observed on sialon tools. In one research program on cutting of a nickel-based alloy, the limit to rate of metal removal was reached when the tool edge deformed and fractured. Wear by attrition at low speed, and by diffusion and interaction with the Si_3N_4 phase at high speed, were observed, and also a type of notch wear, where the outer edge of the chip crossed the tool edge.[6] (The high resistance to interaction with nickel at high temperatures is shown by the very low rate of attack when immersed in molten nickel.) The interactions when machining this alloy, although they were a cause of wear, did not prevent sialon being the best material for machining nickel alloys on an industrial scale, at speeds three or four times higher than with carbide tools.

In the case of machining another nickel-based high temperature alloy, industrial experience demonstrated that, for this alloy, long tool life can be achieved using the Al_2O_3/SiC whisker ceramics.[16] When machining engineering steels, however, sialons are generally not used on a industrial scale because of rapid wear, attributed to interaction between tool and work materials.

The conditions at the tool edge when cutting different alloys are so complex that more research will be required to achieve an understanding of the performance of different ceramic tool materials. At this stage optimum ceramic tooling is selected by empirical testing.

8.4.2 Summary of ceramics, alumina-based composites and sialon

The advances in alloying of ceramic tools today, typified by sialons, Al_2O_3/ZrO_2, and Al_2O_3/SiC whiskers, demonstrate that numerous variations in composition, and even heat treatment, can be introduced to modify structure and properties for specific cutting operations on different work materials. A universal ceramic tool to meet all requirements is unlikely and in the next period, various ceramic types will be competing for use in particular niches. Improvements in rigidity of machine tools and in tool design can be expected to enable ceramic tools to be used in more operations. It is of interest that, in Japan, the ceramic share of indexable inserts is estimated at 8 to 10%, while in the U.S. the proportion is estimated to be lower at about 3 to 4%.[17]

8.5   CUBIC BORON NITRIDE (CBN)

8.5.1   General properties of CBN

Cubic boron nitride (CBN) is not found in nature and it was theoretical considerations which led to the creation of this new substance. Like the synthetic diamonds described next in Section 8.6, cubic boron nitride consists of two interpenetrating face-centered cubic lattices, but one of the face-centered sets consists of boron atoms and the other of nitrogen atoms. Like diamond this is a very rigid structure, but in this case not all the bonds between neighboring atoms are covalent.

It has been stated that 25% of the bonding is ionic.[9] The resultant cubic boron nitride is the next hardest substance to diamond and has many similar, but not identical, properties. The hardness varies with the orientation of the test surface relative to the crystal lattice - between 40 and 55 GPa (4000 and 5500 HV). It is thus much harder than any of the metallic carbides (*Table 7.1*).

As with diamond, cubic boron nitride particles are consolidated in ultra-high pressure equipment.[11] A small percentage of metal or ceramic is blended with the boron nitride to achieve full density. Examples of commercially available CBN products are 'BZN'® (GEC) and 'Amborite'® (De Beers).

Polycrystalline cubic boron nitride is available in two forms i) thin layers consolidated onto a cemented carbide substrate, with a thickness less than 5 mm, ii) solid cubic boron nitride indexable inserts. Both particle size and the proportion and type of second-phase material can be varied for specific applications. Consolidated blanks are usually cylindrical discs from which tool tips of required shape can be cut: using a laser for solid cubic boron nitride tips, or EDM for the layered blanks.

Properties and performance of the tools depend mainly on the very hard cubic boron nitride but the second phase plays an important role. DeBeers reports that Amborite with a higher proportion of second phase gives a longer tool life when used as a finishing grade (at low feeds, e.g. 0.1 mm/rev) and for small depths of cut, while for roughing cuts optimum tool life is achieved by tools with a low content of second phase.[19] Considerable progress with these materials can be expected. The process of production is expensive and, as with diamond, tool tips cost in the region of 5 to 10 times the price of cemented carbide tools.

The room temperature hardness of the polycrystalline cubic boron nitride is given by the producers as: 'BZN' 35 GPa and 'Amborite' 28 GPa. As with the hardness of diamond there is likely to be considerable difference in the values measured by different laboratories and these differences are probably not significant.

To show the influence of temperature on hardness, the values given by Brookes[9] for both diamond and cubic boron nitride in single crystal form and as polycrystalline aggregates 'Syndite' and 'Amborite' - are shown in the graph, *Figure 8.4*.

All values were obtained with a Knoop indenter at a load of 1 kg. As with diamond, the hardness of cubic boron nitride decreases with increments in temperature, but its hardness remains higher than other tool materials over the whole temperature range. The thermal conductivity of 'Amborite' is 100 W/m °C.

A major advantage of cubic boron nitride compared with diamond is its greater stability at high temperatures in air or in contact with iron and other metals. It is stable in air for long periods at temperatures over 1000°C and its behavior as a cutting tool for machining steel at high speed suggests that it does not react rapidly with steel at higher temperatures.

8.5.2 Machining performance of cubic boron nitride

Polycrystalline cubic boron nitride tools have now been evaluated for some years in industrial machining operations. Both industrial trials and laboratory tests indicate a demand for these materials for specific operations where their superior performance makes them economic in spite of the high cost. In particular, they can be used for cutting both hardened steel and hard cast iron at high speeds without the rapid wear which prevents the use of diamond.

Hardened steel rolls (60-68 RC) can be machined at speeds of 45 to 60 m min^{-1} (150 to 200 ft/min) and feeds of 0.2 to 0.4 mm/rev. Chilled cast iron rolls are reported as being machined in the same range of speed and feed. The rate of metal removal is several times greater than that possible with cemented carbide tools.

Tool life is long so that rolls may often be machined to a dimensional tolerance and surface finish which avoid the necessity for a grinding operation. Hardened tool steels, including high speed steel, can be machined in this same speed and feed range.

The ability of CBN to cut such hard materials at high speeds is due to the retention of strength to higher temperatures than other tool materials, combined with excellent abrasion resistance and resistance to reaction with the ferrous work materials.

Figure 8.4 shows the increase in cutting speed achieved by cubic boron nitride tools compared with a K10 cemented carbide when machining a very hard cast iron. These polycrystalline boron nitride tools have very good toughness. Usually they are used with negative rake and with chamfers on the edge similar to those on ceramic tools (*Figure 8.2*). With the correct geometry, they can be employed for taking interrupted cuts on hardened steel - for example, turning bars with slots or holes, or in use as milling cutters.

No observations of deformation of the tool edge have been reported. The wear takes the form of flank and crater wear, the crater starting at or very close to the edge (*Figure 8.5*). Detailed analysis of the mechanisms of wear have not been widely reported and more research in this area would be justified.

There are probably differences in machining behavior of competitive commercial materials, for which somewhat different recommendations are made. It is uncertain how successful they are in machining the highly creep resistant aerospace nickel-based alloys, in which they compete with sialons and Al_2O_3/SiC whisker ceramics.

It seems certain that there are other special fields in which polycrystalline cubic boron nitride tools will play an important part as one of the types of tool materials.[9-19]

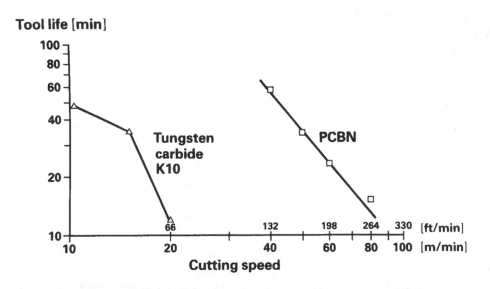

FIGURE 8.4   Tool life *vs* cutting speed for polycrystalline cubic boron nitride tools compare with K10 tungsten carbide-cobalt alloy, at a feed of 0.3 mm/rev with no coolant, when machining hard, martensitic cast iron. (Courtesy of DeBeers Industrial Diamond Division (PTY) Ltd and ASM International[19])

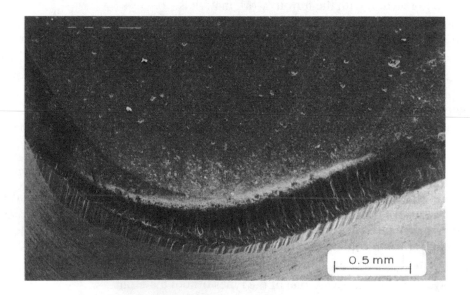

FIGURE 8.5   SEM showing flank and crater wear on chamfered edge of polycrystalline cubic boron nitride tool after cutting hardened tool steel (650 HV) at 75 m min^{-1} (250 ft/min). (Courtesy of de Beers Industrial Diamond Division Ltd)

8.6   DIAMOND, SYNTHETIC DIAMOND, AND DIAMOND COATED CUTTING TOOLS

8.6.1   General properties of diamonds

The hardest of all materials, diamond, has long been employed as a cutting tool although its high cost has restricted its use to operations where other tool materials cannot perform effectively. Because of their very high hardness, all types of diamond tools show a much lower rate of wear and longer tool life than carbides or oxides under conditions where abrasion is the dominant wear mechanism.

The extreme hardness of diamond is related to its crystal structure. This consists of two interpenetrating, face-centered cubic lattices arranged so that each carbon atom has four near neighbors to which it is attracted by co-valent bonds.[9] Diamond crystals are very anisotropic. The hardness and resistance to abrasive wear of any surface are very dependent on the orientation of this surface to the crystal lattice. Hardness measurement is difficult because the indenters must also be diamond but, using Knoop indenters, values for hardness have been shown to vary between 56 and 102 GPa (equivalent to approximately 6,000 to 10,000 HV) in different crystallographic directions. This compares with maximum values of about 1,800 HV for cemented carbide or alumina. On different faces the rate of abrasive wear can vary by as much as 1:80.[10]

The structure of all types of diamond shows stacking faults and voids in the lattice, which create the sites for microcrack growth. Then, because the structure is so rigid, opportunities for crack bluntening by dislocation movement are nil or minimal.

Single crystal, natural diamonds have been used in many industrial applications, for example as dies for drawing fine wire. For cutting operations, large natural diamonds are used as single point tools in specialist fields. The optimum orientation is selected and they are lapped to the required shape and mounted in tool holders. The tool edges can be prepared to quite exceptional accuracy of form and edge perfection and are capable of producing surfaces of extremely high accuracy and finish. They are used for this purpose in production of optical instruments and gold jewelry. Diamonds are deficient in toughness - sharp edges are easily chipped - and this limits the range of operations in which they are used.

8.6.2   Fabrication of synthetic diamond (and CBN) by the HT-HP method

Since the early 1950s, synthetic diamonds have been produced by heating graphitic carbon with a catalyst at temperatures over 1500°C and at ultra-high pressures. This is usually called the HT-HP process for the high temperatures and pressures involved. This synthesis of diamond was made possible by the engineering development of ultra-high pressure processing units in which temperatures and pressures of the order of 1500 °C and 8 GPa could be maintained for a sufficient time to transform carbon into the diamond structure and to grow diamond crystals of usable size.

These processing units also made possible the transformation of cubic boron nitride (CBN) - from a hexagonal form to a structure akin to diamond (*see* Section 8.5).

8.6.3   Synthetic diamond

Synthetic diamonds are small - usually a few tenths of a millimeter - and a range of particle shapes can be produced. In many industrial applications the synthetic diamond grit has replaced natural diamond and it is used very extensively, bonded with metal or polymers, to form grinding wheels. Synthetic diamonds are not made in sizes large enough to make single point tools, but techniques have been developed of consolidating fine diamond powder into blocks of useful size.[11]

The consolidation is a hot-pressing operation carried out at ultra-high pressures within the same HT-HP equipment employed for initial synthetic diamond production. Densification is accelerated by including a metallic or ceramic bonding material - usually metallic for metal cutting tools. A range of polycrystalline diamond tools is available, with diamond grain size from 2 to 25 μm. The proportion of diamond to bonding agent can also be varied for different applications.[18]

Figure 8.6 shows the structures of two polycrystalline diamond tools.[19] Many commercially available tool tips are in the form of laminated bodies. A layer of consolidated diamond, usually 0.5 to 1 mm thick, is bonded to a cemented carbide substrate to form a tool tip usually about 3 mm thick.

The consolidated blanks are usually in the form of thin cylinders up to about 70 mm in diameter. These can be cut to useful shapes, usually employing electrical discharge machining (EDM). Precision cutting edges can be formed by EDM and are often polished. The composite tools can be clamped or brazed to shanks. They can be reground when worn, but the grinding taking longer than with carbide tools. The tools are expensive, costing typically 20 to 30 times the equivalent carbide tool.

These polycrystalline diamond tools are aggregates of randomly oriented diamond particles, which behave as an isotropic material in many applications. It is not possible to achieve as extreme a perfection of cutting edge as with natural diamond, but their behavior in cutting operations demonstrates that the edges are less sensitive to accidental damage, while maintaining exceptional resistance to wear. It is their ability to maintain an accurate cutting edge for very long periods, which has made them successful competitors in specific areas of machining.

8.6.4   Machining performance of synthetic diamond tools

Polycrystalline diamond tools are recommended for machining aluminum alloys. For hypereutectic aluminum-silicon alloys they are particularly useful because carbide tools are very rapidly worn at speeds over 100-150 m min^{-1} (300-450 ft/min), while a very long tool life can be obtained with diamond tools at speeds over 500 m min^{-1} (1500 ft/min). Wear rates are many times lower than on carbide tools. Flank wear occurs and *Figure 8.7* shows flank wear as a function of time for Syndite diamond tools of three grain sizes. Wear rate is dependent on grain size but the differences are relatively small.

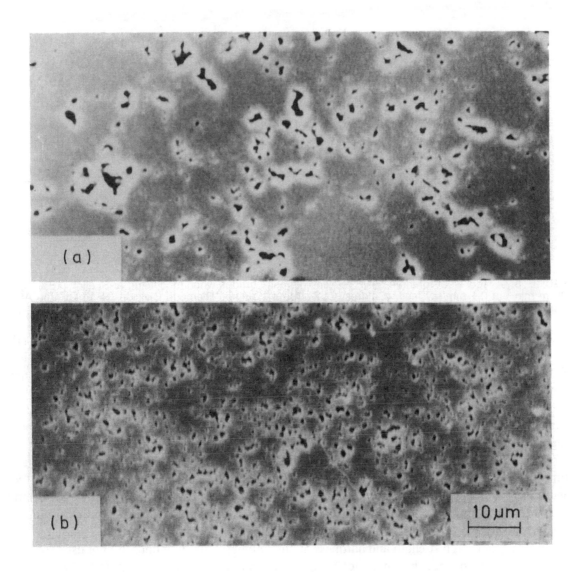

FIGURE 8.6   Etched microstructures of two grades of polycrystalline diamond tool: (a) 25 μm grain size (b) 2 μm grain size (Courtesy of DeBeers Industrial Diamond Division (Pty) Ltd)

Diamond tools are now being used for milling, turning, boring, threading and other operations in the mass production of many aluminum alloys because the very long tool life without regrinding can reduce costs. They are also used for machining copper and copper alloys at speeds over 500 m min^{-1} - copper commutators are an example. Cemented carbides in the soft (pre-sintered) condition are machined with diamond tools and even fully sintered carbide tools of the softer (higher cobalt) grades are regularly machined. Publications concerning polycrystalline diamond tools are available from a number of companies such as General Electric Co., USA, who market 'Compax'® tools and De Beers Industrial Diamond Division who produce 'Syndite'®.

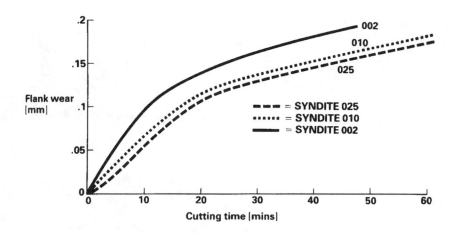

FIGURE 8.7 Tool wear for three polycrystalline diamond grades when machining Al-Si alloy (18% Si) at 100 m min^{-1} at 0.1 mm/rev feed. (Courtesy of DeBeers Industrial Diamond Division (PTY) Ltd and ASM International[19])

The former give room temperature hardness values for 'Compax' in the range 6500-8000 HV (64-78 GPa) while typical values for 'Syndite' are given as 4400-4700 HV (43-46 GPa). In view of the lack of standardization of hardness testing for these materials this does not suggest any significant difference in hardness between the two products. The hardness is somewhat lower than that of single crystal natural diamond but much higher that of other tool materials. High temperature hardness tests have shown that, as with other hard materials, the hardness of both single crystal and polycrystalline diamond decreases as the temperature increases.

Figure 8.8 shows data presented by Brookes[9] compared with values for cemented carbide and alumina. The high temperature hardness, also, of diamond is much higher than that of other tool materials. The thermal conductivity of diamond is very high - 500 to 2000 W/m°C for natural diamond, depending on orientation, and 560 W/m°C for the polycrystalline 'Syndite' (compare 400 W/m°C for copper).[19]

In spite of their high strength and hardness at high temperature, diamonds are not used for high speed machining of steel because tool wear is very rapid. The tools are smoothly worn by a mechanism which appears to involve transformation of diamond to a graphitic form and/or inter-action between diamond and iron and the atmosphere. Diamond is not the stable form of carbon at atmospheric pressure. Fortunately, it does not revert to the graphitic form at temperatures below 1500 °C. *In contact with iron, however, graphitization begins just over 730 °C[12] and oxygen begins to etch a diamond surface at about 830°C.*

It is also disappointing that diamond tools are rapidly worn when cutting nickel alloys. Generally, they have not been recommended for machining high melting point metals and alloys where high temperatures are generated at the interface.

By contrast, major use for polycrystalline diamond tools is in machining non-metallic materials of an abrasive nature - for example, silica flour filled resins, fiber reinforced plastic, printed circuit board laminates, wood and wood-based products, un-fired and some fired ceramics.

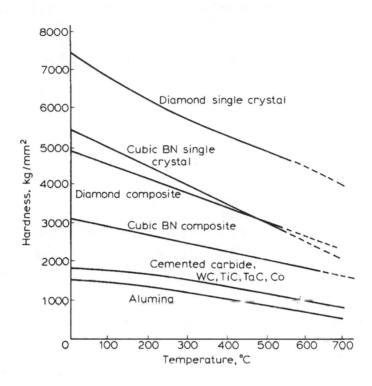

FIGURE 8.8 Hardness *vs* temperature for diamond and cubic boron nitride (After Brookes and Lambert[9])

8.6.5 Fabrication of synthetic diamond by the CVD process

In the period of the 1990s there was great interest in the synthesis of diamond materials by a variety of chemical vapor deposition processes (CVD). Windischmann and colleagues have demonstrated that diamond wafers that are 250 mm diameter and 0.3 millimeters thick can be produced, crack free, with proper stress control.[20,21] Diamond's outstanding thermal properties mean that such wafers, when cut into small rectangles, might be an excellent base-support in the packages for integrated circuits (ICs) - cost permitting. Thicker (e.g. 1.5 mm), polycrystalline diamond slabs have also been deposited, cut-up to size, and then brazed onto a cemented carbide substrate. Norton, Crystallume, Kennametal, General Electric, Sumitomo Electric, Toshiba Tungaloy, Mitsubishi Metal Corporation, Asahi Diamond Company, and Sandvik are some of the many companies that are competing to make these products viable for the metal cutting industry. Microwave CVD, hot filament CVD, plasma torch, and combustion synthesis are some of the techniques that have been used. Busch and Dismukes[22] indicate that the DC arc-jet deposition method is likely to be the most effective CVD process if measured by deposition cost per square centimeter.

Rather than grow the thicker slabs and braze them onto carbides, an alternative procedure is to directly deposit by CVD thin (2 to 5 μm) coatings onto the surface of carbide cutting tools. This excites the research community even more because it holds the promise of quickly depositing a highly wear resistant coating on a relatively cheap substrate. The advantage of direct-CVD is that no polishing and honing of the tool is required, thus saving finishing costs. Originally, it was this reduction in finishing costs that allowed the thin (1 to 10 μm) coatings on cemented carbide tools to be economical.

Metallurgical sections through the larger substrates (250 mm in diameter and 0.3 millimeter thick) show fine grained equiaxed grains on the lower surface where nucleation first takes place in the growing process onto an inert surface in a hydrogen atmosphere.[20] Columnar grain growth then occurs vertically towards the upper surface. Within the columnar grains, stacking faults, voids and dislocation sites are present. These become the source for microcrack growth. With the nil or limited ability for dislocations to flow in the diamond lattice, the lack of crack-tip blunting is the cause of early failure. Similarly, this gradation can be seen in the coated CVD tools: fine-grained equiaxed structures at the interface with the carbide substrate, growing in a columnar fashion to the tool's free surface. These structures also contain the flaws discussed. These are almost certainly the source of the early failures that occur today in research laboratories, as scientists hope to machine the nickel and titanium aerospace alloys with the CVD diamond coated tools.

There is a further problem with CVD coated carbides: To form a good metallurgical bond at the interface in a coated tool, the characteristics of the coating and the substrate should be matched as closely as possible. This includes matching of thermal expansion coefficients, lattice parameters, and surface chemistry. Unfortunately, the adhesion of the diamond coating to the cobalt in the tool substrate is difficult. Efforts are being made to selectively etch away the cobalt at the substrate interface before the coating is applied. In this way the diamond has more of the WC or the mixed (WC+TiC+TaC) to bond onto.[20]

8.6.6   Summary of diamond, synthetic (HT-HP) diamond and coated diamond

In summary, as a metal cutting engineer in a production setting, one should be "on guard" about any advertised claims in which diamond tools and coated diamond tools give phenomenal performance! There seems to be no other tool material that is so "work material dependent".

On the one hand, diamond tools give outstanding performance with aluminum alloys, especially the aluminum-silicon casting alloys containing 17-23% Si. In these, the silicon content is above the eutectic composition and the structures contain large grains of silicon, up to 70 μm across, in addition to the finely dispersed silicon of the eutectic structure. The large silicon crystals greatly increase the wear rate, even when using carbide tools and thus the diamond HT-HP inserts have met with great success. Since cutting forces with such aluminum alloys are still relatively low (despite the abrasive silicon) the CVD diamond coated tools produced by such companies as Crystallume and Kennametal have also met with success. Similarly, the cutting forces are low with non-metallic resins, printed circuit board e-glass and fiber reinforced plastics - these are also nicely machined with CVD coated diamond tools.

On the other hand, when cutting steel, diamond is quickly dissolved away as described in detail at the end of section 8.6.4. Also, the authors' own laboratory attempts to use CVD diamond coated tools for aerospace alloys have met with only limited success. Nickel, titanium (and steel) alloys create much higher shear stresses on the rake face. The evidence seems to indicate that today's CVD diamond coatings still have too many stacking faults, voids and poor adhesion to cobalt, so that these higher shear stresses quickly propagate microcracks eventually leading to the complete stripping away of the CVD diamond coating. Despite today's unreliable performance, the machining performance of aerospace alloys with CVD diamond tools prepared with a variety of CVD processing methods, still seems a fruitful area of research.

8.7   GENERAL SURVEY OF ALL TOOL MATERIALS

8.7.1   100 years of cutting tool performance

At the beginning of Chapter 6 the evolutionary character of the development of tool materials was emphasized. The evolutionary progress is shown by the following table listing the speeds commonly used for cutting medium carbon steel with different classes of tool material:

TABLE 8.2   Evolution of tool materials as measured by cutting medium carbon steel for a cutting time of approximately 20 minutes. Note that diamond tools are not listed in this table. They are not recommended for cutting steel because iron dissolves diamond.

Tool Material	Cutting Speed	
	$(m\ min^{-1})$	(ft/min)
Carbon Steel	5	16
High speed steel	30	100
Cemented carbide	150	500
Ceramic	600	2000

The table suggests a dramatic but over-simplified history of tool materials. One might suppose that a more efficient tool material would quickly and completely eliminate a less efficient one. *However, because of the enormous variety of metal cutting conditions, a range of tool materials remains in use, and will continue to do so.* It has been an objective of these chapters to offer some explanation for the continuing demand for a variety of tool materials.

By no means have carbide or harder tools taken the place of high speed steel for day-to-day end-milling, drilling, tapping and broaching operations in a typical machine shop. Indeed in a small "job-shop" where the machinists are making "one-of-a-kind" prototypes, the focus is on fixturing, setups, and inspection, rather than increasing the cutting speed when cutting itself starts. In these the environments, high speed steels are often favored over carbides even for regular turning operations. Such factors explain why carbon and high speed steels still account for approximately half of all tools sold - as shown in *Table 8.3*.

This is because tool life is determined by a large number of different mechanisms which come into play in different ways depending dramatically on the conditions of use. There is no simple relationship between tool performance and any single property of the tool material such as hardness or wear resistance. Emphatically so, the general phrase "wear resistance" is not a unique property of a tool material, but is a complex interaction between the tool and work materials very dependent on the cutting conditions. Therefore, the subject of tool performance is bound to be complex.

8.7.2   Annual expenditures on cutting tools

High speed steels and cemented carbides were the most extensively used tool materials in the decades of the 1980s and 1990s, accounting for U.S. sales of about $2 to 2.5 billion dollars per year.[23,24] It is now of interest to revisit some of the data in Chapter 1. The U.S. consumption of new machine tools (CNC lathes, milling machines, etc.) is $7.5 billion dollars per year. *Metcut Research Associates*, estimates that the annual labor and overhead costs of machining are about $300 billion dollars per year (this excludes work materials and tools). For comparison purposes a ratio is noted of {2.5 → 7.5 → 300} billion dollars for {disposable cutting tools → fixed machinery → labor investments}.

TABLE 8.3   Estimated Breakdown of Cutting Tool Costs (shown in billions of dollars) in the Decades of the 1980s and 1990s [23, 24]

Material	Cost, $
High carbon, low alloy, HSS	$1 - 1.25 \times 10^9$
Cemented carbides	0.75×10^9
Ceramics	0.25×10^9
Diamond, CBN	$0.25 - 0.50 \times 10^9$
TOTAL	approx. $2 - 2.5 \times 10^9$

Of the approximately $2 to 2.5 billion per year (in the U.S.) of consumable tool materials, high speed steel accounts for about $1-1.25 billion, cemented carbides account for about $750 million, and ceramics and diamond about $25 million each (*Table 8.3*). Many manufacturers keep sales data confidential - thus such estimates in *Table 8.3* are bound to be approximate. The growth of *carbide* cutting tools in the next few years has been predicted and *Figure 8.9* shows the estimates (for both U.S. and globally) by the Kennametal company.[26] The predictions indicate that PVD coated carbides will grow in use but that CVD coatings will continue to have the highest share of sales. The required trends towards dry cutting will promote further use of ceramics and diamond tools but their proportional sales will still remain less than carbides. The share of uncoated carbide tools will decline.

8.7.3   An informal, closing analogy

Note that *Figure 8.9* excludes high speed steel and carbon steels, which in fact still account for over half of the annual sales of cutting tools. The following question may still linger in the reader's mind: Why is an older tool material still so dominant?

In a lecture situation where informality is permitted and indeed desirable, the authors often answer this question with the following analogy - We might all like to own a Lamborghini or a Ferrari and speed down the highway undisturbed by other drivers (and the police). But even wealthy people that own such cars rarely use them for those mundane weekend tasks at the shopping mall. The owners use them even less for car pooling their children to soccer, or dragging supplies back from the lumber yard. A rough and tumble mini-van is best for those tasks. This analogy holds up quite well for cutting tools. Ceramic and diamond tools are happiest when speeding down a long bar in turning, or across a disc brake in a facing operation. But for day-to-day operations, where set-up times are long, where the tool might be used in a variety of ways, and where some unexpected knocks and accidents might occur on the tool edge, high speed steels are equivalent to the rough and tumble mini-van. Especially in a job-shop they give more reliable performance. And as pointed out earlier, the job-shop is a place where set-up times and fixturing might be very time consuming. Even if a few seconds or minutes can be saved by using a diamond coated tool once cutting has started, it will not make much difference to the overall economics.[†]

As a result the average tool-room will always contain all the tools listed above. Over the coming years the percentage of diamond and ceramic tools will grow. Coated carbides will continue to displace high speed steel. However, high speed steel will continue to hold an important place as reflected in the table above. Given this economic reality, the authors suggest that the research community would do well to re-evaluate some of their efforts.

Rather than being over-infatuated with the promise of exotic diamond tools, large benefits can still come from improving the performance of TiN, and other coatings, for high speed steel and carbides. These issues are revisited in Chapter 9 in the context of deoxidized steels (*see* Section 9.13.2). And the conclusions in Chapter 13, reconsider the overall economics of CVD diamond coatings.

---

†. The reader might be interested to use a search-engine such as <www.hotbot.com> and type in "metal-cutting" or "cutting-tools" in the search-window. Of the many thousands of locations that come up, it is interesting to see how many small to medium sized job shops and suppliers are "out-there" and that the work they do will favor the all-purpose tool steels. These comments are not meant to downplay the importance of carbide, ceramic or diamond tools - merely to keep things in economic proportion vis-a-vis all tool material types.

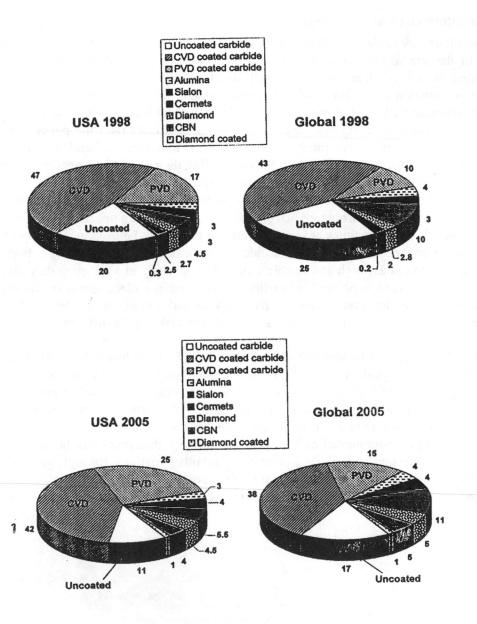

FIGURE 8.9 U.S. and global markets for carbide, ceramic, diamond and other hard tooling in 1998 and 2005. Courtesy of the Kennametal Company[25] Figure excludes high speed steel.

8.8 REFERENCES

1. King, A.E. and Wheildon, W.M., *Ceramics in Machining Processes*, Academic Press, New York (1960)
2. Dearnley, P.A. and Trent, E.M., *Metals Technol.*, **9**, (2), 60 (1982)
3. Kim, C.H., et al., *J. Appl, Phys.*, **44**, (11), 5175 (1973)
4. Lewis, M.H, Bhatti, A.R., Lumby, R.J and North, B., *J. Mat. Sci*, **15**, 103 (1980)
5. Hartley, P., *Engineering.* (Sept. 1980)
6. Jawaid, A., Bhattacharyya, S.K., Lewis, M.H. and Wallbank, J., *Met. Technol.*, **10**, 482 (1983)
7. Jack, K.H., *Metals Technol.*, **9**, (7), 297 (1982)
8. Hatschek, R.L., *American Machinist*, 110 (Jan 1983)
9. Brooks, C.A. and Lambert, W.A., *Ultrahard Materials Application Technology.* (*De Beers*), p.128 Hornbeam Press Ltd (1982)
10. Wilkes, E.M. and Wilks, J., *J. Phys. D: Appl. Phys-* **5**, 1902 (1972)
11. Hibbs, L.E., Jr., and Wentorf, R.H., Jr., *8th Plansee Seminar*, Paper No 42 (1974)
12. Hitchener, A.P., Thornton, A.G. and Wilks, *J. Wear of Materials*, p. 728 A.S.M.E., (1981)
13. Tabuchi, N. et al., Samimoto Electric Technical Review. **18** 57-65 (1978)
14. International Scminar, *Superhand Materials*, Kiev. (1981)
15. Ezugwu, E.O. and Wallbank, J., *J. Mat. Sci. & Tech.*, **3**, 881 (1987)
16. Bhattacharyya, S.K., *et al., Proc. 6th Int. Conf. Prod. Eng.*, Osaka, 176 (1987)
17. North, B., *Tech. paper No MR86-451*, Soc. Mfg. Eng. (1986)
18. Heath, P.J., VDI/CIRP Conf., *'Cutting Materials'*, Dusseldorf, (Sept. 1989)
19. Heath, P.J. *A.S.M. Handbook*, 9th ed., Vol. 16, 105
20. Windischmann, H., Information from *Saint-Gobain Industrial Ceramics* Inc. Norton Diamond Films, Northboro, MA. *See* <www.saintgobain.com>
21. Howard, W.N., Spear, K.E., and Frenchlach, M., *Appl. Phy. Lett.*, **63**, (19), 26 (1993)
22. Busch, J.V., and Dismukus, J.P., *Diamond and Related Materials*, **3**, 295 (1994)
23. Klutznick, P.M., *Industrial Outlook for 200 Industries with Projections for 1985*, U.S. Department of Commerce, Washington D.C., 1981
24. Komanduri, R., *Tool Materials*, Encyclopedia of Chemical Technology, Fourth Edition, Volume 24, 390 (1997) John Wiley and Sons Inc.
25. *See* <www.kennametal.com>
26. Huston, M.F., and Knobeloch, G.W., *Cutting Materials, Tools and Market Trends,* in the Conference on High-Performance Tools, Dusseldorf, Germany, p. 21, (1998)

CHAPTER 9

MACHINABILITY

9.1   INTRODUCTION

9.1.1   Defining "machinability"

The machinability of an alloy is similar to the palatability of wine - easily appreciated but not readily measured in quantitative terms

The term machinability is used in innumerable books, papers and discussions. It might therefore be taken to imply that there is a property or quality of a material which can be clearly defined. In fact, there is no unique and unambiguous meaning to the term machinability. To the active practitioner in machining, engaged in a particular set of operations, the meaning of the term is clear. For that particular person's experiences, the machinability of a work material can often be measured in terms of the numbers of components produced per hour; the cost of machining the component; or the quality of the finish on a critical surface.

Problems arise because there are so many practitioners. Each one of them carries out a wide variety of operations, each with a different criteria of machinability. A material may have good machinability by one criterion, but poor machinability by another. Also, relative machinability may change when a different type of operation is being carried out - turning versus milling - or when the tool material is changed.

To deal with this complex situation, the approach adopted in this chapter is to discuss the behavior of a number of the main classes of metals and alloys during machining. Also, there are explanations of their behavior in terms of their composition, structure, heat treatment and properties. Despite the lack of a universal metric, "machinability" may be assessed by one or more of the criteria below:

(1) *Tool life*. The amount of material removed by a tool, under standardized cutting conditions, before the tool performance becomes unacceptable or the tool is worn by a standard amount.

(2) *Limiting rate of metal removal*. The maximum rate at which the material can be machined for a standard short tool life.
(3) *Cutting forces*. The forces acting on the tool (measured by a dynamometer, under specified conditions) or the power consumption.
(4) *Surface finish*. The surface finish achieved under specified cutting conditions.
(5) *Chip shape*. The chip shape as it influences the clearance of the chips from around the tool, under standardized cutting conditions.

9.1.2   Machinability and temperature effects

Chapter 5 emphasized that the temperatures generated in tools are a major influence on the rate of tool wear and on the limit of the rate of metal removal. Temperatures at the tool work interface vary with the composition of the work material. Throughout this chapter the metals and alloys discussed are compared in terms of the temperatures generated. This has not yet been investigated in the deepest manner required for a conclusive survey, but significant relationships have emerged.

Temperature distribution at the tool/work interface was determined by changes in microhardness or microstructure in the heat affected regions of carbon and high speed steel tools (Chapter 5). For comparison of different work materials a standard tool geometry was adopted (+ 6° top rake being the most essential feature). The feed and depth of cut were constant at 0.25 mm and 1.25 mm, respectively, and cutting time was generally 1 min; no coolant was used. Temperatures were determined as a function of cutting speed, the accuracy of temperature measurement being about +/- 25°C. The parameter used as the criterion is the maximum temperature observed at any position on the interface.

The metals and alloys with the best machinability are discussed first.

9.2   MAGNESIUM

Of the metals in common engineering use, magnesium is the easiest to machine, scoring a top ranking by almost all the criteria. Rates of tool wear are very low because magnesium does not alloy with steel, and the metal and its alloys have a low melting point (650°C). Temperatures at the tool-work interface are low even at very high cutting speed and feed rate. Turning speeds may be up to 1,350 m min^{-1} (4,000 ft/min) in roughing cuts, and finishing cuts can be faster with good tool life. Magnesium alloys behave, in this respect, much like the pure metal.

The tool forces when cutting magnesium are very low compared with those when cutting other pure metals, and they remain almost constant over a very wide range of cutting speeds,[1] *Figure 9.1*. Both the cutting force (F_c) and the feed force (F_t) are low and the power consumption is considerably lower than that when cutting other metals under the same conditions. The low tool forces are associated with the low shear yield strength of magnesium, and, more important, associated with the small area of contact on the rake face of the tool over a wide range of cutting speeds and rake angles. This ensures that the shear plane angle is high and the chips thin - only slightly thicker than the feed.

Steel or carbide tools can be used and the surface finish produced is good at both low and high cutting speeds. The chips formed are deeply segmented and easily broken into short lengths, so that chip disposal is not difficult even when cutting at very high speed. The hexagonal structure of magnesium is probably mainly responsible for the low ductility. This leads to the fragile segmented chips and the short contact length on the tool rake face. The worst feature of magnesium is that the fine chips can be ignited. This causes considerable fire risk.

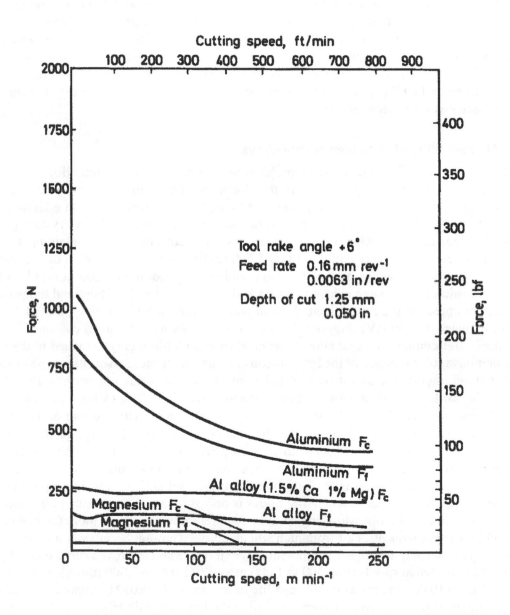

FIGURE 9.1 Tool forces *vs* cutting speed - magnesium and aluminum (From data of Williams, Smart and Milner[1])

9.3   ALUMINUM AND ALUMINUM ALLOYS

9.3.1   Introduction

Alloys of aluminum in general also rate highly in the machinability table by most of the criteria. As with magnesium, the melting points of aluminum (659°C) and its alloys are low. Consequently, the temperatures generated during cutting are never high enough to be damaging to the heat-treated structures of high speed steel tools. Good tool life can be attained up to speeds of 600 m min^{-1} (2,000 ft/min) when using carbide tools, and 300 m min^{-1} (1,000 ft/min) with tools of high speed steel. Speeds as high as 4,500 m min^{-1} (15,000 ft/min) have been used for special purposes. (Further discussion of high speed, and very high speed machining of aluminum alloys appears in Chapter 11.) Wear on the tool takes the form of flank wear, but no detailed study of the wear mechanism has been reported.

9.3.2   Machinability of aluminum-silicon alloys

High tool wear rates become a serious problem with only a few aluminum alloys. In aluminum-silicon castings containing 17-23% Si, the silicon content is above the eutectic composition. The structures thus contain large grains of silicon, up to 70 μm across. In addition, they contain a finely dispersed silicon of the eutectic structure. The large silicon crystals greatly increase the wear rate, even when using carbide tools.[2] The eutectic alloys, containing 11-14% Si can be machined at 300-450 m min^{-1} (1,000-1,500 ft/min) with good carbide tool life, but the presence of large silicon grains may reduce the permissible speed to only 100 m min^{-1} (300 ft/min). The drastic effect of large silicon particles is the result of the high stress and temperature which these impose on the cutting edge. Silicon particles have a high melting point (1,420°C) and high hardness (> 400 HV). *Figure 9.2* shows a section through the worn cutting edge of a carbide tool after cutting a 19% Si alloy.[3] The layer of silicon (dark gray) attached to the worn surface demonstrates the action of the large silicon crystals which cause an attrition type of wear.

The photomicrographs demonstrate that the wear of tools depends not only on the phases present in the work material, but also on their size and distribution. Small silicon particles in the eutectic bypass the cutting edge. Either they do not make contact with the tool at all, or they make only rubbing contact with one of the flat surfaces. However, the large primary silicon grains cannot pass by the cutting edge but are divided when they make contact with it. Part of the silicon passes away with the chip and part continues on the new machined surface. At the same time, a layer of sheared silicon is strongly bonded to the tool surface, causing rapid attrition wear. The machining of hypereutectic Al-Si alloys is one of the most important applications for polycrystalline diamond tools. Most engine manufacturers now use these tools for machining pistons and other components, cast from high silicon alloys. Turning, boring and milling operations are carried out at much higher speed with longer tool life and improved surface finish. Pistons have been turned at speeds from 300 to 1,000 m min^{-1} (1,000 to 3,000 ft/min), and at a feed of 0.125 mm/rev (0.005 in/rev) with tool edge life of the order of 100,000 components. The diamond tools have proved so advantageous for cutting the hypereutectic alloys that they are used for turning and milling other aluminum alloys even where carbide tools give reasonable tool life. Polycrystalline diamond tools can be kept in use for months. This gives economies in spite of the higher tool cost.

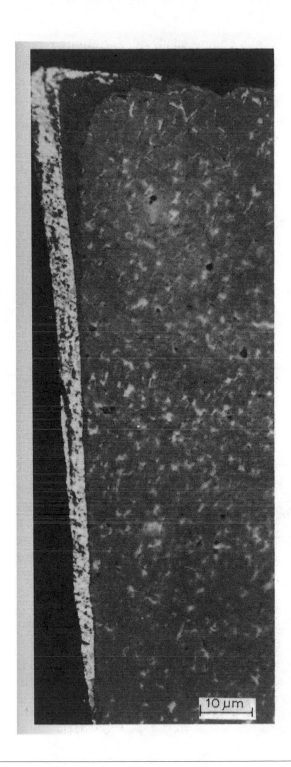

10 μm

FIGURE 9.2 Section through cutting edge of cemented carbide tool used to cut 19% Si-Al alloy at 122 m min^{-1} (400 ft/min)[3]

9.3.3 Aluminum alloys versus commercially pure aluminum

In general, tool forces when cutting aluminum *alloys* are low, and tend to decrease slightly as the cutting speed is raised (*Figure 9.1*). High forces occur, however, when cutting *commercially pure* aluminum, particularly at low speeds. In this respect aluminum behaves differently from magnesium, but in a similar way to many other pure metals.

The area of contact on the rake face of the tool is very large when machining commercially pure aluminum. As explained in Chapter 4, this leads to several effects: a high feed force (F_t); low shear plane angle; very thick chips; consequently high cutting force (F_c); and high power consumption. The effect on pure aluminum of most alloying additions or of cold working, is to reduce the tool forces, particularly at low cutting speed. In general most aluminum alloys, both cast and wrought, are easier to machine than pure aluminum, in spite of its low shear strength.

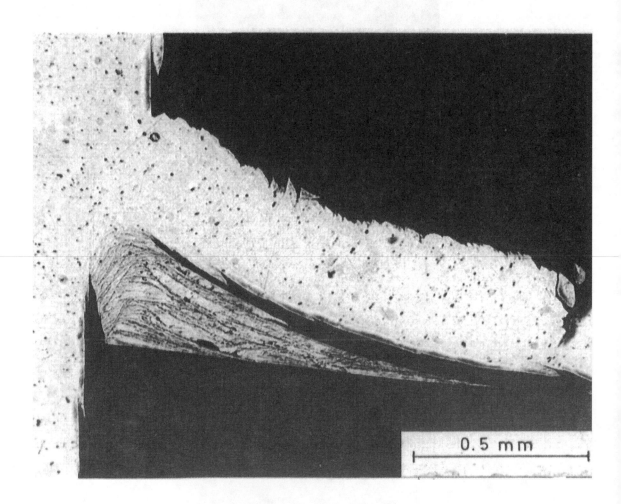

FIGURE 9.3   Built-up edge when cutting an aluminum alloy at 38 m min^{-1} (125 ft/min)

A built-up edge is not present when cutting *commercially pure aluminum*. However, the surface finish tends to be poor except at very high cutting speed. Most *aluminum alloys* have structures containing more than one phase, and with these a built-up edge is formed at low cutting speeds, *Figure 9.3*. At higher speeds, e.g. above 60-90 m min^{-1} (200-300 ft/min), the built-up edge may not occur. Tool forces are low where a built-up edge is present, and the chip is thin, but the surface finish tends to be poor. The built-up edge may be reduced or even eliminated by use of diamond tools.

9.3.4 Some industrial/practical constraints

One of the main machinability problems with aluminum is in controlling the chips. Extensive plastic deformation before fracture occurs with aluminum alloys. This is because aluminum has a face-centered cubic structure, rather than the hexagonal structure of magnesium.

When cutting aluminum and some of its alloys, the chips are continuous, rather thick, strong and not readily broken. The actual form of the chip varies greatly, but it may entangle the tooling. It often requires interruption of the operation to clear the chips. In drills, taps and cutters of many types, it may clog the flutes or spaces between the teeth. Thus, modified designs of tools are often required for cutting aluminum. The cutting action can be improved by modifications to rake and approach angles. The introduction of chip breakers, or curlers, which deflect the chips into a tight spiral is also useful. Another approach is to modify the composition of the alloys to produce chips which are fragmented or more easily broken. "Free machining" aluminum is discussed in the next sub-section. Further on in this chapter, the role of "free machining" additions in brasses and steels is also described.

9.3.5 Free machining aluminum alloys

The standard aluminum specifications now include "free machining" alloys containing additions of lead, or of {lead + bismuth}, or tin, or antimony in proportions up to about 0.5%. How these additions function is not completely certain, but the chips are more readily broken into small segments.

The low melting point metals do not go into solid solution in aluminum. Therefore, they are present in the structure as dispersed fine globules. They act to reduce the ductility of the aluminum as it passes through the primary shear plane to form the chip. There is also evidence that the "free machining additions" are drawn out in the secondary shear zone. This provides a form of "internal lubrication". The main purpose of the "free-cutting" additives in aluminum and its alloys is improvement in the chip form for CNC and automatic machines. Better tool life and an increase in the metal removal rate are an additional bonus.

9.3.6 Summary on aluminum alloys

The excellent machinability of aluminum alloys makes them ideal work materials to be shaped in CNC or multi-spindle automated machine tools - provided the chips can be broken up for continuous operation. Completely automatic production of certain classes of shapes can be introduced with confidence because long tool life and consistent performance can be guaranteed even at high rates of metal removal. This is still true even when using multi-spindle automatic machines where a great variety of operations is involved, such as turning, milling, drilling, tapping and reaming.

Attempts to use such unattended automated methods on other classes of work material, have not been as successful as when machining aluminum alloys. As a result, consumer products and automobile products that need to be of reasonably high strength, but produced cheaply and in quantity, are often designed and made from aluminum alloys.

9.4 COPPER, BRASS AND OTHER COPPER ALLOYS

9.4.1 Introduction

Copper is another highly ductile metal with a face-centered cubic structure, like aluminum, but it has a higher melting point (1,083°C). In general, copper-based alloys also have good machinability for the same reason as aluminum alloys.[4] Although the melting point is higher, it is not high enough for the temperatures generated by shear in the flow-zone to have a very serious effect on the life or performance of cutting tools. Both high speed steel and cemented carbide tools are employed. Good tool life is obtained, the wear on the tools being in the form of flank wear or cratering or both.

The most important field of machining operations on copper-based alloys is in the mass production of electrical and water fittings using high speed automatic machines. These are mostly very high speed lathes, but fed with brass wire of relatively small diameter, so that the maximum cutting speed is limited to 140 - 220 m min^{-1} (450 - 700 ft/min), although the tooling is capable of good performance at much higher speeds if required.

A built-up edge does not occur when cutting high-conductivity copper. There is a flow-zone at the tool work interface over a wide range of cutting speed. The tool forces are very high, particularly at low cutting speed (*Figure 9.4*). As with aluminum, this is essentially due to the large contact area on the rake face, resulting in a small shear plane angle and thick chips.

For this reason, high conductivity copper is regarded as one of the most difficult materials to machine. In drilling deep holes, for example, the forces are often high enough to fracture the drill. Additional problems in the machining of pure copper are poor surface finish, particularly at low speeds. At higher speeds, cutting forces are lower and surface finish improves, but tangled coils of continuous chips are difficult to clear.

9.4.2 Machinability of brasses

The machining qualities of copper are somewhat better after cold working and are greatly improved by alloying. *Figure 9.4* shows the reduction in cutting force (F_c) as a result of cold working. This reduces the contact area, giving a larger shear plane angle and a thinner chip.

The tool forces are lower for the 70/30 brass, which is single-phased. There is an even greater reduction when cutting the two-phased 60/40 brasses. The forces are low over the whole speed range, with thin chips and small areas of contact on the tool rake face.

The forces are lowest in alloys of high zinc content where the proportion of the β phase is greater. These low tool forces and power consumption with the α-β brasses, together with low rates of tool wear, are a major reason for classifying them as of high machinability.

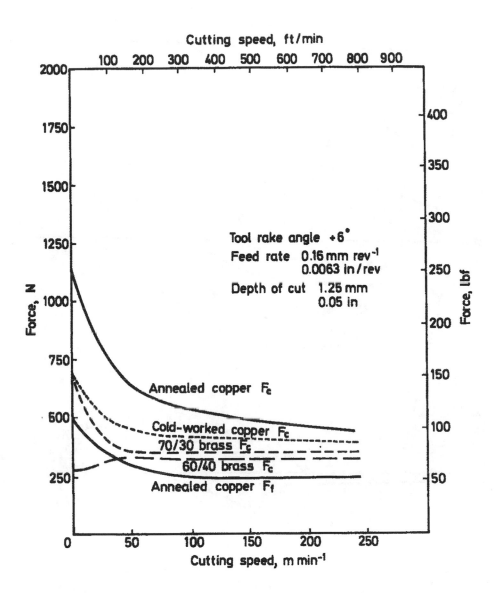

FIGURE 9.4   Tool forces *vs* cutting speed-copper and brass (From data of Williams, Smart and Milner[1])

Figure 9.5 shows the temperature distribution in tools used to cut high conductivity copper at three speeds, using carbon steel tools. The temperatures were much lower than those measured when cutting steel (see Chapter 5). Temperature gradients were much smaller. Even at 530 m min^{-1} (1,750 ft/min), the highest measured temperature was in the range 500-550°C. The temperature at the cutting edge was not measurably (+/-25°C) lower than the highest on the rake face.

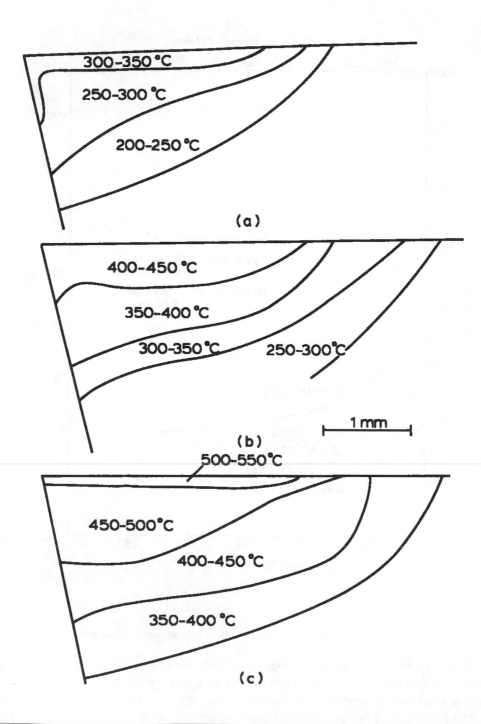

FIGURE 9.5 Temperature distribution in tools used to cut high-conductivity copper at a feed of 0.22 mm/rev at (a) 120 m min^{-1} (400ft/min); (b) 240 m min^{-1} (800ft/min); and (c) 530 m min^{-1} (1,750 ft/min)

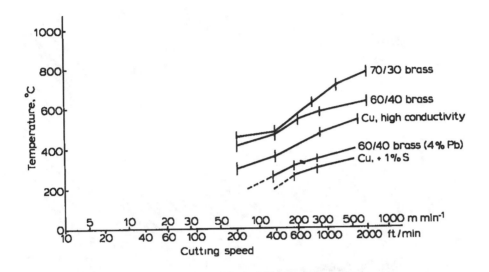

FIGURE 9.6   Maximum interface temperature vs cutting speed

Figure 9.6 shows the maximum interface temperature for copper and a number of its alloys as a function of cutting speed. For convenience the speed is plotted on a logarithmic basis. With high conductivity copper the maximum temperature was 300°C at a cutting speed of 70 m min^{-1} (230 ft/min), rising to 560°C at 700 m min^{-1} (2,300 ft/min).

The effect of alloying with zinc was to raise the temperature for any speed. Temperatures were higher for 70/30 brass than for 60/40 brass, particularly at speeds over 200 m min^{-1} (650 ft/min) and reached 800°C at 700 m min^{-1} (2,300 ft/min). With copper and 60/40 brass, the temperature was nearly uniform (within +/- 25°C) over the contact area.

However, with 70/30 brass there was a higher temperature region about 1 mm from the tool edge, as when cutting steel. With copper and brass the heat source is a flow-zone as shown for 60/40 brass in *Figure 9.7a*. Thus, with copper and brass, very high speeds can be used when cutting with high speed steel tools. It is probably only with 70/30 brass that high temperatures, and consequent high wear rates, limit the rate of metal removal with high speed steel tools.

9.4.3   Free machining brass

A more important limitation on speed is the problem of clearing the continuous chips from the working area. This problem is most acute with automatic tooling. To deal with it, "free-machining brass" was developed many years ago. It can be made by addition of lead in the proportion of 2 to 3%, and sometimes higher. [4] Lead is soluble in molten brass but is rejected during solidification, precipitating particles usually between 1 and 10 microns in diameter, which should be uniformly dispersed to achieve good machinability. Environmental issues now prevent the use of leaded brass. However, there has been much research on understanding how the lead additions help machinability (*Figure 9.7b*). Since engineers must now search for other materials to take the place of lead, it makes sense to analyze its beneficial function in machinability.

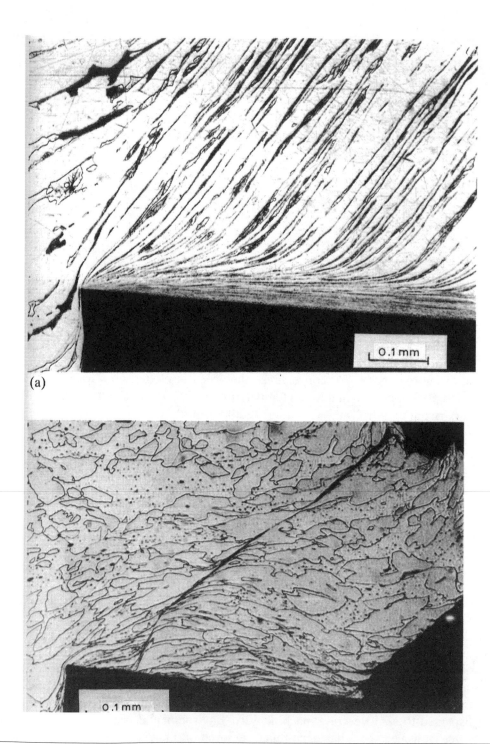

FIGURE 9.7   Quick-stop (a) Continuous chip with flow-zone after cutting 60/40 brass at 120 m min^{-1} (400 ft/min); (b) segmented chip without flow-zone after cutting leaded 60/40 brass at 120 m min^{-1} (400 ft/min)

When machining leaded brass, thin chips are produced, not much thicker than the feed. These are fragmented into very short lengths which are readily disposable. Leaded brass also reduces the tool wear rate in comparison with other types of brass. Free-machining brass can be cut for long periods on automatic machines without requiring shut-down to replace tools or to clear chips. Many small parts are economically made from free-machining brass because of the low machining cost, in spite of the high price of copper.

The addition of lead greatly reduces tool forces (*Figure 9.8*), which become almost independent of cutting speed. No definite flow-zone is formed on the rake face of the tool. The absence of a flow-zone, the greatly reduced strain in chip formation, and the segmented chips which result from the addition of lead, can be seen by comparing *Figures 9.7a* and *9.7b*. With the segmented chips the contact area on the tool rake face is reduced. Lower ductility of the brass, as a result of lead addition, may be partly responsible for the segmentation of the chips in the primary zone. However, the main reason for the reduction in tool forces is the action of lead at the tool/work interface.

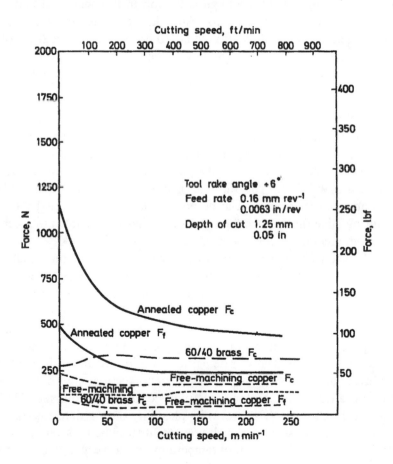

FIGURE 9.8   Tool forces *vs* cutting speed - free cutting copper and brass (From data of Williams, Smart and Milner[1])

Seizure between the brass and the tool seems largely to be eliminated by the lead which is concentrated at the interface. *Figure 9.9a* shows the contact area on the rake face of a high speed steel tool after a quick-stop during cutting a leaded brass at 180 m min^{-1} (600 ft/min). A concentration of lead can be seen, which had been molten during cutting and had wetted the steel tool. Confirmation of this action comes from observations of monatomic layers of lead on the under surface of leaded brass chips. In addition, there are thicker layers, in striations, in the cutting direction.[5]

Figure 9.9 also shows the results of Wolfenden and Wright[6] using the equations in Chapter 5. These calculations confirm that the lead is molten at the end of the chip tool contact area. This concentration of lead at the interface reduces the tool forces. It also facilitates chip fracture by reducing the compressive stress acting on the shear plane. Molten lead accumulates at the interface because it wets the steel tool, as shown by the small contact angle of the lead droplets on the tool. This also demonstrates that the tool surface at the interface is freed from oxide and other surface films by the action of the work material. Lead does not wet oxidized steel surfaces.

Another result of the action of lead is a large reduction in the energy of cutting both on the shear plane and at the tool/work interface. This results in a considerable reduction of tool temperature, *Figure 9.6*. The effect of lead addition to 60/40 brass is to reduce the tool temperatures below those for high conductivity copper. The lower temperatures reduce tool wear. The main advantages are the elimination of chip control problems and the reduction in tool forces.

9.4.4   Environmental concerns

Unfortunately for "machinability" alone, lead has become recognized as a serious health hazard. Possible dangers arise both in melting and machining of leaded brass and also in the use of leaded brass as fittings for water supply. With the possibility of regulations restricting the use of leaded brass, producers now have to consider the machining of unleaded 60/40 brass on automatic machines. Continuous brass chips can be broken into short lengths by forming grooves parallel with the cutting edge on the rake face of tools. It has been shown that certain shapes of groove are successful in breaking chips of unleaded 60/40 brass.[7]

Additions are made also to high conductivity copper to improve its machinability. Additives have to be confined to those which do not appreciably reduce electrical conductivity, or cause fracture during hot working. About 0.3% sulfur is usually added. This forms plastic non-metallic inclusions of Cu_2S. The effect is to reduce greatly the tool forces, particularly at low speeds, *Figure 9.8*. Also it produces thin chips which curl and fracture readily. The surface finish is greatly improved. The action of Cu_2S can probably be attributed to the reduction of seizure between copper and the rake face. The sulfide particles are plastically deformed in chip formation, and very thin layers of sulfide are observed on the contact area after quick-stops. The flowzone, normally present after cutting high conductivity copper, is eliminated. The essential features of Cu_2S as a free-machining phase are: i) its plasticity during deformation in cutting, and ii) its strong adhesion to the tool surface, which prevents it from being swept away. One result of the sulfur addition is a large reduction in tool temperature up to very high cutting speed (*Figure 9.6*). The temperatures are even lower than those when cutting leaded brass.

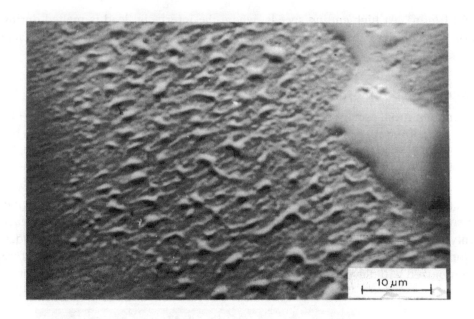

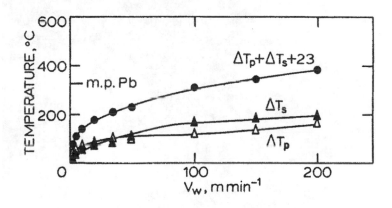

FIGURE 9.9 a) Lead on rake contact area of high speed steel;[5] quick-stop after cutting leaded brass at 180 m min^{-1} (b) calculations confirming melting of lead at end of contact zone[6]

9.4.5 Gun metal

Wise and Samandi have investigated machining behavior of other copper-based alloys and some of their findings are summarized next.[7,8] Gun metals are similar to zinc brasses but contain a solution - for example, 5% Zn, 5% Sn, 1-5% Pb with the balance copper. They are normally used as castings with lead added to improve machinability. Almost all criteria of machinability are improved by lead. Cutting forces are low and are nearly constant over the whole cutting speed range, particularly the feed force. Lead is concentrated at the interface and semi-continuous chips are formed with periodic cracks through the chips. There is some continuity on the under side but no flow-zone. Tool interface temperatures are low, e.g. 350°C at a speed of 140 m min^{-1} (460 ft/min). At high cutting speeds, where the interfacial lead is melted during cutting, the rate of wear is increased compared with that at lower speeds where the interfacial lead is solid.

9.4.6   Aluminum bronze

Two aluminum bronzes have also been studied.[7,8] The first contained 10% Al, 5% Fe and 5% Ni. The second, of improved machinability, contained 6% Al, 2% Si and 0.5% Fe. In both alloys, the balance was copper. Both alloys are basically solid solutions of Al and Cu, but also contain numerous fine precipitated particles containing Fe, Si and Al dispersed in the matrix. The two alloys showed similar machining behavior, very different from that of brasses. At low speeds, the chips are completely segmented and discontinuous while at high speed they are segmented, but the segments are joined together by a continuous flow-zone at the tool/work interface (*Figure 9.10*). Only slight plastic deformation takes place in the body of the chip before sudden fracture occurs along the line of the shear plane. This leads to large fluctuations of cutting force. These were synchronized with the periodic fracture, and a characteristic high noise level during cutting.

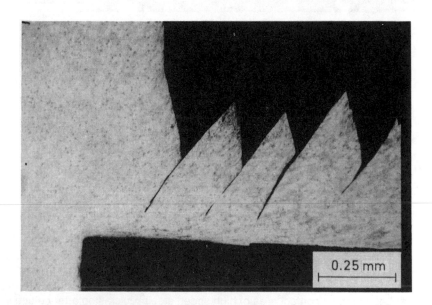

0.25 mm

FIGURE 9.10   Quick-stop of aluminum bronze (10% Al); speed 75 m min^{-1} (Courtesy of M. Samandi[8])

The cutting force and the feed force are both lower when cutting the 6% Al alloy. Because the chip is segmented, the contact length on the tool rake face is very short and the maximum temperature occurs less than 0.5 mm from the tool edge. The maximum temperature as a function of cutting speed is shown in *Figure 9.11*. The 10% Al alloy generates much higher temperatures, and the 6% Al alloy can be cut at much higher speeds for this reason. Aluminum bronze can be machined with high speed steel tools, but cemented carbide (WC-Co alloys) can be used with much longer tool life and at higher cutting speed.

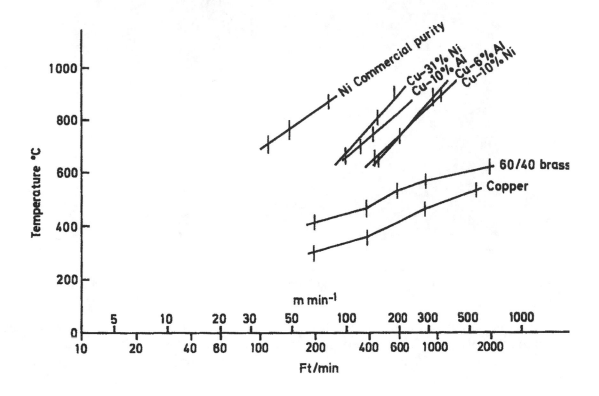

FIGURE 9.11   Maximum interface temperature vs cutting speed

9.4.7   Cupro-nickel alloys

Cupro-nickels are highly ductile, single-phase alloys forming a continuous series of solid solutions from copper (m.pt. 1,083°C.) to nickel (m.pt. 1,453°C.). The machining investigations were carried out on two alloys with 10% and 31% nickel.[7,8] The behavior in machining is, in many ways, the direct opposite to that of aluminum bronzes. Continuous chips are formed with no segmentation over the whole cutting speed range. The work material is strongly bonded to high speed steel or carbide tools over the whole contact area.

At low cutting speed, the cutting and feed forces were very high and the chips very thick with a very small shear plane angle. As the cutting speed was raised chips became thinner and cutting and feed forces dropped rapidly, as when cutting pure iron, nickel and copper. No built-up edge was formed at low speed. At speeds of 25 m min^{-1} and higher, a clearly defined flow-zone was formed, where the maximum temperature was 500°C.

At higher speeds the temperature gradients were of a similar pattern as for iron and steel with a maximum about 1.2 mm from the cutting edge (*Figure 9.12*). The maximum temperatures as a function of cutting speed are shown in *Figure 9.11* for the two cupro-nickel. These are higher than for any other copper-based alloy. Temperatures for the 31% Ni alloy were much higher than for the 10% Ni alloy at all speeds.

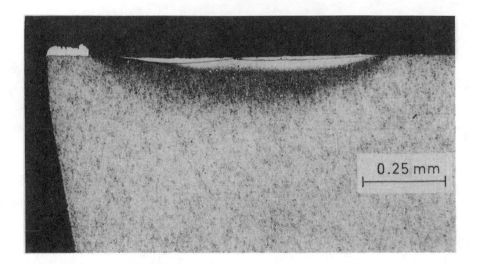

FIGURE 9.12 Section through high speed steel tool used to cut cupro-nickel (31% Ni) at 350 m min^{-1}. Etched to show temperature distribution

Flank wear occurred on high speed steel tools, and also cratering in the high temperature region on the rake face. Rapid cratering occurred at a speed of 200 m min^{-1} with the 31% Ni alloy. It occurred at 350 m min^{-1} with the 10% Ni alloy, when the interface temperature exceeded about 900°C.

9.4.8   Summary on copper, brass and copper based alloys

The copper-based alloys considered here, demonstrate some of the ways in which alloying of a ductile metal influences its machinability. When an alloying metal goes into solid solution in the base metal - e.g. Zn or Ni in Cu - a strong continuous chip is formed. The most important factor influencing machinability is that the temperature generated in the flow-zone is raised at any cutting speed. This is because the *alloying elements increase the energy expended in the work material in the thermo-plastic shear zone at the tool/work interface.* This increases the rate of tool wear and limits cutting speed.

Lead is insoluble in solid copper and is present as dispersed particles. This influences the behavior of alloys in metal cutting, because the lead becomes concentrated at the tool/work interface. Also, it becomes strongly bonded to the steel tools. It provides a layer of very low shear strength and greatly reduces energy expenditure. This further reduces tool temperatures, permitting higher cutting speeds. It also causes segmentation of the chips, solving many chip control problems. This action is seen both with free machining brass and with gun metal.

The machining behavior of the aluminum bronze alloys studied was very different from that of the other copper alloys. Discontinuous chips were formed with very short contact length on the tool surface. Thus, high temperatures were generated close to the cutting edge. Further research is required to determine whether this behavior is caused by aluminum in solid solution, or whether the dispersed particles of hard phases in these alloys caused the periodic fracture on the shear plane.

9.5   COMMERCIALLY PURE IRON

Commercially pure iron, like commercially pure copper and aluminum, is in general a material of poor machinability. At room temperature the structure is body-centred cubic (α iron), transforming to face-centered cubic (γ iron) a just over 900°C. In both conditions, it has relatively low shear strength but high ductility. The tool forces are much higher than for copper (*Figure 4.12*), particularly at low cutting speed. However, the forces decrease rapidly as the speed is raised.

The high tool forces are associated with a large contact area on the rake face of the tool, and thick chips. No built-up edge is observed at any speed, and quick-stops show a flow-zone about 25-50 µm (0.001-0.002 in) thick seized to the rake surface of the tool (*Figures 3.14* and *5.4*). This is the main heat source raising the temperature of the tool.

Temperature distributions in cutting tools are discussed in Chapter 5. The problem is considered in relation to the cutting of a very low carbon steel, which is in effect a commercially pure iron. *Figures 5.10* to *5.18* demonstrate the temperature pattern characteristic of tools used to cut this material. *This pattern is very important in relation to the machinability of iron and of steel.* The upper limit of the rates of metal removal, using high speed steel and carbide tools, is determined by tool wear and deformation mechanisms controlled by these temperatures.

In general, this same type of temperature pattern occurs in tools used to cut carbon and alloy steels, including austenitic stainless steels. A high temperature is generated on the rake face *well back from the cutting edge,* leaving a low temperature region near the edge.[†] The temperature increases as cutting speed and feed are raised.

9.6   STEELS: ALLOY STEELS AND HEAT-TREATMENTS

9.6.1   Overview

It is in the cutting of iron, steel and other high melting-point alloys that the problem of machinability becomes of major importance in the economics of engineering production. With these higher melting-point metals, the heat generated in cutting becomes a controlling factor. It imposes constraints on the rate of metal removal and the tool performance, and hence on machining costs. This main section reviews low-carbon, medium-carbon and general steels of various heat treatment. "Free cutting steels" and stainless steels warrant main sections of their own (Sections 9.7 and 9.8).

9.6.2   Influence of alloying elements on the forces and stresses on the tool

Alloying elements in steel (carbon, manganese, chromium, etc.) increase its strength. This influences both the stresses acting on the tool and the temperatures generated. As with copper

---

[†]. It will be seen later that titanium alloys machine in such a way that the maximum temperature is much closer to the cutting edge. Indeed the delicate cutting edge is, relatively speaking, much hotter than when machining steels. This difference in temperature pattern is the major difference between iron/steel on the one hand, and titanium/titanium alloys on the other.

and aluminum, the effect of alloying additions to iron is often to reduce the tool forces as compared with pure iron (*Figure 4.13*). However, the stress required to shear the metal on the shear plane to form the chip is greater. And when cutting steel, as opposed to iron, the chip is thinner, the shear plane angle is larger and the area of the shear plane is much smaller.

It is clear that the *cutting force is reduced by the addition of alloying elements*. However, the created contact area is much less. *Hence, the average compressive stress on the tool is always higher by adding the alloying elements*. How the average stress correlates with the maximum on the delicate cutting edge has been investigated. The evidence reviewed in Chapter 4 demonstrates that stress is at a maximum near the edge, generally 1.5 to 2 times the average value that might be measured by a dynamometer. Numerical values of this stress when cutting steel, and how it is affected by alloying additions, have been determined for the very limited number of cases shown in *Tables 4.7* and *4.8*. More comprehensive testing and analysis are desirable.

The yield stress of steels is influenced by both composition and heat treatment. When steels are heat treated to high strength and hardness, the compressive stress which they impose on tools during cutting may become high enough to deform the cutting edge (*Figure 6.15*) and destroy the tool. When using high speed tools, the machining of steels with hardness higher than 300 HV becomes very difficult; even at low speeds where the tool is not greatly weakened by heat.

Cemented carbide tools can be used to cut steels with higher hardness. Even here, tool life becomes very short and permissible cutting speeds very low when the hardness exceeds 500 HV. *Figure 9.13* shows the speeds and feeds at which carbide tools were found to deform severely when cutting steels of different hardness.[9]

For cutting fully hardened steel, the tool materials must retain their yield strength to higher temperatures. Ceramic tools can be used to machine steel hardened to 600 to 650 HV. Higher rates of metal removal and longer tool life can be achieved with cubic boron nitride tools on fully-hardened tool steel (see Chapter 8). Because they are less tough than cemented carbides, the range of operations on which they can be used is restricted. Additionally, tool costs are very high.

9.6.3   Relative machinability of steel

A very high percentage of all cutting operations on steel is carried out using high speed steel or cemented carbide tools. Guidance for selecting the optimum speed and feed for a particular cutting operation involves many factors. Nominal cutting speeds and feeds for machining different steels are often proposed by tool manufacturers and in books and papers on metal cutting.[10,11] Most commonly, these relate the nominal cutting speed to the hardness of the steel.

Table 9.1 gives typical recommended cutting speeds for single point turning operations. Such recommendations can be only a starting point for trials in practical machining operations. However, *Table 9.1* is included here to give some idea of the range of speeds used in machining carbon and low alloy steels of different strength and hardness.

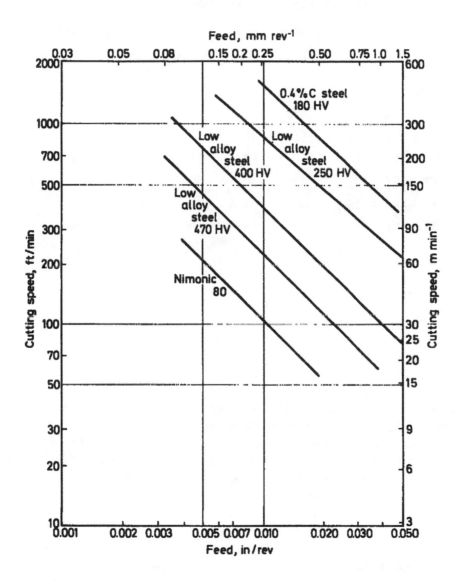

FIGURE 9.13 Conditions of deformation of cemented carbide tools when cutting steels of different hardness[9]

9.6.4   Combined influence of alloying, heat treatment, stress and temperature

To permit higher metal removal rates, steel work-materials are often heat treated to reduce the hardness to a minimum. The heat treatment for medium or high carbon steel often consists of annealing just below the transformation temperature (about 700°C). This "spheroidizes" the cementite - the form in which it has least strengthening effect, *Figure 9.14*. For some operations

TABLE 9.1   Typical maximum turning speeds for steel

Hardness		High speed steel tools				Cemented carbide tools			
		Feed, 0.5 mm/rev		Feed, 0.25 mm/rev		Feed, 0.5 mm/rev		Feed,0.25 mm/rev	
(HV)	(Rockwell)	(m min⁻¹)	(ft/min)	(m min⁻¹)	(ft/min)	(m min⁻¹)	(ft/min)	(m min⁻¹)	(ft/min)
90-125	48-69 RB	40	130	55	180	180	600	205	680
125-160	69-82 RB	34	110	46	150	155	510	180	690
160-210	82-93 RB	27	90	35	115	130	420	160	530
210-250	13-22 RC	21	70	30	100	115	380	140	460
250-300	22-30 RC	18	60	24	80	100	330	130	420
300-350	30-35 RC	15	50	18	60	85	280	115	370
350-400	35-41 RC					70	230	85	280
400-450	41-45 RC					45	150	60	200
450-500	45-49 RC					35	120	45	150

a coarse pearlite structure is preferred. This structure is obtained by a full annealing treatment in which the steel is slowly cooled from above the transformation temperature.

In the machining of low carbon steel containing a lot of pro-eutectoid ferrite, slow cooling from the annealing temperature is essential. This ensures that the carbon is not present in solution in the ferrite, nor as very finely dispersed particles. The heat treatment is necessary to make sure that the carbon is not redissolved with the natural heating of the cutting process - thereby accidentally returning the steel to a more-difficult-to-machine state.[12] When machining low carbon steels with high speed steel tools, failure by deformation of the tool edge occurs at a much lower speed if the heat treatment is incorrect. This is because carbon is available in the steel workpiece to strengthen the ferrite and increase its rate of strain hardening.[13]

Thus, the rate of metal removal when cutting steel alloys is strongly limited by the high stress imposed on the tool edge. The yield stress of the steel work material, and its rate of strain hardening, are the main factors in its machinability.

Of equal or greater importance however, is the influence of the alloying elements in the steel work material on the temperatures and temperature gradients generated in the cutting tools. These cannot at present be predicted from the mechanical properties because of the extreme conditions of strain and strain rate in the flow-zone, which is the heat source. This is discussed in Chapter 5.

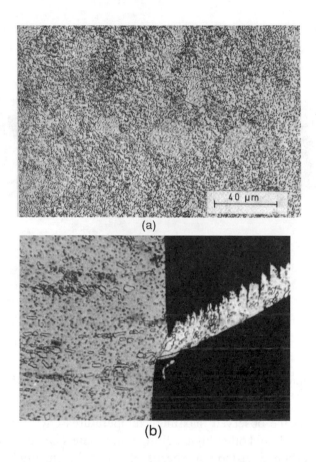

FIGURE 9.14   a) Spheroidized steel; b) quick stop of Spheroidised tool steel

Experimental study of the influence of alloying elements on the temperatures in the tools is a more effective way of investigating this aspect of machinability. The studies carried out so far, show that the introduction into iron of strengthening alloying elements has two main effects:

(1) The same characteristic temperature gradient is maintained, but...

(2) with alloying, a lower cutting speed is required to generate the same temperature.

Figure 9.15 shows an etched section through a high speed steel tool used to cut a 0.4%C steel at 61 m min^{-1} (200 ft/min) at a feed of 0.25 mm/rev.[14] The temperature gradient is similar in character to that when cutting the very low carbon steel (*Figure 5.11*).

The highest temperature, 850-900°C occurs at more than 1 mm from the tool edge. The temperature at the tool edge is 600-650°C. This temperature is observed at the cutting edge over a wide range of cutting conditions. It is associated with the temperature at which the thermoplastic shear band in the secondary shear flow-zone, is initiated in most engineering steels. This includes carbon, low alloy and stainless steels (see Chapter 5, *Figure 5.19*).

With the 0.04%C steel, this temperature was not reached until the speed was 152 m min^{-1} (500ft/min). When cutting the 0.4% steel, a crater is formed by shearing of the surface layers of the tool. However, this occurred at a much lower speed than when cutting the 0.04%C steel, and at a lower temperature.

In summary, the two steels with differences in carbon contents (0.04% versus 0.4%) show the same characteristic temperature gradient and same tool wear patterns. However, the increased carbon content creates all these effects at a much lower speed.

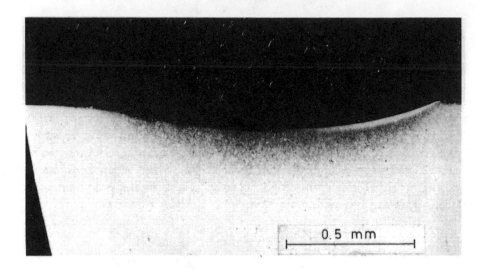

FIGURE 9.15 Temperature distribution in high speed steel used for cutting 0.4% C steel at 61 m min⁻¹ (200 ft/min) (After Dines[14])

Figure 9.16 compares the observed maximum temperatures on the rake face of high speed steel tools. A range of steels and other high melting point metals and alloys is shown. The values for copper cut with carbon steel tools at speeds of up to 300 m min⁻¹ (900 ft/min) are also included. This graph should be compared with *Figure 9.6*. It contains comparable data for tools used to cut copper-based alloys. In the limited range of steels tested, the slopes of the lines are similar but not identical.

Increasing the carbon content in the steel lowers the cutting speed required to achieve any temperature above 650°C. The addition of other alloying elements (Cr, Ni, Mo) into low-alloy engineering steel and the two stainless steels tested, also creates higher temperatures for a given cutting speed. The lowest speed required to reach a temperature of 800°C was 24 m min⁻¹ (80 ft/min) when cutting a molybdenum containing austenitic stainless steel. This compares with 131 m min⁻¹ (430 ft/min) for the lowest carbon steel.

Cratering wear, when it was observed, was always in the heat-affected region on the tool rake face (*Figure 9.15*). The lowest temperature at which it occurred was generally about 700°C. At this temperature, wear was by a diffusion/interaction mechanism (*Figure 6.19*). Wear by a superficial shearing mechanism occurred only when the temperature reached 800°C or higher, depending on the strength of the work material.

Ranking the steel work-materials by order of the temperatures which they generate in the tools, indicates the relative maximum cutting speed which can be used. These temperatures are towards the rear of the contact length. Further research on the precise temperatures at and very close to the cutting edge, will provide an even more useful criterion for this aspect of machinability of steels.

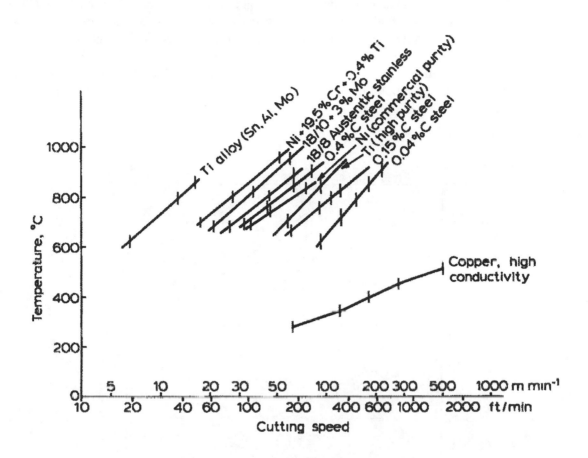

FIGURE 9.16 Maximum interface temperature *vs* cutting speed

9.6.5 Built-up edge effects

Steels containing more than about 0.08% carbon, have an appreciable amount of pearlite in their structure. Thus, at lower cutting speeds, a built-up edge is formed which has a major influence on all aspects of machinability (*Figure 3.21*). As cutting speed is increased, a limit is reached above which a built-up edge is not formed. This limit is dependent also on the feed. The conditions under which a built-up edge is formed are shown for two steels in the machining charts of *Figures 7.29* and *9.17*, for one standard tool geometry. The tool material was a steel cutting grade of cemented carbide.

The influence of cutting speed on the built-up edge can be demonstrated by taking a facing cut on a bar of steel rotating at a constant spindle speed. Imagine that the cut is started from a small hole in the center and the tool is fed outward. A built-up edge will be present at first, but the cutting speed continuously increases as the tool moves outward.

Thus at a critical speed, the shape of the chip changes and the surface finish improves as the built-up edge disappears. There are also conditions at very low rates of metal removal where a built-up edge is absent and discontinuous chip segments form. However, when machining steel with a normal, low sulfur content, using high speed steel or cemented carbide tools, this is usually below 1 m min^{-1} (3 ft/min). Practical experience shows that there are conditions where the built-up edge is much smaller or is eliminated. An example is when cutting in a normal speed range using CVD-coated cemented carbide tools, ceramic tools or cubic boron nitride tools. This results from lower frictional bonding at the interface. It again emphasizes the economic importance of CVD and PVD coatings.

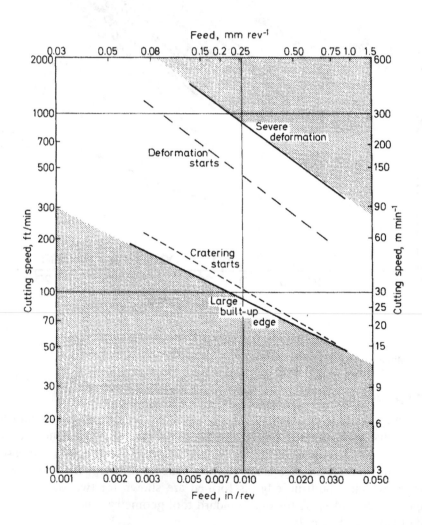

FIGURE 9.17 'Machining chart' for steel cutting grade of carbide used for cutting Ni-Cr-Mo steel (Hardness = 258 HV)

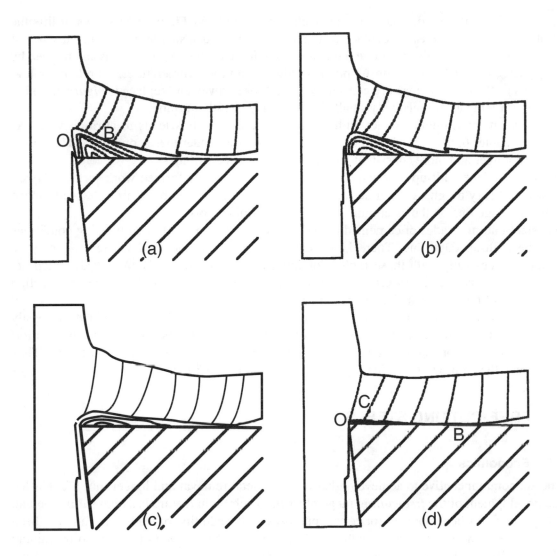

FIGURE 9.18 Transition from built-up edge to flow-zone with increasing cutting speed[15]

The built-up edge, which is formed on both high speed steel and cemented carbide tools, con-sists of steel, greatly strengthened by extremely severe strain, the pearlite being much broken up and dispersed in the matrix (*Figures 3.21*). Hardness as high as 600/700 HV has been measured on a built-up edge. This is considerably harder than steel wire of the highest tensile strength. The built-up edge can therefore withstand the compressive and shearing stresses imposed by the cut-ting action.

When the cutting speed is raised, temperatures are generated at which the dispersed pearlite can no longer prevent recovery or recrystallization. The built-up edge structure is weakened until it can no longer withstand these stresses. The built-up edge then collapses and is replaced by a flow-zone. The transition can be pictured diagrammatically as in *Figure 9.18*.[15]

In effect, the built-up edge alters the geometry of the tool. As *Figure 3.21* shows, it lifts the chip off the rake face. Thus, the contact area to be sheared is much smaller than in the absence of a built-up edge. This results in a large reduction in the forces acting on the tool as can be seen by the dip in *Figure 4.13*. Power consumption is reduced and tool temperatures are relatively low. Fragments of the built-up edge are constantly being broken away and replaced (*Figure 3.21*) but usually the fragments are relatively small. Tool life may be rather erratic.

Especially if intermittent contact with the tool edge occurs, this leads to attrition wear. As shown in Chapter 6, high speed steel tools are generally used under these conditions. They often give much longer and more consistent tool life than cemented carbides.

Fragments of the built-up edge which break away on the newly formed work surface (*Figure 3.21*) leave it very rough. A better surface finish is usually produced by cutting at speeds above the built-up edge line on the machining charts, using carbide tools.

The steel-cutting grades of cemented carbide are employed most efficiently using conditions above the built-up edge line. There is a wide range of speed and feed where steel may be machined successfully with these tools. Continuous chips are generally produced, which are often strong and not easily broken. The form of the chip depends not only on the composition and structure of the steel, but also on the speed, feed and depth of cut.

It is important, particularly on automatic machines, that chips should be of a form easily cleared from the cutting area. Manufacturers of indexable carbide inserts have put much effort into designing "chip breakers". These are grooves in the rake face behind the cutting edge. They curl or break the chips over a wide range of cutting conditions (*Figure 7.36*).

9.7   FREE-CUTTING STEELS

9.7.1   Economics

The economic incentive to achieve higher rates of metal removal and longer tool life, has led to the development of the *free-cutting* range of steels. Their main feature is a high sulfur content, but they can be further improved for certain purposes by the addition of lead. Tellurium has been added to steel as a replacement for sulfur and evidence has been given of improved machining qualities. Tellurium has certain toxic properties, however. It involves a hazard for steel makers, and the use of tellurium steels is unlikely to become widespread.

9.7.2   The role of MnS additions in "free cutting" steels

Typical free-cutting steel compositions are given in *Table 9.2*. The manganese content of these steels must be high enough to ensure that all the sulfur is present in the form of manganese sulfide (MnS). Steel makers pay attention not only to the amount but also to the distribution of this constituent in the steel structure.

TABLE 9.2   Typical compositions of free-cutting steels

Steel type	Percentage by weight			
	C	Mn	S	P
Low Carbon	0.15	1.1	0.2 -0.3	0.07 max
	0.15	1.3	0.3 -0.6	0.07 max
28 Carbon	0.28	1.3	0.12-0.2	0.06 max
36 Carbon	0.36	1.2	0.12-0.2	0.06 max
44 Carbon	0.44	1.2	0.12-0.2	0.06 max

Figure 9.19 shows the MnS in a typical free-cutting steel. Control of the MnS particle shape, size and distribution is achieved during the steel making. It is influenced both by deoxidation of the molten steel before casting the ingots and by the hot rolling practice. Supplementary lead additions are, usually, about 0.2 - 0.3 per cent. Lead is insoluble in molten steel, or nearly so, and good distribution is difficult to achieve. In order to disperse the lead, it is added in the form of "lead-shot" to the steel as it is tapped from the ladle into the ingot molds. When lead is added to high sulfur steels it is usually found attached to the MnS particles, often as a tail at each end (*Figure 9.19*).

Free machining varieties of a wide range of engineering steels are produced. Those produced in the largest quantity are the low-carbon, plain carbon steels. The cost of a free-cutting steel is higher than that of the corresponding steel of low sulfur content. This must be justified by reducing the machining cost. An increase in speed from 20 to 100 per cent over that used for the steel of low sulfur content (see *Table 9.1*) has been demonstrated.[17]

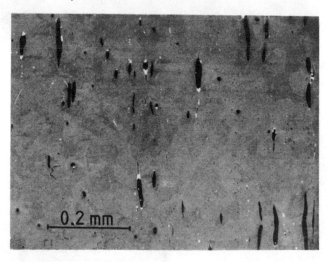

FIGURE 9.19 SEM of leaded free-cutting steel, showing MnS inclusions (dark) and Pb inclusions (white) (After Milovic)[16]

The free-cutting steels are used extensively for mass production of parts on CNC and auto-matic machine tools. They permit the use of higher cutting speeds, give longer tool life, good surface finish, lower tool forces and power consumption, and produce chips which can be more readily handled.[18,19,20] Above all, they are used because they can be relied on to perform more consistently than the non-free-cutting steels in the automatic machine cycle. In spite of extensive investigations into the mechanisms by which MnS acts to improve machinability, the under-standing is still incomplete.

MnS particles dispersed in steel are plastically deformed when the steel is subjected to metal working processes. Those in *Figure 9.19* had been elongated during hot rolling of the bar from the ingot. In this respect they behave differently from many carbide and oxide particles which rigidly maintain their shape, or are fractured, while the steel matrix flows around them. There have been laboratory studies of the deformation of MnS particles, the extent of which depends on the amount of strain and the temperature.[21] Micro-examination of quick-stop sections shows that, on the shear plane, the sulfides are elongated in the direction of the shear plane (*Figure 9.20*). In the flow-zone adjacent to the tool surface, the elongation is very much greater.

Figure 9.21 shows sulfide particles that are extensively drawn-out in the flow-zone. Their thickness is on the limits of resolution of the optical microscope. Some may be too thin to be seen, i.e. less than 0.1 μm. In the flow-zone, on the under surface of free-cutting steel chips, a very high concentration of these thin ribbons of sulfide can be seen in scanning electron micro-scope pictures after etching in nitric acid (*Figure 9.22*).[14] The steel in this example contained 0.4% S. It is possible that, in this form, they may provide surfaces of easy flow where work done in shear is less than in the body of the metal. The contact length on the rake face of the tool is shortened by the presence of MnS.[14] Separation of the chip from the tool to which it is bonded, requires fracture. The weak interfaces between the sulfide ribbons and the steel create nuclei for this fracture. The shorter contact length results in thinner chips and lower tool forces, and lower power consumption (*Figures 9.23* and *9.24*).[22]

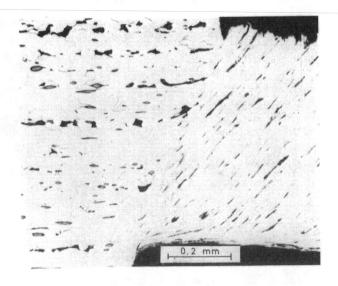

FIGURE 9.20 Deformation of MnS inclusions on the shear plane when cutting free-cutting steel

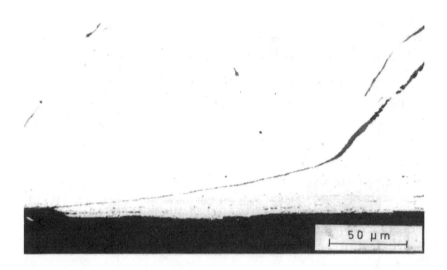

FIGURE 9.21 Deformation of MnS inclusions in the flow-zone of free-cutting steel chip (After Dines[14])

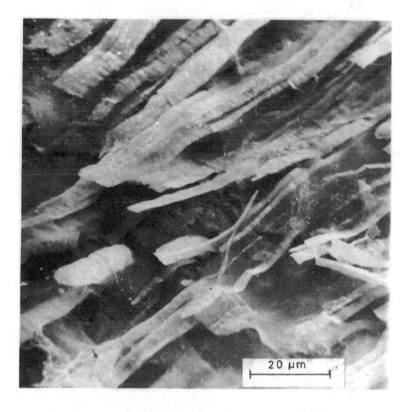

FIGURE 9.22 SEM of MnS in flow-zone of chip after etching deeply in HNO_3 (After Dines[14])

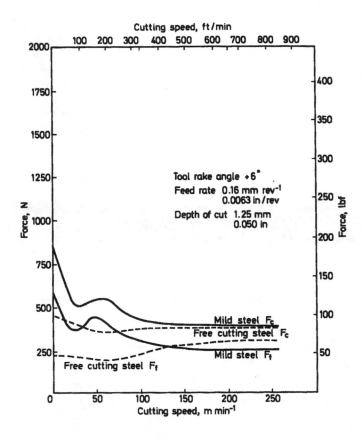

FIGURE 9.23 Tool force *vs* cutting speed: carbon and free-cutting steels (after data of Williams, Smart and Milner[1])

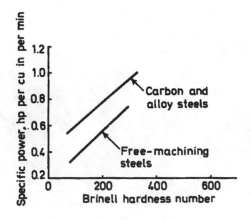

FIGURE 9.24 Power consumed during cutting: carbon and free-machining steels[22] (By permission from Metals Handbook, Volume 1, Copyright American Society for Metals)

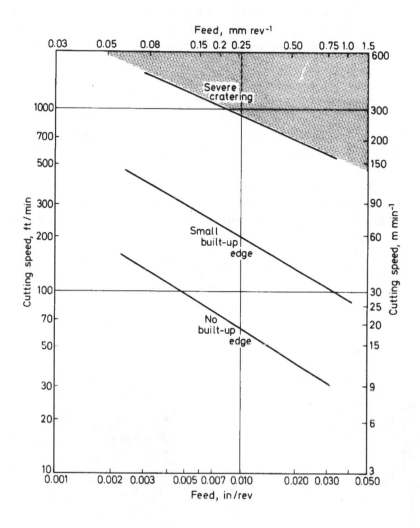

FIGURE 9.25 'Machining chart' for steel-cutting grade of carbide used for cutting a low carbon free-cutting steel

9.7.3 MnS behavior on the *steel-cutting grades* of carbide

After cutting, layers of MnS are often found covering parts of the tool surface. When using the steel cutting grades of carbide tool, sulfide is found covering the contact area on the tool rake face. This may prevent almost entirely the formation of a built-up edge at low cutting speed (see machining chart, *Figure 9.25*). On the steel-cutting grades of carbide, at high speed, MnS seems to act as a lubricating layer - interposed between tool and work material on the rake face. The presence of the sulfide layers can be investigated by electron probe analysis. They are also readily demonstrated by "sulfur prints" of the tool rake surface.

Figure 9.26a shows a photomicrograph of the contact area on the rake surface of a steel-cutting grade of carbide tool after cutting a free-cutting steel. *Figure 9.26b* is the sulfur print of this region, which shows a high concentration of sulfide over the whole contact area. The mechanism by which the MnS particles are deposited on the tool surface is not certain. It has been suggested that they are extruded onto the tool surface from their "sockets" in the steel, like toothpaste from a tube.

Although a large built-up edge is not observed on steel-cutting grades of carbide tool when cutting high sulfur steels, a very small built-up cap wraps itself around the tool edge and is present even at high speeds[16,23,24] (*Figure 9.27*). This is either commensurate in size with the feed or smaller. Beyond it, the chip flows over the tool surface with much sulfide at the interface.

This small build-up may be an important part of the mechanism by which sulfides are continuously deposited on the rake surface. It does not cause the deterioration in surface finish which accompanies a large built-up edge. By restricting the rate of flow at the tool edge, it appears to be responsible for a reduction in the rate of flank wear. Milovic[16,24] has shown that the rake face temperature is lower when cutting a high sulfur steel compared with the corresponding steel of low sulfur content.

The retention of a sulfide layer at the interface - with the mixed crystal, steel-cutting grades of carbide tool - is most probably the result of a bond formed between the MnS and the cubic carbides of the tool material. The bond is strong enough to keep it anchored to the tool surface. It resists the flow stress imposed by the chip. To be effective however, enough sulfide must be present in the work material. This is needed to replace those parts of the sulfide that are, from time-to-time, carried away by the under surface of the chip.[15]

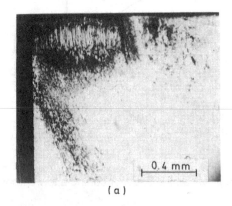
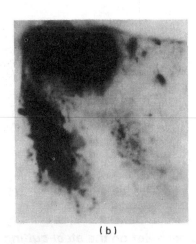

0.4 mm

(a)

(b)

FIGURE 9.26 (a) Sulfides on rake face of steel-cutting grade carbide tool after free-cutting steel at high speed; (b) sulfur print of the same tool[15]

9.7.4   MnS behavior on the *straight WC-Co grades* of carbide

By contrast, when using high speed steel tools or "straight" cemented carbides, the sulfides are not found on the contact area of the tool face except at very low cutting speeds. However, they often cover areas of the tool beyond the contact region. A built-up edge always seems to prevail when high sulfur steels are machined at relatively low speeds with high speed steel and the "straight" WC-Co tools. This may persist to even higher speeds than when cutting normal low sulfur steels.

From these last two sub-sections a central message of this chapter becomes even more established. "Machinability" is not a unique property of a work material. The above results show an important point with MnS based "free cutting steels": It is only when using the steel-cutting carbide grades, or tools coated with TiC or TiN, that the formation of a large built-up edge is prevented. In this limited regard, the authors emphasize that "machinability" is more dependent on tool material than work material.

9.7.5 The role of lead additions in "free cutting" steels

The role of lead has been less studied than that of sulfur. Lead is used as additions of about 0.25% to both high sulfur steel, and steel of normal sulfur content. Lead permits even higher cutting speeds. It gives both better surface finish and better control of chips without serious detriment to the mechanical properties of the steel. The percentage of lead in steel is about one tenth of that in free-machining brass. Discontinuous chips, a feature of leaded brass, are not formed. Examination of quick-stopped tools shows local deposits of lead on the contact area or nearby, but not the high concentration observed with leaded brass (*Figure 9.9*). One investigation of the under surface of chips from leaded free-cutting steel, used Auger electron spectroscopy. It showed a uniform layer, about one atom thick, of lead, together with a concentration of sulfur.[25]

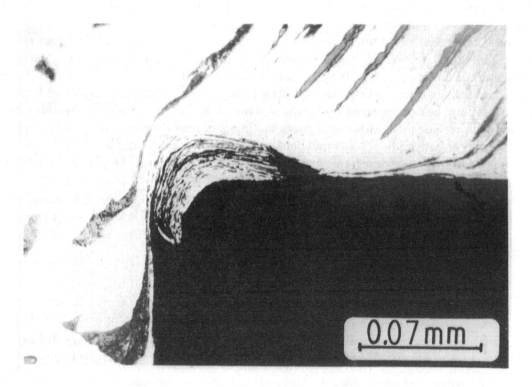

0.07mm

FIGURE 9.27 Built-up 'cap' on steel-cutting grade of carbide tool with high sulfur steel- 63 m min[-1] [19]

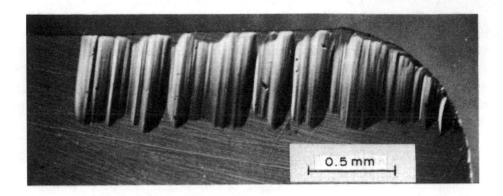

FIGURE 9.28 Deep grooves in rake face of WC-Co tool after cutting high-sulfur leaded steel at 46 m min^{-1} for 10 min (After Milovic[19])

At high cutting speed, sections through quick-stops often show a modest secondary shear zone. Flow is retarded at the under surface of the forming chip, but without a definite flow-zone. The tool rake face temperatures are often considerably lower than when cutting the corresponding lead-free steels. These observations suggest that, under some cutting conditions, lead at the interface prevents complete seizure between the steel work-material and the tool. This effect depends on the tool material.

The reduction of seizure, allowing sliding at the interface, may in fact cause an acceleration of wear despite the lower interface temperatures. With high speed steel tools, or WC-Co tools, the rake face of the tool may be rapidly worn when cutting leaded steel. Deep, "horseshoe-shaped", localized craters are often observed on high speed steel tools.[24] They start very close to the tool edge. These may be accompanied by localized flank wear. The horseshoe configuration is not observed with carbide tools. However, craters starting at or near the edge may develop rapidly in WC-Co tools (*Figure 9.28*).[19] This rapid damage to the tool is most common at low or medium cutting speed. It may disappear if the speed is raised. Some CVD coated tools appear to be very resistant to this form of wear.

In summary, the particles of lead in steel are plastically deformed during machining. This is seen clearly where it is associated with MnS inclusions. The action of lead as a free-cutting additive is very complex.[†] Because it is complex, the value of "free cutting" steels as a solution to particular production problems must be investigated for each case.

---

†. Today's demands of "dry cutting" - see end of Chapter 10 - are connected to the effects described in the last few pages. The results from the authors' laboratory experiments show a surprising range of possible "internal lubrication" effects from the various "free cutting" additions. The variations between the "straight" and "steel-cutting" grades of carbides are perhaps the most surprising of all, even to many practitioners - here the "machinability" of the "free cutting" steels is totally changed by the bonding effects in the secondary shear zone. Dry cutting will therefore be very easy with the "steel-cutting" grades and less so with the "straight, WC-Co" grades.

9.7.6 Variable machinability of <u>non</u> free-cutting steel

9.7.6.1 Sulfur content

High sulfur content in steel promotes machinability, but very low sulfur content makes machining more difficult. The rates of flank wear increase with decreasing sulfur content, *Figure 9.29*, as has been demonstrated in a variety of tests.[26] Steels which have been cleaned of non-metallic inclusions by electro-slag remelting (ESR) are reported to be more difficult to machine. In some cases small amounts of sulfur have been reintroduced into these steels, e.g. 0.015%S, to reduce costs of machining.

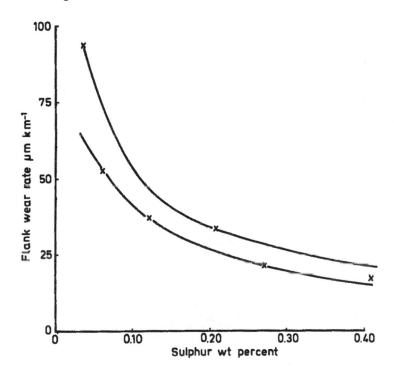

FIGURE 9.29 Influence of sulfur in steel on the rate of flank wear (Courtesy of Naylor, Llewellyn and Kean,[26] British Steel Corporation. Swinden Laboratories)

9.7.6.2 Variability between batches

When machining non free-cutting steels the tool wear rates and the permissible rates of metal removal, are found to vary widely for different batches of steel conforming to the same standard specification.[26] *Figure 9.30* shows flank wear rate *vs* cutting speed curves for bars of a medium carbon steel from different heats. The differences may be very large. This poses severe problems in the engineering industry for those involved in the planning of machining, particularly on transfer lines and highly automated operations. Not all the causes of the different behavior have been determined. When machining low carbon steels, the very large differences in the life of high speed steel tools are caused by variations in the amount of interstitial carbon and nitrogen in the ferrite. On a practical level, quality control of heat treatment is vital. It seems that although steel making became more quality control oriented in the 1990s, more rigor is desirable. [25-31]

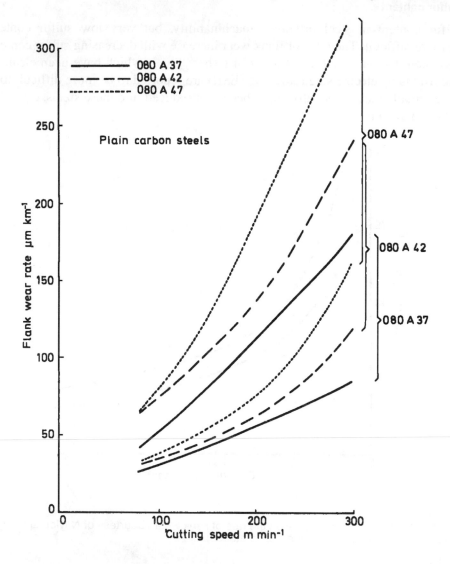

FIGURE 9.30 Variability of flank wear rate on carbon steels of the same specification (Courtesy of Naylor, Llewellyn and Kean,[26] British Steel Corporation, Swinden Laboratories)

9.7.6.3 Damaging aspects of Al_2O_3 and hard inclusions

Major variability in tool wear rate can be attributed to non-metallic inclusions.[31] Until recently, most attention has been paid to acceleration of wear by the abrasive action of very hard, rigid inclusions, particularly Al_2O_3. Evidence for abrasive action by Al_2O_3 has been largely statistical, associated with rapid wear of tools when cutting steels de-oxidized with aluminum. The final stage of steel making requires deoxidation and a variety of de-oxidizing agents are used as well as aluminum. These include manganese, silicon and calcium. Thus a range of oxide inclusions is observed in commercial steels. Many of these remain undeformed in the work material during machining and are not detected on the worn tool surfaces.

9.7.6.4 Beneficial aspects of inclusions that naturally coat the tool

Steel-cutting grades of carbide tools are often covered with layers of oxide. They contain aluminum, silicon and calcium. Calcium is an essential element in these layers, which have a glassy appearance. They become strongly bonded to the worn rake face and, sometimes, the flank surfaces. The layers are plastically deformed in the direction of movement of the work material, as with MnS which may be also present.[15,17,27] *Figure 9.31* shows such a silicate layer on the rake face of a carbide tool. *Figure 9.32* shows a rather thick layer. *Work materials containing inclusions which form such layers can be shown to give much longer tool life and permit higher cutting speeds.* This is especially true when using steel-cutting grades of carbide tooling containing TiC and TaC. It is also true with CVD-coated tools with coatings of TiC or TiN.

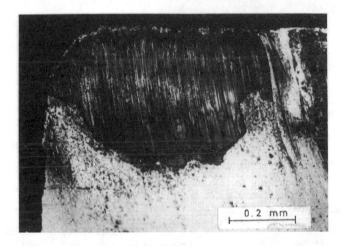

FIGURE 9.31 Silicate layer on take face of steel cutting grade of carbide used to cut steel at high speed

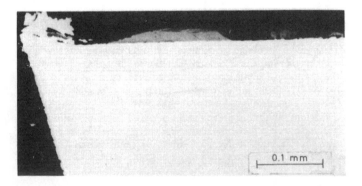

FIGURE 9.32 Section through silicate layer on rake face of tool[11]

9.8   AUSTENITIC STAINLESS STEELS

Austenitic stainless steels are generally regarded as more difficult to machine than carbon or low alloy steels. They bond very strongly to the tool, and chips often remain stuck to the tool after cutting. When the chip is broken away it may bring with it a fragment of the tool. This is particularly evident when cutting with cemented carbides, giving poor, erratic tool performance.

The forces are not greatly different from those found when cutting normalized medium carbon steel (*Figure 9.33*). The *temperature pattern* imposed on the tool is of the same general character as when cutting other steels, with a cool region at the cutting edge. With an 18% Cr, 8% Ni alloy the *temperature value* at any speed is rather higher than when cutting a medium carbon steel (*Figure 9.16*). Since the carbon content of 18-8 stainless steel is very low, the higher temperature must be attributed to the strengthening effect of nickel and chromium. This raises the temperature in the flow-zone. With austenitic stainless steel containing 3% Mo, the temperatures are higher still (*Figure 9.16*).

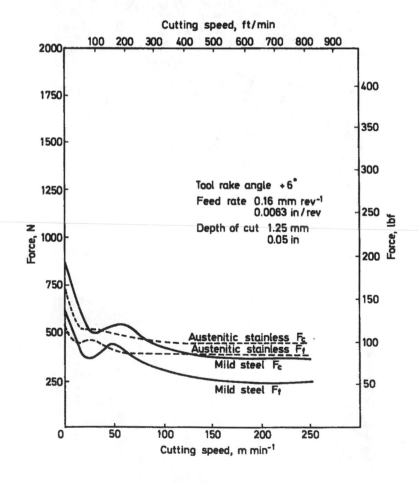

FIGURE 9.33   Tool force vs cutting speed: austenitic stainless steel

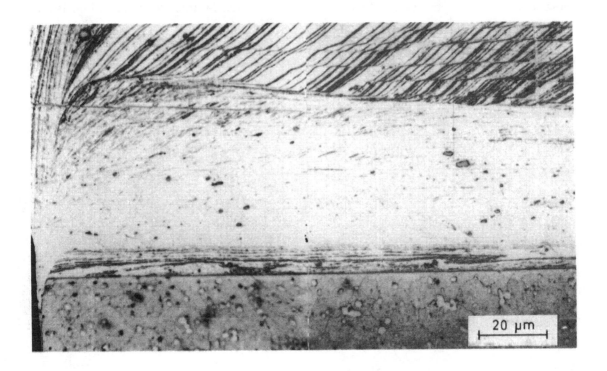

FIGURE 9.34 Section through built-up edge on tool used to cut austenitic stainless steel

A built-up edge is formed in a cutting speed range somewhat lower than with carbon steels, and has rather a different character, being more like an enlarged flow-zone (*Figure 9.34*).

Cratering of high speed steel tools occurs in the high temperature region on the rake face by diffusion and by superficial shear. At interface temperatures over 700°C wear by diffusion occurs (*Figure 6.18*). At 800°C this diffusion wear becomes rapid. Above 800°C, wear by superficial plastic deformation may be the dominant wear mechanism in cratering (*Figure 6.13*).

Cratering by the hot shearing mechanism (*Figure 6.24; mechanism 1*) occurs at speeds lower than those for medium carbon steels. The relatively high temperatures restrict the rates of metal removal. For example, at a feed of 0.25 mm (0.01 in) per rev, cutting speeds with high speed steel tools are generally lower than 25 - 30 m min^{-1} (80 - 100 ft/min). Flank wear when cutting austenitic stainless steels is characteristically very smooth. This is true of both high speed steel and carbide tools. Flank wear increases regularly as cutting speed is raised.

The cutting speed and feed, for a steel-cutting carbide, are shown in the machining chart, *Figure 9.35*. There is evidence that the rate of crater wear, when cutting austenitic stainless steel with WC-Co grades of cemented carbide, becomes very slow after the first short time of cutting. This is possibly due to intermediate phases formed at the interface. WC-Co alloys are thus frequently used to cut austenitic steels.

Austenitic stainless steels are strongly work-hardening. Therefore, particular problems arise when cutting into a severely work-hardened surface, such as that left by a previous machining operation with a badly worn tool. The use of sharp tools and a reasonably high feed rate are two recommendations for prevention of damage to tools caused by this work-hardening. It means that the roughing vs finishing sequences might be different from regular steels. With stainless steels, the goal is to "not overly work harden" the surface by too many repeated roughing cuts.

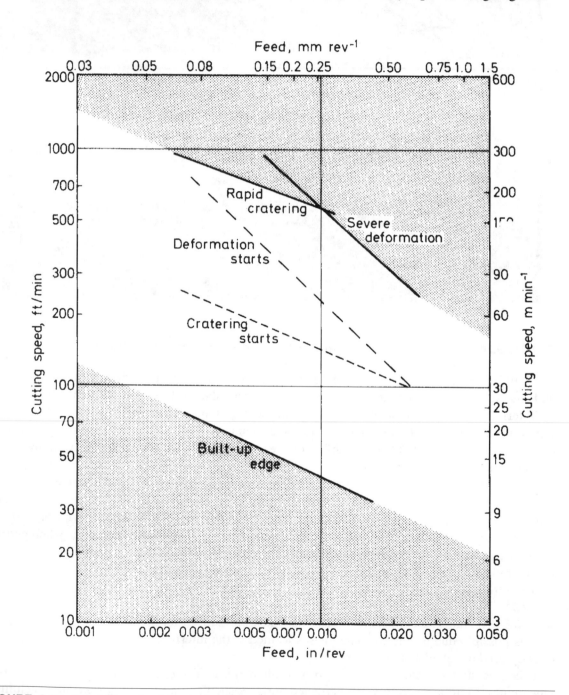

FIGURE 9.35 'Machining chart' for steel cutting grade of carbide used for cutting austenitic stainless steel

9.8.1 Free cutting stainless steels

Free-cutting austenitic stainless steels increase tool life and metal removal rates. These have high sulfur contents. As with the free-cutting ferritic steels, their improved machinability is associated with the plastic behavior of the sulfides in the flow zone.

The use of free-cutting stainless steels is restricted, because the introduction of large numbers of sulfide inclusions reduces corrosion resistance under some conditions. An alternative free-cutting additive is selenium, which forms selenide particles which behave like sulphides during machining. Selenides are less harmful to corrosion resistance. Improved machinability also occurs when using calcium deoxidation to produce protective layers at the tool/work interface. Importantly, these may well improve machinability without loss of corrosion resistance.[32]

9.9 CAST IRON

Flake graphite cast irons are considered to have very good machining qualities. In fact, a major reason for the continued large-scale use of cast iron in engineering is not only the low cost of the material and the casting process, but also the economics of machining the finished component.

By nearly all criteria cast iron has good machinability: low rates of tool wear, high rates of metal removal, relatively low tool forces and low power consumption. The surface of the machined cast iron is rather matt in character, but ideal for many sliding interfaces. The chips are produced as very small fragments which can readily be cleared from the cutting area even when machining at very high speeds. It is a somewhat dirty and dusty operation, throwing a fine spray of graphite into the air, so that protection for factory personnel may be required.

As when cutting other materials, there is a great difference between the behavior of cast iron when sheared i) on the shear plane and ii) at the tool work interface. The most important characteristic is that fracture on the shear plane occurs at very frequent intervals, initiated by the graphite flakes. The chip is thus composed of very small fragments a few millimeters in length. Because the chips are not continuous, the length of contact on the rake face is very short. The chips are thin and the cutting force and power consumption are low. The cutting force is low also because graphite flakes are very weak. One flake may extend an appreciable way across the shear plane.

Table 9.3 shows values for the cutting force (F_c) for a typical pearlitic cast iron in comparison with steels. This aspect of machinability is influenced by the grade and composition of the cast iron. Low strength cast irons, the structure of which consists mainly of ferrite and graphite, are the most machinable, permitting the highest rates of metal removal. Permissible speeds and feeds are somewhat lower for pearlitic irons, and decrease as the strength and hardness are raised. As with steel, there are extensive published data for cutting speeds when machining gray cast iron with high speed steel and cemented carbide (WC- Co) tools.[33,36] Recently, there is much data for ceramic tools.[37] *Table 9.4* shows typical recommendations for turning gray cast irons. These are classified according to hardness, when using the three classes of tool material.

The use of ceramic (alumina and sialon) tools for machining cast iron has increased greatly in recent years, mainly in mass production turning, boring and milling operations. Most cast iron brake drums, clutch faces and flywheels are, today, being machined with ceramic tools. The excellent surface finish achieved often eliminates the need for a grinding operation. Because the chips are fragmented, very high cutting speeds can be used without problems of chip control.

TABLE 9.3   Cutting forces: pearlitic, flake-graphite cast iron versus mild steel. Feed: 0.16 mm rev^{-1} (0.0063 in/rev) - Depth of cut: 1.25 mm (0.05 in)

Cutting speed		Cast iron forces				Mild steel forces			
		F_c		F_f		F_c		F_f	
m min^{-1}	(ft/ min)	(N)	(lbf)	(N)	(lbf)	(N)	(lbf)	(N)	(lbf)
30	100	222	50	232	52	520	115	356	80
61	200	245	55	285	64	490	110	364	82
91	300	245	55	320	72	445	100	325	75
122	400	267	60	338	76	422	95	313	70

By contrast, the highly alloyed irons and chilled irons, with very little graphite and containing large amounts of iron carbide (Fe_3C) and other metal carbides, become very difficult to machine. Chilled iron rolls may be machined only at speeds of the order of 3 to 10 m min^{-1} (10-30 ft/min) with cemented carbide tools. Higher speeds can be used with ceramic or cubic boron nitride tools. Alumina tools are recommended for use on chilled cast iron of hardness (430 HV) at about 50 m min^{-1} (150 ft/min). Turning operations are being carried out with cubic boron nitride tools on chilled iron with hardness 55-58 RC (600-650 HV) at about 80 m min^{-1} (240 ft/min) at feeds up to 0.4 mm/rev.

TABLE 9.4   Typical maximum turning speeds for flake graphite cast iron

Hardness	High speed steel tools		Cemented carbide (WC-Co) tools		Ceramic (alumina or sialon) tools	
	Feed, 0.5 mm/rev.		Feed, 0.5 mm/rev.		Feed, 0.25 mm/rev.	
(BHN)	($m\ min^{-1}$)	(ft/min)	($m\ min^{-1}$)	(ft/min)	($m\ min^{-1}$)	(ft/min)
115-200	40	130	120	400	450	1500
150-200	25	80	90	300	400	1300
200-250	20	65	70	230	250	900
250-300	12	40	55	180	180	600

The majority of engineering cast irons are of the ferritic or pearlitic types. Their behavior on the shear plane during cutting can be predicted quite well from their strength and lack of ductility as measured in standard laboratory tests. Their behavior at the tool/work interface in the secondary shear zone is less "conventional". Graphite might be expected to act as a lubricant and to inhibit seizure at the tool/work interface, but there is no evidence that it acts in this way.

When cutting with carbide or high speed steel tools, a built-up edge is formed which persists to higher cutting speeds than with steels. *Figure 7.18* is a machining chart for a pearlitic iron cut with a WC-Co tool. It shows the region in which the built-up edge persists. The built-up edge changes shape as speeds and feeds are raised. It eventually disappears and a form of cratering appears on the tool.

Figure 9.36 shows a section through a built-up edge seized to the rake face and to the worn flank. This consists of fine fragments of the metallic parts of the cast iron structure, severely plastically strained and welded together. The cementite and other constituents are usually so highly dispersed that they cannot be resolved with an optical microscope. The top part of *Figure 3.10* shows the structure observed in a polished but unetched section. Graphite is probably present, broken up into very thin, fragmented layers, since a black deposit is formed when the built-up edge is dissolved in acid.

Thus, under the compressive stress and conditions of strain at the tool surface, flake graphite ferritic and pearlitic cast irons behave as plastic materials. The feed force (F_t) is thus often higher than the cutting force (F_c) as shown in *Table 9.3*, and nearer to the value of F_t when cutting steel.

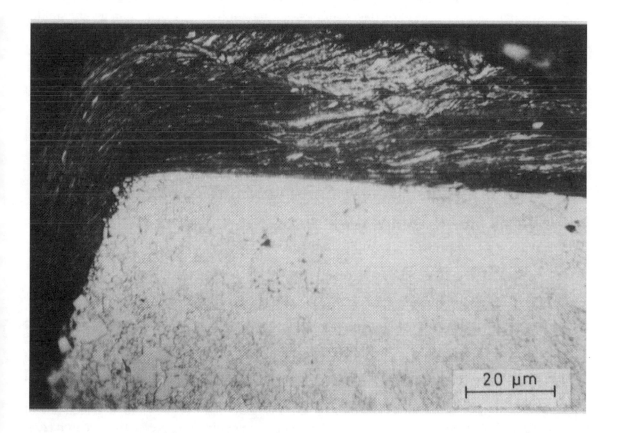

FIGURE 9.36 Section through built-up edge on carbide tool used to cut pearlitic flake-graphite cast iron

It is common practice when machining with either high speed steel or cemented carbide tools, to cut cast iron under conditions where a built-up edge is formed, and very good tool life can be achieved. With a discontinuous chip, the built-up edge is more stable and is less frequently detached from the tool even when an interrupted cut is involved. The wear is by attrition (*Figures 3.10* and *3.11*) and the longest tool life is achieved with tungsten carbide cobalt tools of fine grain size.

At higher rates of metal removal the built-up edge disappears. Then, to resist cratering and diffusion wear on the flank, a fine grained steel-cutting grade of carbide tool should be used. These contain small amounts of TiC and TaC - the M grades in *Table 7.3*. CVD-coated tools give low wear rates and permit the use of even higher cutting speeds.

Investigations show that the temperature distribution in tools used to cut cast iron is different from that when cutting steel. In the absence of a continuous chip, the highest temperatures are observed in the region of the cutting edge. With high compressive stress and high temperature at the edge, the upper limit to the rate of metal removal occurs when the tools are deformed at the edge, (*See* machining chart, *Figure 7.18*).

Spheroidal graphite (SG) irons have better mechanical properties than flake graphite irons, and in recent years have replaced them in many applications. In the SG irons the graphite is present as small spheres instead of flakes. In fact, during cutting, they behave in a very similar way to the flake graphite irons, and can generally be machined using very similar techniques. The graphite spheres act to weaken the material in the shear plane and initiate fracture, but are rather less effective in this respect than flake graphite. The SG chips are formed in rather longer segments, but these are weak, easily broken and much nearer in character to flake graphite iron, than to steel chips.

One problem which is sometimes encountered is that, with ferritic SG iron, the flow-zone material is extremely ductile and may cling to the flank face of the tool when cutting at high speeds. This causes high tool forces, high temperatures and poor surface finish. This problem can be largely overcome by using a high flank angle on the tool.

9.10 NICKEL AND NICKEL ALLOYS

9.10.1 Commercially pure nickel

Nickel has a lower melting point (1,452 °C) than iron (1,535 °C). The metal and its alloys are, in general, more difficult to machine than iron and steel. Nickel is a very ductile metal with a face-centered cubic structure and, unlike iron, it does not undergo transformations in its basic crystal structure up to its melting point.

Commercially pure nickel has poor machinability on the basis of almost all the criteria. Tool life tends to be short and the maximum permissible rate of metal removal is low. The tools fail by rapid flank wear plus deformation of the cutting edge, at relatively low cutting speeds. With high speed steel tools, a recommended turning speed is 50 m min^{-1} (150 ft/min) at a feed rate of 0.4 mm (0.015 in) per rev. Tool forces are higher than when cutting commercially pure iron (*Figure 9.37*). The contact area on the rake face is very large, with a small shear plane angle and very thick chips.

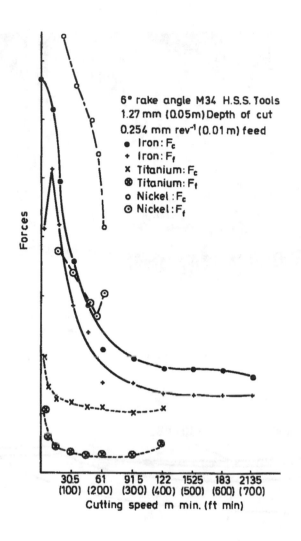

Forces

6° rake angle M34 H.S.S. Tools
1.27 mm (0.05m) Depth of cut
0.254 mm rev⁻¹ (0.01 m) feed
● Iron: F_c
+ Iron: F_f
× Titanium: F_c
⊗ Titanium: F_f
○ Nickel : F_c
⊙ Nickel : F_f

30.5   61   91.5   122   152.5   183   213.5
(100) (200) (300) (400) (500) (600) (700)

Cutting speed m min. (ft min)

FIGURE 9.37   Tool force *vs* cutting speed: comparison of iron, nickel and titanium[35]

As with iron and other pure metals, no built-up edge is formed, and the tool forces decrease steadily as the cutting speed is raised. The contact area becomes smaller and the chip thinner. However, over the whole speed range, the forces are relatively high. The high temperatures generated in the flow-zone lead to high rates of tool wear. *In particular, with commercially pure nickel, there is a characteristic adverse distribution of temperature in the tools, which is very different from that when cutting pure iron.*[35]

Figure 9.38a shows an etched section through a high speed steel tool used to cut commercially pure nickel at 45 m min⁻¹ (150 ft/min) at a feed of 0.25 mm (0.010 in) per rev feed. During manual disengagement of the tool, separation took place along the shear plane, leaving the chip very strongly bonded to the tool. The flow-zone, which is the heat source, is very clearly delineated adjacent to the tool rake face. The derived temperature gradient from this tool is in *Figure 9.38b*.

(a)

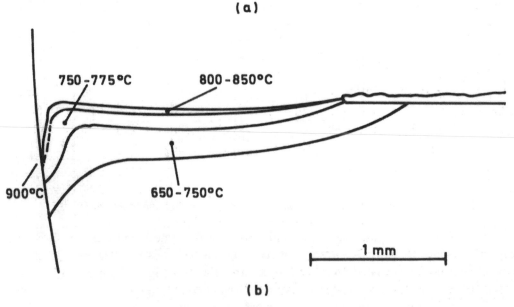

(b)

FIGURE 9.38 (a) Section through high speed steel tools used to cut commercially pure nickel, etched to show temperature distribution (b) temperature contours derived from (a)[26]

Figures 9.38a and *b* should be compared with *Figures 5.11a* and *b* to contrast the characteristic temperature distributions in tools used to cut commercially pure nickel and commercially pure iron. There are two major differences:

(1) Temperatures over 650 °C appear at much lower speeds when cutting nickel.

(2) The relatively cool region at the tool edge is not present when cutting nickel.

The difference in temperature distribution is seen also if the rake surfaces of tools used to cut nickel and iron are compared (*Figures 9.39* and *5.13*). Temperature is seen to be high along the main cutting edge when cutting nickel. Consequently, tools used for cutting commercially pure nickel tend to be deformed along the main cutting edge, where both compressive stress and temperature are high even at relatively low cutting speeds. Once the tool edge has deformed and a wear land has been started, a new heat source develops at the flank wear land and may result in rapid collapse of the tool.

Cemented carbide tools have a higher compressive strength at high temperature. They can therefore be used for cutting nickel and its alloys at much higher speeds than high speed steel tools. Carbide tools wear mainly on the flank by a diffusion or deformation mechanism, cratering not being a major problem.

Cemented carbide tools are not, however, generally recommended for cutting commercially pure nickel. The very strong bonding of the nickel chips to the tool surface often leads to damage to the tool when the chips are removed.

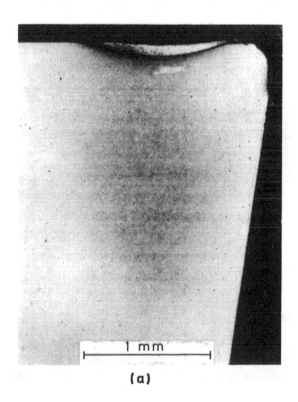

(a)

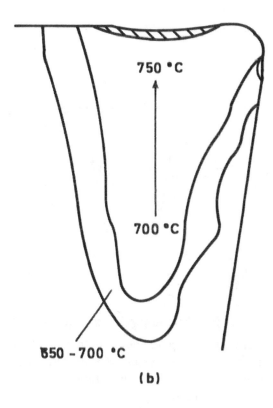

750 °C

700 °C

650 – 700 °C

(b)

FIGURE 9.39 (a) Rake face of tool used to cut nickel, etched to show temperature distribution; (b) temperature contour derived from (a)[35]

9.10.2 Lightly alloyed nickel work materials

Small amounts of alloying elements in commercially pure nickel affect its machining qualities in ways similar to those discussed for iron *vs* steel. Even when the alloying additions result in considerable strengthening of the nickel, the cutting forces are often reduced. This is because the contact length on the rake face is smaller, the shear plane angle is larger and the chip thinner.

The temperature distribution in tools used to cut the alloys of nickel has shown a temperature pattern similar in character to that in tools used to cut steel *and unlike that when cutting commercially pure nickel.* As might be expected, the temperatures with nickel alloys are higher at any given speed, as compared with iron and steel.

For example, *Figure 9.40* shows the isotherms in a high speed steel tool used to cut a nickel-based alloy. The alloy contained 19.5% chromium and 0.4% titanium. It was machined at 23 m min $^{-1}$ (75 ft/min) and at a feed of 0.25 mm/rev. The relatively cool cutting edge, and the high temperature region about 1 mm from the tool edge, are similar to the features observed when cutting steel.

Figure 9.16 shows the maximum temperature *vs* cutting speed curve for the above alloy and for commercially pure nickel. The speed required to generate a temperature of 800 °C, under the standard conditions, was 21 m min $^{-1}$ (70 ft/min) for the alloy. It was only 52 m min $^{-1}$ (170 ft/min) for commercially pure nickel.

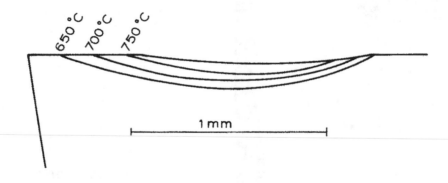

FIGURE 9.40 Temperature distribution in high speed steel tool used to cut Ni-Cr-Ti alloy at 23 m min $^{-1}$ (75 ft/min)

9.10.3 Highly creep-resistant nickel alloys used in aerospace industries

The example given above is for one of the least creep-resistant nickel based alloys. The highly creep-resistant alloys used in the aerospace industry, are some of the most difficult materials to machine. These alloys are strengthened by a finely dispersed second phase, as well as by solid solution hardening.

A built-up edge is formed when cutting these two-phased alloys at low cutting speeds (*Figure 9.41*). As the speed is raised, the built-up edge disappears but very high temperatures are generated even at relatively low speeds in the flow-zone at the tool/work interface. The temperatures are often high enough to take into solution the dispersed second phase in the nickel alloy, and may be well over 1,000°C.

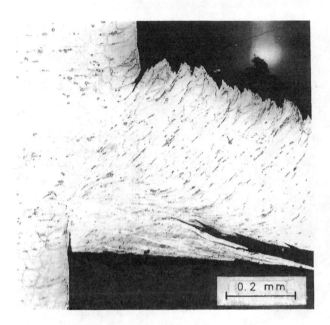

FIGURE 9.41 Built up edge when cutting creep resistant two-phased nickel alloy, quick-stop

Because these creep-resistant alloys are metallurgically designed to retain high strength at elevated temperatures, the stresses in the flow-zone are very high. The result is a destruction of the cutting edge under the action of shear and compressive stresses acting at high temperature.

Figure 6.14 is a section through the cutting edge of a high speed steel tool used to cut one such wrought alloy at 10 m min^{-1} (30 ft/min) and shows the tool material being sheared away. For many operations high speed steel must be used - for drilling and tapping small holes, for broaching and for most milling operations.

Cemented carbides, usually WC-Co alloys of medium to fine grain size, are used for turning, facing, boring and sometimes in milling operations and for drilling large holes. When they are used, carbide tools are more efficient because of the higher speeds and longer tool life. Even so, it is rare to find carbide tools operating at a speed as high as 60 m min^{-1} (200 ft/min).

The steel-cutting grades of carbide are usually worn more rapidly than the WC-Co grades. Coated carbides have been found to offer some advantages but not radically so. When cutting the most advanced of the "aerospace alloys", however, the inadequacy of cemented carbide tools becomes apparent. The tearing apart of a carbide tool used to cut one of the most creep-resistant cast nickel-based alloys is shown in *Figure 9.42*. The cutting speed in this case was only 16 m min^{-1} (50 ft/min).

The cost of machining the nickel-based aerospace alloys is very high. Metal removal rates are limited by the ability of conventional tool materials to withstand the temperatures and stresses generated. Much effort is now being put into employing ceramic tools to increase the efficiency of these operations.

Using both sialon and Al$_2$O$_3$/SiC whisker ceramics, cutting speeds up to 250 m min^{-1} (800 ft/min) are now employed for the machining of nickel-based gas turbine discs. Much effort has to be put into the machine tools, tooling and the details of the operation to achieve success.

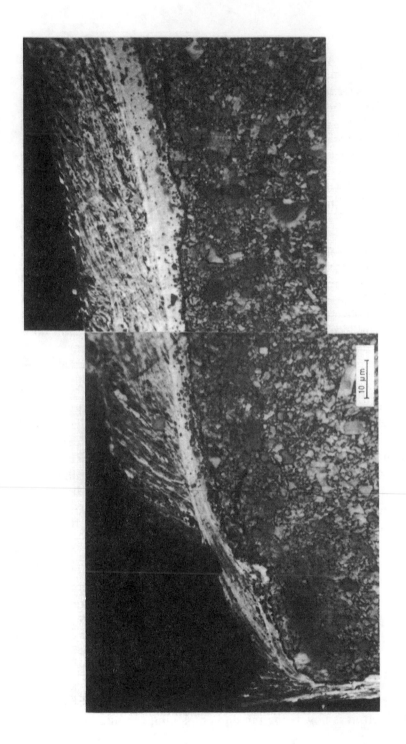

FIGURE 9.42 Section through cutting edge of carbide tool used to cut cast creep-resistant nickel alloy at 17 m min^{-1} (50 ft/min) cutting speed[3]

9.11   TITANIUM AND TITANIUM ALLOYS

9.11.1   Commercially pure titanium

Titanium and its alloys are generally regarded as having poor machinability. The melting point of Ti is 1,668°C. It is a ductile metal with a close-packed hexagonal structure at room temperature, changing to body-centered cubic at 882°C. The commercially pure metal is available in a range of grades depending on the proportion of carbon, nitrogen and oxygen. The hardness and strength increases and the ductility decreases as the content of these elements is raised.

The machining characteristics of titanium are different in several respects from those of the other pure metals so far considered. The tool forces and power consumption are considerably lower than when cutting iron, nickel or even copper. This is especially the case in the low-speed range, as shown in *Figure 9.37*. However, these low tool forces are associated with a much smaller contact area on the rake face of the tool than when cutting any of the other metals discussed (except magnesium). This means that the compressive *stresses* on the delicate cutting edge are very high. Consequently, tool life is short: it is terminated by flank wear and/or deformation of the tool. *The rates of metal removal for a reasonable tool life are much lower than when cutting iron.* As a final note, because of the small contact area, the shear plane angle is large and the chips are thin, often not much thicker than the feed.

9.11.2   Segmented chip formation in commercially pure titanium

Titanium chips are continuous but are typically segmented. With titanium alloys, the segmentation becomes very marked. Narrow bands of intensely sheared metal being separated by broader zones only lightly sheared (*Figure 9.43*). The intensely sheared layers are thermoplastic shear bands, to which titanium is particularly susceptible because of its thermal properties, especially its low thermal conductivity.

In a typical "segmentation cycle" each period of thermoplastic shear is very short-lived and relieves the stress. Next, compressive strain continues by dislocation movement until the next thermoplastic shear band is initiated. At the tool surface, the flow-zone is continuous and bonded very strongly to high-speed steel or carbide tools.

With titanium and its alloys, the flow-zone is very thin - usually less than 12 μm thick. No built-up edge is formed when cutting commercially pure titanium. During a quick-stop, the chip often remains bonded to the tool. Or, a layer of titanium is left bonded to the tool, the chip having separated by ductile fracture within itself, rather than at the interface. During normal disengagement of the tool, the chip frequently remains attached.

The main problems of machining titanium are that the tool life is short. Permissible rates of metal removal are low, in spite of the low tool forces. It is the high temperatures and unfavorable temperature distribution in tools used to cut titanium which are responsible for this.[34,35] The temperatures in the flow-zone are higher than when cutting iron at the same speed.

For example the maximum temperature on the rake face of a tool was 900 °C after cutting a commercially pure titanium at 91 m min^{-1} (300 ft/min). It was only 650 °C after cutting iron at this speed under the standard cutting conditions. The maximum temperature *vs* cutting speed relationship - when cutting a commercially pure titanium with low carbon and nitrogen content - is shown in *Figure 9.16*. Temperature gradients in tools used to cut titanium are shown in *Figure*

9.44. These should be compared with those for cutting iron (*Figure 5.10*) and nickel (*Figure 9.38*).

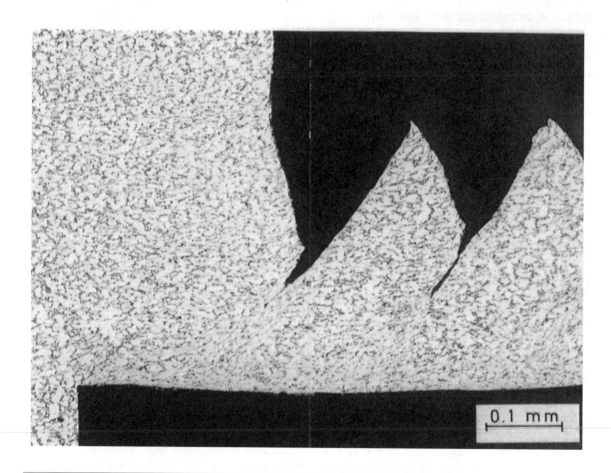

FIGURE 9.43 Section through forming titanium alloy chip, quick-stop (After Freeman[25])

The temperature distribution is like that when cutting iron, but the cool zone close to the edge is very narrow, and the high temperature region is much closer to the tool edge.

The flow zone in titanium alloys is initiated at a temperature close to 0.5 of the melting point in degrees Kelvin of titanium. At this initiation, the temperature at the tool edge is just under 700°C. The 700°C contour (*Figure 9.44b*) is within 0.1 mm of the tool edge, compared with 0.4 mm for iron and steel (*Figure 5.11*). The center of the high temperature region when cutting titanium and its alloys is usually about 0.5 mm from the tool edge. The total contact length is very short and the heated region does not extend far along the rake face.

Thus, when machining commercially pure titanium, although the tool forces are low, the stress on the rake face is high. Furthermore, the highly stressed region near the tool edge is at a very high temperature. This is the worst possible combination! It leads to deformation of the tool edge and rapid failure, with the formation of a new heat source on the deformed and worn flank. Frequently, failure is initiated at the nose radius of the tool.

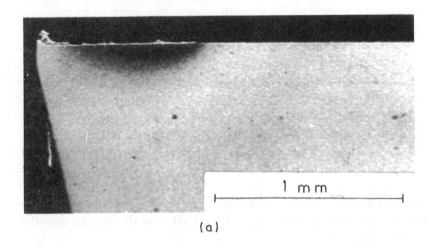

(a)

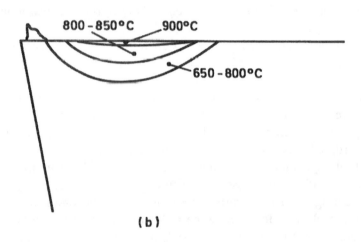

(b)

FIGURE 9.44 (a) Section through high speed steel tools used to cut commercially pure titanium, etched to show temperature distribution; (b) temperature contours derived from (a)

9.11.3 Titanium alloys

The temperature gradients in tools used to cut titanium alloys are similar in character to those found when cutting the commercially pure metal.[34] In general, the effect of alloying additions is to raise the temperature for any set of cutting conditions, and therefore, to reduce the permissible cutting speed.

When cutting commercially pure titanium, the influence of increasing amounts of the interstitial impurity elements, carbon, nitrogen and oxygen, is very pronounced. In one series of experiments,[34] an increase in oxygen content from *only* 0.13% to 0.20% reduced the cutting speed required to produce a temperature of 900°C in the tool from 91 m min^{-1} (300 ft/min) to 53 m min^{-1} (175 ft/min).

With alloys containing a second phase, the temperature increase for any cutting speed is much more marked. Under the standard test conditions, tools used to cut a titanium alloy containing 6% Al and 4% V were heated on the rake face to over 900 °C at a cutting speed of 19 m min^{-1} (60 ft/min).

The tool temperature *vs* cutting speed relationship for this alloy is shown in *Figure 9.16*. When cutting a commercial alloy with 11% Sn, 2.25% Al and 4% Mo, high speed steel tools failed due to stress and temperature after cutting for only 30 seconds at a speed of 12 m min^{-1} (40 ft/min). In the failure of the high speed steel tools, not only was the edge deformed downward under compressive stress. In addition, the heated high speed steel was sheared away to form a crater on the rake face, as was observed when cutting steel.

Apart from deformation, diffusion wear seems to be the main process, responsible for the wear both of high speed steel and of carbide tools when cutting titanium alloys. With cemented carbide tools, longer life is achieved with the use of the WC-Co alloys than with the steel-cutting grades containing TiC and TaC.

The introduction of TiC into the cutting tool, which is so strikingly successful in combating diffusion wear when cutting steel, has an adverse effect in relation to diffusion wear when machining titanium and its alloys.

There is evidence that the cubic carbide grains containing TiC are lost more rapidly by diffusion into titanium flowing over the tool surface than are the WC grains.[34] Resistance to diffusion wear and resistance to deformation at high temperatures make the WC-Co grades of carbide useful for cutting titanium alloys. Even with these, the cutting speeds which can be used for machining the more creep-resistant alloys are low, e.g., 30 m min^{-1} (100 ft/min).

The strong adhesion of the chip to the tool may also cause problems when the machining operation involves interrupted cuts. These lead to the breaking away of the adherent chip, removing fragments of the tool edge, and causing inconsistent tool life.[36-39] Ceramic tools based on alumina are worn more rapidly by processes of attrition.[36]

Even the CVD coatings applied to carbide tools are not as successful in reducing rate of wear when cutting titanium alloys. However, some reduction in wear is achieved. The mechanism controlling the crater wear of cutting tool materials in the machining of titanium alloys has been studied in detail by Kramer and colleagues.[37-39] Despite the drawbacks of the intense chip/tool adhesion described above, it is suggested that crater wear itself is reduced. This is because the adhesion maintains a reaction layer at the interface. Crater wear is then limited by the rate of dissolution through the reaction layer into the titanium work material.

Another strategy for decreasing tool wear is to identify materials that have high chemical stability with respect to titanium. Some rare earth metals are known to have relatively high enthalpies of solution in titanium and their compounds are expected to have corresponding low solubilities.

Thermochemical analysis of compounds of the rare earth elements and other selected compounds has identified candidate materials that may be chemically stable with respect to titanium. Several of these, including scandium carbide, scandium nitride and a number of rare earth and transition metal borides, have sufficient hardness. Eventually, they may be applied as coatings or incorporated in composite materials to provide sufficient toughness to be considered for industrial machining applications.[38,39]

9.12 ZIRCONIUM

The machining behavior of commercially pure zirconium is very similar indeed to that of titanium. The contact area on the rake face of the tool is short, the shear plane angle high and the chips are thin. The same sort of temperature pattern is imposed on the tool as when cutting titanium. These are the only two high melting-point metals investigated so far, for which the temperature gradients in the cutting tools are similar in character. The very close similarity of these two metals in structure and properties is paralleled by their machining qualities. There is some hazard in machining zirconium because fine chips may ignite.

9.13 CONCLUSIONS ON MACHINABILITY

9.13.1 General observations

In this chapter the machining qualities of some of the more commonly used metals and alloys have been described and discussed.

On the one hand, it is clear that machining behavior is complex, not easily or meaningfully evaluated by a single measurement.

On the other hand, some useful *ad hoc* tests can be specified for prediction of tool life; for rates of metal removal; or for power consumption under particular sets of operating conditions. However, these cannot be regarded as evaluations of machinability, valid for the whole range of operations encountered industrially. The results of such tests should *always be accompanied by a statement of the machining operation* used, precise tooling materials and the test conditions.

The changes that take place in a material as it passes through the primary shear plane, and the very different changes that occur in the secondary shear zone at the tool/work interface, give deeper insights into the concept of "machinability". The former can, to some extent, be directly related to properties measured by standard laboratory mechanical tests. However, the behavior in the secondary shear zone can be investigated only by observations of the machining process, since it cannot readily be simulated by model tests.

It is with the higher melting-point metals and alloys that the temperature and temperature distribution in the secondary shear zone (the flow-zone) play an important role in almost every aspect of machinability. The evidence of laboratory experiments shows that each of the major metals imposes on the tools a characteristic temperature pattern which differs greatly for different metals.

The reason why a particular temperature pattern is associated with a particular metal is an interesting subject for research. Alloying elements which increase the strength also raise the flow-zone temperature for any cutting speed and, in this way, reduce the maximum rate of metal removal when cutting high melting-point alloys.

"Free-cutting" alloying additives such as sulfur and lead may lower the interface temperature by reducing the energy expended in the flow-zone, but their action is complex and not yet fully understood. This is a most difficult region to study because of the small size and inaccessibility of the critical volume of metal. However, a better understanding of behavior in the flow-zone is an essential prerequisite for comprehension of machinability. These investigations have also theoretical interest in relation to the behavior of materials subjected to extreme conditions of strain and strain rate.

Improved appreciation of machinability also requires an understanding of the interactions of tool and work material at the interface. Rates of tool wear can be reliably predicted and controlled only when an understanding of bonding, diffusion and interaction at the interface has improved. This is particularly important with the introduction of new and expensive tool materials. It is also required for rational development of "free-machining" work materials. The concentration of useful phases at the tool/work interface is very dependent on strong bonding between these phases and the tool. Also, the plastic behavior of such inclusions under the extreme conditions of stress, temperature and strain at the interface is important.

9.13.2 Important observations on machinability variations in steel

Many of the sections in this chapter show good examples of how "specific" machinability can be. In particular, when machining steel, calcium inclusions, in the extreme conditions of the flow-zone, become attached to the cubic carbides of the steel-cutting grades of carbide. Two important metallurgical issues are emphasized:

- Without the extreme conditions of temperature and stress in the secondary shear zone the "internal lubrication" effects of the silicate layers might not be activated
- Without the specific reaction - between the silicate layers in the calcium deoxidized steels and the mixed carbide crystals in the steel-cutting grades of carbide - the effect does not occur with such intensity and benefit

When cutting such deoxidized steels at high speed, the performance of the tools is greatly increased. *This is of great economic importance for industry.*

Apart from increased rate of metal removal and longer tool life, surface finish may be improved and chip shape altered beneficially. Once again, the authors emphasize that the same work materials may show no improvement in machinability when cutting with a WC-Co grade of carbide - or with high speed steel tools - or at low cutting speeds where a built-up edge is formed with any type of tool material.

"Machinability" is not a unique property of a material! It is a mode of behavior of the material during cutting. All other aspects of the system must be concurrently included in an assessment of a material's machinability. Tool material and cutting speed are perhaps the two most important parameters to include. Specifying the work material's heat treatment and trace inclusions is also vital. (Many research papers exclude such information - thus invalidating any cross-comparisons with other research findings). Assessments of machinability should always specify the specific conditions of cutting, for which they have validity.

The deoxidation practice in the final stages of steel making varies considerably in different steel works and for production of steels for different purposes.[17] This accounts for much of the variability in machining quality of carbon and low alloy steels.

For example, the rapid wear rate of carbide tools that occurs when cutting steels that have been de-oxidized with aluminum alone, *may have as much to do with the <u>absence</u> of other inclusions which form protective layers as to the <u>presence</u> of abrasive Al_2O_3*. There are many studies in progress in many countries related to the most effective inclusions to form protective layers.[28-41] Many of these focus attention on calcium-aluminum silicates in a range of compositions of a mineral called anorthite.

There is some evidence that these inclusions need not be drawn out into long thin ribbons in the flow-zone, as demonstrated for MnS inclusions. They may remain rigid in the flow-zone (Region *cd* and *ef* in *Figure 5.5*) but be plastically deformed as they are smeared on the tool surface (Region *OXY* in *Figure 5.5*). The use of calcium in deoxidation seems essential, and the product resulting from its use is often referred to as calcium de-oxidized steel.

It has required many years for these initial observations to be translated into commercial practice. Steels with "improved machinability" or "inclusion modification" were first developed in Japan, Germany and Finland, and are now produced in most industrial countries. Data on industrial performance suggest that a major advantage is more uniform machinability: low wear rate, reduced cutting forces and improved chip form.

However, good machinability is rarely the primary quality required of a steel. De-oxidation practice has to be designed to ensure correct response to heat treatment. This heat treatment must also produce appropriate qualities in the final product - such as yield strength, creep resistance, fatigue resistance, and fracture toughness. These objectives may demand other types of inclusion - and these may favor or impede "machinability".

Whether or not the "special deoxidation" practices can be generally adopted, it is certain that there is much scope in the future for adjustment of steel making practice. *The ultimate goal is to produce steels capable of more consistent performance during machining at high rates of metal removal.* (And without resorting to the addition of large percentages of sulfur or lead).

9.14 REFERENCES

1. Williams, J.E., Smart, E.F. and Milner, D.R., *Metallurgia*, **81**, (3), 51, 89 (1970)
2. Sully, W.J., *I.S.I. Special Report*, **94**, 127 (1967)
3. Trent, E.M., *I.S.I. Publication*, **126**, 15 (1970)
4. Davies, D.W., *Inst Metallurgists Autumn Review Course 3*, No 14, p. 176 (1979)
5. Stoddart, C.T.H., *et al.*, *Metals Technol.*, **6**, (5), 176 (1979)
6. Wolfenden, A., and Wright, P.K., *Metals Technol.*, **6**, (8), 297 (1979)
7. Samandi, M. and Wise, M.L.H., *International Copper Research Association Project Report*, University of Birmingham (1989)
8. Samandi, M., *Ph.D. Thesis*, University of Birmingham (1990)
9. Trent, E.M., *Proc. Int. Conf. M. T.D.R., Manchester 1967*, p. 629, (1968)
10. A.S.M. Handbook, 8th ed., Volume 3 on Machining (1976)
11. *Machining Data Handbook*, 3rd ed., Vol 1, Machinability Data Center, Cincinnati
12. Hau-Bracamonte, J.L. and Wise, M.L.H., *Metals Technol.*, **9**, (11), 454 (1982)
13. Trent, E.M. and Smart, E.F., *Metals Technol.*, **9**, (8), 338 (1982)
14. Dines, B.W., *Ph.D. Thesis*, University of Birmingham (1975)
15. Trent, E.M., *I.S.I. Special Report*, **94**, 77 (1967)
16. Milovic, R., *Ph.D. Thesis*, University of Birmingham (1983)
17. Opitz, H., Gappisch, M. and König, W., *Arch. für das Eisenhüttenwesen*, **33**, 841 (1962)
18. Marston, G.J. and Murray, J.D., **208**, *J.I.S.I.*, 568 (1970)
19. Shaw, M.C., Smith, D.A. and Cook, N.H., *Trans. A.S.M.E.*, **83B**, 181 (1961)
20. Moore, C., *Proc. Int. Conf M.T.D.R. Manchester, 1967*, p. 929 (1968)

21. Baker, T.J. and Charles, J., *J.I.S.I.*, **210**, 680 (1972)
22. A.S.M. Handbook, 8th ed., Vol 1, (1976)
23. Wilber, W.J., *et al., Proc 12th Int. Conf M.T.D.R.*, p. 499 (1971)
24. Milovic, R. and Wallbank, J., *J. Appl. Metalworking*, **2**, (4), 249 (1983)
25. Stoddart, C.T.H., *et al., Nature*, **253**, 187 (1975)
26. Naylor, D.J., Llywellyn, D.T. and Keane, D.M., *Metals Technol.*, 3, (5,6), 254 (1976)
27. Pietikainen, J., *Acta Polytechnica Scandinavica*, No. 91 (1970)
28. Helle, A.S. and Pietikainen, J., *Behaviour of non-metallic inclusions during machining steel*, Inst. Metals Conf. (Nov. 1988)
29. Subramanian, S.V. and Kay, D.A.R., *Inclusion engineering for improved machinability*, Inst. Metals Conf. (Nov. 1988)
30. Nordgren, A. and Melander, A., *Inclusion behaviour in turning Ca treated steel*, Inst. Metals Conf. (Nov. 1988)
31. Ramalingam, S. and Wright, P.K., *Trans. ASME, Journal of Engineering Materials and Technology*, **103,** (2), 151 (1981)
32. Bletton, O., Duet, R. and Pedarre, P., *Influence of oxide nature on machinability of stainless steel*, Inst. Metals Conf. (Nov. 1988)
33. Opitz, H. and König, W., *I.S.I. Special Report*, **94**, 35 (1967)
34. Freeman, R., *Ph.D. Thesis*, University of Birmingham (1975)
35. Smart, E.F. and Trent, E.M., *Int. J. Prod. Res.*, **13**, (3), 265 (1975)
36. Dearnley, P.A. and Grearson, A.N., *Mat. Sci. ct Tech.*, **2**, 47 (1986)
37. Hartung, P.D. and Kramer, B. M., *Annals of the CIRP,* **31**, (1) 75 (1982)
38. Kramer, B.M., Viens, D., and Chin, S., *Annals of the CIRP,* **42**, (1) 111, (1993)
39. Kramer, B.M., *Thin Solid Films,* **108,** 117 (1983)
40. Fripan, M., and Schneider, J., *Ceramic Cutting Materials,* in the Conference on High-Performance Tools, Dusseldorf, Germany, p. 117, November 1998
41. Yang, X., and Liu, C.R., *Machining Science and Technology*, **3**, (1) 107 (1999)

CHAPTER 10	# COOLANTS AND LUBRICANTS

10.1 INTRODUCTION

10.1.1 Functions of cutting fluids: coolants versus lubricants

A tour of most factories will demonstrate that some cutting operations are carried out dry. However, in many other cases, a flood of liquid is directed over the tool, to act as a coolant and/or a lubricant. These cutting fluids perform a very important role and many operations cannot be efficiently carried out without the correct fluid.[1] They are used for a number of objectives:

(1) To prevent the tool, workpiece and machine from overheating and distorting.
(2) To increase tool life.
(3) To improve surface finish.
(4) To help clear the chips from the cutting area.

Many machine tools are fitted with a system for handling the cutting fluids. Such systems include circulating pumps, piping and jets for directing the fluids to the tool, and filters for clearing the used fluid.

A very large number of cutting fluids is available commercially from which the production planners select the one most suitable for a particular application. With very little guidance from theory, both the development of cutting fluids and their selection depend on a vast amount of empirical testing. A successful fluid must not only improve the cutting process in one of the ways specified, but must also satisfy a number of other requirements. It must not be toxic or offensive to the operator; it should not be a fire hazard; it must not be harmful to the lubricating system of the machine tool; it should not corrode or discolor the work material; it should give some corrosion protection to the freshly cut metal surface; and, of course, it should be as cheap as possible.

There are two major groups of cutting fluid:

- water-based or water-miscible fluids
- neat cutting oils

10.1.2 Water-based cutting fluids: overview of their cooling effects

As *coolants*, the water-based fluids are much more effective.[1-7] They consist of an emulsion. This is usually a mineral oil that can dissolve in water. A typical proportion is between 1: 10 and 1: 60 of oil to water. In addition to the mineral oil they contain emulsifiers and inhibitors. These prevent corrosion and the growth of bacteria and fungi. To increase the *lubricating* properties, animal or vegetable fats and oils are usually introduced. Also it is common to add "extreme pressure" lubricating substances containing chlorine and/or sulfur. Recently "synthetic fluids" have been developed to complement the emulsions. These are oil-less organic chemicals which may contain surface-active molecules or chlorine additives.

10.1.3 Neat cutting oil fluids: overview of their lubricating effects

Neat cutting oils are usually mineral oils supplied in a range of viscosities suitable for different applications. Like the water-based emulsions, the *lubricating* properties can be improved by addition of fatty oils, chlorine and sulfur. Chlorine is usually added as chlorinated paraffins. Sulfur may be added to mineral oil as elemental sulfur. This is known as "active sulfur" because it may be responsible for staining the machined work material, particularly if this is a copper-based alloy. Sulfur may also be introduced as sulfurized fat, where the sulfur is strongly bonded and not readily released. This avoids the staining problem.

10.1.4 Selection for different processes and work material types

Selection of the optimum cutting fluid from all the commercially available grades is a difficult problem. However, it is one which may be very important in many practical machining operations. The choice is influenced by many parameters including the work material, the cutting conditions and the machining operation involved. The latter is of particular importance. *Table 10.1*, prepared at the Machine Tool Industry Research Association[2] gives a guide to the selection of cutting fluids for a number of work materials and a range of machining operations.

10.1.5 Dry versus water-based versus oil-based fluids: an overview

There are many applications where cutting is carried out dry, in air, with no advantage being found in the use of a cutting fluid. For example, many turning and facing operations using carbide or ceramic tools are carried out dry, the cost of a fluid being avoided. Very many operations on cast iron require no cutting fluid.

Single point turning, planing and shaping, and drilling of shallow holes are among the operations where simple water-based coolants may be the only fluids required.

Lubricating requirements are most exacting with difficult operations such as broaching, lapping, thread cutting, reaming, trepanning of deep holes, and the hobbing of gears. For such operations, workshop trials must be the criterion for the optimum cutting fluid.

Many laboratory tests on operations such as drilling, tapping and reaming have been carried out under controlled conditions to determine the effect of lubricants. Although there is much scatter in individual tool life test results, the use of the correct lubricating additives to a cutting oil has been shown to give advantages in tool performance well outside any possible error in measurement.[6-17]

TABLE 10. 1   Guide to the selection of cutting fluids for general workshop applications[1-18].

Machining operation	Workpiece material			
	Free-machining and low-carbon steels	**Medium-carbon steels**	**High-carbon and alloy steels**	**Stainless and heat resistant alloys**
Grinding	Clear type soluble oil, semi-synthetic or chemical grinding fluid			
Turning	General-purpose, soluble oil, semi-synthetic or synthetic fluid		Extreme-pressure soluble oil, semi-synthetic or synthetic fluid	
Milling	General-purpose, or fatty, soluble oil, semi-synthetic or synthetic fluid	Extreme pressure soluble oil, semi-synthetic or synthetic fluid	Extreme pressure soluble oil, semi-synthetic or synthetic fluids (neat cutting oils may be necessary)	
Drilling	Fatty or extreme pressure, soluble oil, semi-synthetic or synthetic fluids			
Gear shaping	Extreme pressure soluble oil, semi-synthetic or synthetic fluid		Neat-cutting oils preferable	
Hobbing	Extreme pressure soluble oil, semi-synthetic or synthetic fluid (neat cutting oils may be preferable)			Neat cutting oils preferable
Broaching	Extreme pressure soluble oil, semi-synthetic or synthetic fluid (neat cutting oils may be preferable)			
Tapping	Extreme pressure oil, semi-synthetic or synthetic fluids (neat cutting oils may be necessary)		Neat cutting oils preferable	

Note: Some entries deliberately extend over two or more columns, indicating a wide range of possible applications. Other entries are confined to a specific class of work material.

10.2 COOLANTS

It is in connection with the machining of steel and other high melting-point metals that the use of coolants becomes essential. Their use is most important when cutting with high speed steel tools, but they are often employed also with carbide tooling. Coolants must always be used on automatic lathes where several tools are used simultaneously or in quick succession to fabricate relatively small components.

In Chapter 5 the two main sources of heat in a cutting operation are discussed - i) on the primary shear plane and ii) at the tool-work interface (especially in the flow-zone on the tool rake face). The work done in shearing the work material in these two regions is converted into heat, while the work done by sliding friction makes a minor contribution to the heating under most cutting conditions.

Coolants cannot prevent the heat being generated, and do not have direct access to the zones which are the heat sources. Heat generated in the primary shear zone is mostly carried away in the chip and a minor proportion is conducted into the workpiece. Water-based coolants act efficiently to reduce the temperature both of the work-piece and of the chip after it has left the tool. The cooling of the chip is of minor importance, but maintaining low temperatures in the workpiece is essential for dimensional accuracy.

The removal of heat generated in the primary shear zone can have little effect on the life or performance of the cutting tools. By contrast, as has been demonstrated, the heat generated at and near the tool/work interface is of much greater significance, particularly under high cutting speed conditions where the heat source is a thin flow-zone seized to the tool.

The coolant cannot act directly on the thin zone which is the heat source. However, the coolant can remove heat from those surfaces of the chip, the workpiece, and the tool which are accessible to the coolant. Removal of heat by conduction through the chip and through the body of the workpiece is likely to have relatively little effect on the temperature at the tool/work interface. This is because both chip and workpiece are constantly moving away from the contact area, allowing very little time for heat to be conducted from the source. For example, when cutting at 30 m min^{-1} (100 ft/min) the time required for the chip to pass over the region of contact with the tool is of the order of 0.005 seconds. The tool is the only stationary part of the system. It is the tool which is damaged by the high temperatures and, therefore, in most cases, cooling is most effective through the tool.

10.2.1 Importance of fluid jet and nozzle location

The tool is cooled most efficiently by *directing the coolant towards those accessible surfaces of the tool* which are at the highest temperatures. These are surfaces from which heat is most rapidly removed, and the parts of the tool most likely to suffer damage. Knowledge of temperature distribution in the tool can, therefore, be of assistance in a rational approach to coolant application. This is illustrated by experimental evidence from the authors' laboratory cutting tests on tools used to cut a very low carbon steel and commercially pure nickel.[4,5]

With steel, nickel and titanium alloys, the tool life is extended in turning if the coolant jet can gain good access to the clearance faces of the tool. Also, Chapter 9 showed that when machining commercially pure nickel the hottest region is very close to the cutting edge.

In most cases it is desirable to direct the jet at the flank face. This can be most easily done in turning operations. In some of the authors' experiments, the tool was held upside down in a toolpost at the back of the lathe. The chip was formed on the rake surface and fell directly down into the bed. This arrangement allowed the cutting fluid to pour directly down into the flank face region and provide consistent cooling action to the face and edge.[5] In other experiments, especially for steels, the authors have directed the jet at the end clearance face. This prevents the hot region from spreading over to the nose radius and causing premature failure.[4]

10.2.2 Machining steel with coolants

Figure 10.1 shows sections through the cutting edge of high speed steel tools used to cut a very low carbon steel at high speed: *a)* dry, *b)* flooded over the chip and tool rake face by a water-based oil-emulsion, and *c)* with a jet of the coolant directed towards the end clearance face of the tool.

The tools were sectioned and etched to show the temperature gradients in the tool by the method described in Chapter 5. *Figure 10.2* shows the temperature contours derived from the structures in *Figure 10.1*. Similarly, *Figure 10.3* shows the temperature contours on the rake faces of tools used for cutting under the same conditions as those in *Figure 10.2*. These illustrate a number of important features relevant to the action of coolants.

First, the coolant application was unable to prevent high temperatures at the tool/work interface. This is because heat continues to be generated in the flow-zone which is inaccessible to direct action by the coolant. Temperatures over 900°C were generated at the hottest part of the rake face of the tool. In terms of the actual *temperature value* it made no difference whether cutting was dry, flooded with coolant, or with a jet directed at the end clearance face.

Second, the action of the coolant reduced the volume of the tool material which was seriously affected by overheating. A jet directed to the end clearance face (*Figure 10.3c*) was much more effective in this respect than flooding over the rake face. The *temperature patterns* within the tool were then much different.

Third, the damage to the end clearance face caused by deformation of the tool when cutting dry (*Figure 10.3a*) or when flooded by coolant from on top, (*Figure 10.3b*) was prevented when the temperature of the end clearance face was reduced by a coolant jet. The wider cool zone at the cutting edge in this tool suggests that the rate of flank wear by diffusion would also be reduced by this method of cooling.

10.2.3 Machining commercially pure nickel with coolants

The cool zone at the cutting edge, which is a feature of tools used to cut steel, is absent when cutting commercially pure nickel, as demonstrated in Chapter 9. The high temperature at the main cutting edge leads to wear at this edge and failure by plastic deformation.

When cutting nickel, therefore, the coolant was found to be very effective when directed as a jet on to the flank face below the main cutting edge. *Figure 10.4a* shows a section through a tool used to cut commercially pure nickel dry, while *Figure 10.4b* shows the corresponding tool after cutting with a jet of coolant on the flank face. The corresponding temperature gradients are shown in *Figure 10.5. These reveal the considerable reduction in temperature and wear near the tool edge, achieved by a coolant directed to the correct part of the tool.*

10.2.4 Practical considerations with coolants

Of course it is not always possible in a factory situation to aim a coolant at precise location as described above. Besides which, the results in the last few pages are for turning where it is possible to aim the jet at different faces of the tool. By contrast, in milling and drilling, specific clearance faces are not individually accessible. (Gun drilling being an exception, where the hole through the drill brings the cutting fluid to the end clearance face of the drill.)

In milling and regular drilling, overall flooding of the whole cutting zone is the only practical solution. Flooding over the forming chip requires larger amounts of coolant to be effective. Many machine tools are equipped with pumps to direct 12 to 25 liters per minute to the cutting area.

Kurimoto and Barrow investigated such flood cooling when machining steel with both high speed steel[6] and cemented carbide tools.[7] They used the e.m.f. from a work/tool thermocouple. The measured temperature was lower when using a water-based coolant than when cutting dry, over the whole cutting speed range.

Figure 5.9 shows results from data presented for a steel-cutting grade of cemented carbide tool, and *Figure 10.6* from tests using high speed steel tools. Coolant was flooded over the rake face of the tools[6,7] - rather than into either of the clearance faces.[4,5]

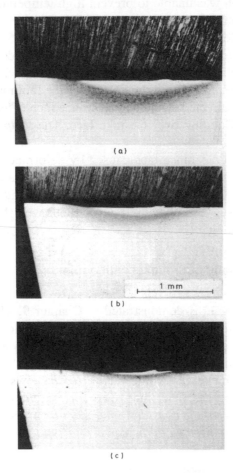

(a)

(b)

1 mm

(c)

FIGURE 10.1 (a) Section through high speed steel tool after cutting iron in air at 183 m min⁻¹ (600 ft/min), etched to show temperature distribution; (b) for tool flooded with coolant over rake face; (c) with jet of coolant directed at end clearance face[4]

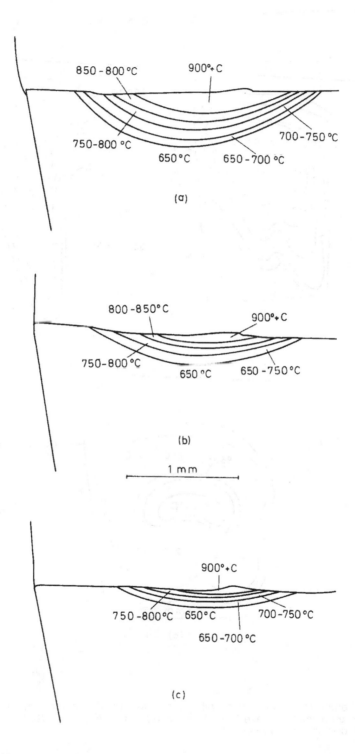

FIGURE 10.2 (a) Temperature contours derived from *Figure 10-1a* (b) temperature contours derived from *Figure 10.1b*; (c) temperature contours derived from *Figure 10.1c*[4]

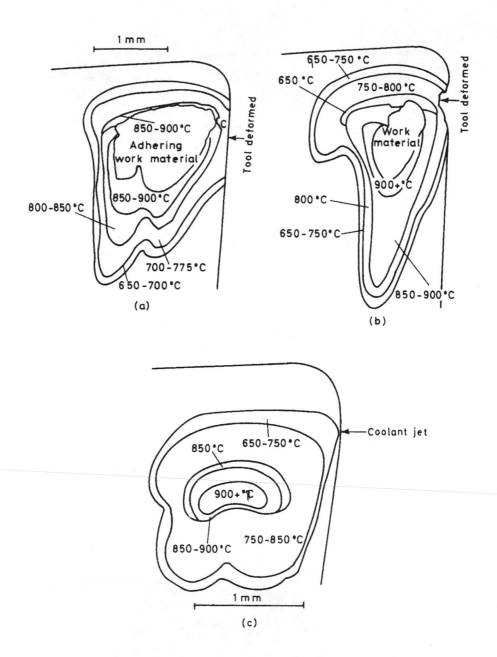

FIGURE 10.3 (a) Temperature contours on rake face of tool used to cut iron in air, conditions as *Figure 10.1a*; (b) temperature contours on rake face of tool, condition as *Figure 10.1b*; (c) Temperature contours on rake face of tool, conditions as *Figure 10.1c*[4]

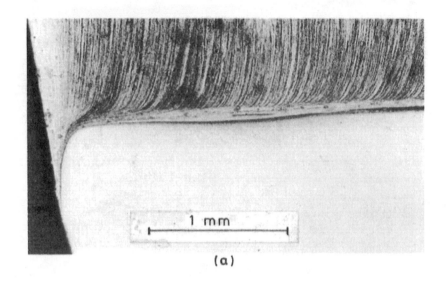

(a)

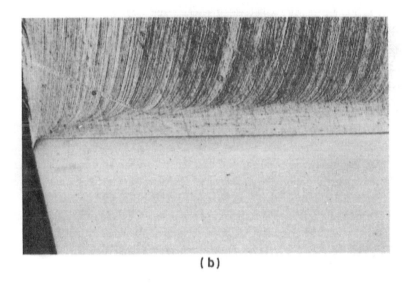

(b)

FIGURE 10.4 (a) Section through high speed steel tool used to cut nickel in air at 46 m min⁻¹ (150 ft/min), etched to show temperature distribution; (b) with jet of coolant on flank-clearance face[4]

The measured reductions in temperature were real and consistent. However, in each case the effectiveness of the coolants decreased as the cutting speed was raised. This was also observed in the experiments by the authors.[4,5]

Water on its own had the greatest cooling power. An emulsion of 1:15 lubricant in water was effective but resulted in measurably higher temperatures than the pure water. The temperature reduction using neat oil was small compared with cutting in air.

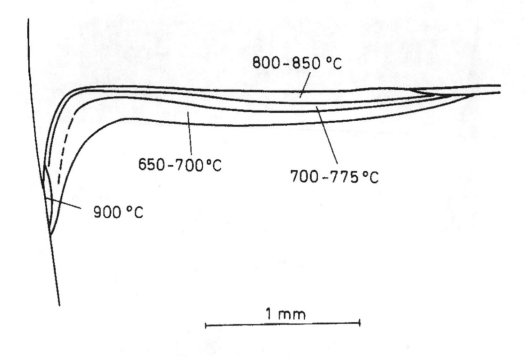

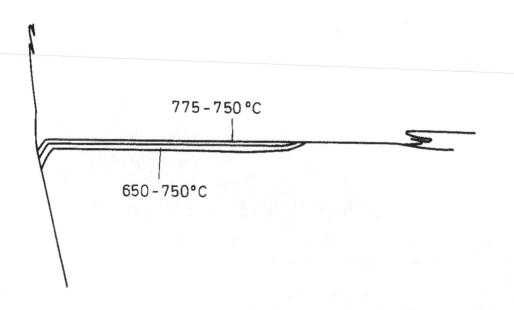

FIGURE 10.5   (a) Temperature contours derived from *Figure 10.4a*; (b) temperature contours derived from *Figure 10.4b*[4]

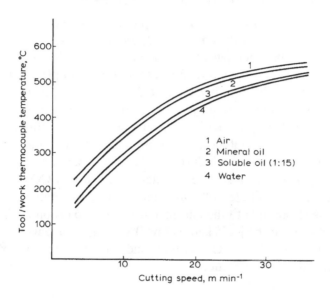

FIGURE 10.6 Tool temperature (tool/work thermocouple). Influence of cutting fluid when the work material is a low alloy engineering steel, feed 0.2 mm/rev. (Data from Kurimoto and Barrow[6])

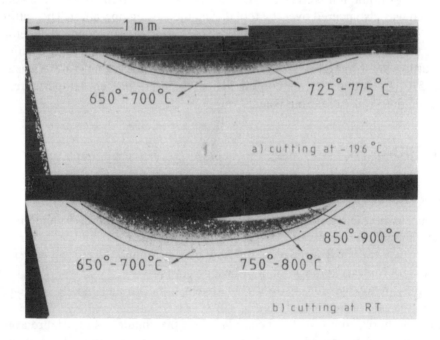

FIGURE 10.7 Section through high speed steel tool used to cut 0.4% C steel at 61 m min^{-1} etched to show temperature contours. (a) Tool and work material cooled to - 196°C in liquid nitrogen; (b) tool and work material at room temperature. (Courtesy of J.L. Hau-Bracamonte[8])

10.2.5 Potential use of high-pressure jets and carbon dioxide

A further potential use of cutting fluids involves employment of very high pressure jets (up to 280 MPa) of soluble oil directed under the chip towards the position where it breaks contact with the tool. It has been reported[9] that such jets can shorten the length of contact on the rake face of the tool and thus greatly reduce cutting and feed forces. In one example when cutting a 0.2% C steel at 180 m min^{-1} and a feed of 0.4 mm/rev, the feed force decreased from 800 N when cutting in air to under 200 N with the high pressure jet. The chip was changed from a continuous one with large curvature to short, curled segments. It seems probable that this is brought about by mechanical stress exerted by the jet rather than by any lubrication action. Whether such high pressure jets can be regularly employed in industrial machining operations is a matter of interest.

Another method which is also very effective and clean is the use of CO_2 as a coolant. CO_2 at high pressure is supplied through a hole in the tool and allowed to emerge from small channels under the tool tip as close as possible to the cutting edge. The expansion of the CO_2 lowers the temperature, and the tool close to the jet is kept below 0°C. Improvements in tool life using this method have been confirmed but both CO_2 cooling (and mist cooling) are expensive to apply and have not been widely adopted in practice.

10.2.6 A special experiment with liquid nitrogen

The potential and limitations to the use of coolants are, perhaps, best illustrated by the results of an extreme experiment.[8] High-speed steel tools were used to cut a bar of 0.4% carbon steel at the temperature of liquid nitrogen (- 196°C). Both tool and work material were cooled in liquid nitrogen which was also poured over the tool during the cutting operation. The cutting speed was 61 m min^{-1} (200 ft/min) at a feed of 0.25 mm/rev for a time of 30 seconds. The temperature gradients of the tool cooled in liquid nitrogen are shown in *Figure 10.7a* and the temperatures for a tool when cutting in air in *Figure 10.7b*. The maximum interface temperature was reduced from 900°C to 775°C. Even this extreme cooling action could not prevent temperatures in the flow-zone high enough to cause cratering wear.

10.3 LUBRICANTS

10.3.1 Conditions of seizure impede the action of external lubricants

The term lubrication in relation to cutting fluids is used here to describe action by the fluid at the interface. The major objectives of the use of lubricants are to:
* improve surface finish (by modifying the flow pattern around the cutting edge)
* reduce the tool forces and the amount of heat generated
* improve the life of the cutting tool (but under some circumstances lubricants cause an increase in the rate of wear)

Throughout the book, emphasis has been placed on the conditions of seizure at the tool/work interface. Seizure is an essential feature. It distinguishes metal cutting from all other metal working processes. In areas of seizure at the interface, especially where the tool and work materials are strongly bonded, there is *no possible access of externally applied lubricants to most of the interface.* Under these conditions, the introduction of substances such as lead, sulfides and plastic silicates by including them *inside the work material* (as described in Chapter 9) is the only practical method of getting a lubricant to the seized part of the interface.

10.3.2 Lubrication at the periphery of contact or during intermittent cutting

To understand the value of cutting lubricants, the emphasis must now be shifted to consideration of i) those areas of the tool/work interface which are not seized, and ii) of those conditions of cutting where seizure is reduced to a minimum or eliminated.

Seizure may be avoided when the cutting speed is very low, as near the center of a twist drill. Seizure may also be avoided when cutting tooth engagement times are very short. Examples include many milling operations - or, the multi-toothed hobs used for gear-cutting - or, when the feed and depth of cut are very small, so that no position on the interface is more than a few tenths of a millimeter from the periphery of the contact area.

Under such conditions metal-to-metal contact may be very localized. Consequently, the tool and work surfaces may be largely separated by a very thin layer of the lubricant which acts to restrict the enlargement of the areas of contact. In other words, what are known as "*boundary lubrication*" conditions may exist. In this situation, a good lubricant, especially one with the extreme pressure additives chlorine and sulfur, will reduce cutting forces, reduce heat generation, and greatly improve surface finish.

Under conditions of cutting where seizure does takes place, there is a peripheral zone around the seized area where contact is partial and intermittent. This is shown diagrammatically in *Figure 3.17* for a simple turning tool, operating under conditions of seizure but without a built-up edge. In the peripheral region, the compressive stress forcing the two surface together is lower than in the region of seizure.

10.3.3 Experiments in controlled atmospheres: an overview

The action of lubricants in this peripheral region, and the character of effective lubrication are shown in experiments on metal cutting in controlled atmosphere carried out by Rowe and others.[10,11] When cutting iron, steel, aluminium or copper in a good vacuum (10^{-3} mbar) or in dry nitrogen, both the cutting force (F_c) and the feed force (F_t) were much higher than when cutting in air. The chip remained seized to the tool over a longer path and was much thicker. The admission into the vacuum chamber of oxygen, even at a very low pressure, resulted in reduction of the contact area and tool forces to those found when cutting in air.

Therefore, the oxygen in the air surrounding the tool under normal cutting conditions acts to restrict the spreading of small areas of metallic contact in the peripheral region into large scale seizure. Unless the cutting speed is very low, oxygen in the surrounding atmosphere cannot completely prevent seizure in the region near the cutting edge; however, it can restrict the area of seizure. The freshly generated metal surfaces on the underside of the chip and on the workpiece are very active chemically and are readily re-welded to the tool even after initial separation.

The role of the oxygen in air, or the chlorine and the sulfur in a lubricant, is to combine with the new metal surfaces and reduce their activity and their affinity for the tool.

10.3.4 Importance of "natural air" as a lubricant

The freshly generated surface of the work material is very clean and chemically active. However, the surfaces of steel or carbide tools, before cutting starts, are contaminated with oxide, with adsorbed organic layers and other accidental oils. When cutting starts, the strength of the bond formed between tool and work material at first depends on the cleanliness of the tool surface. However, the unidirectional flow of work material across the tool surface tends to remove contaminated layers. This action requires a finite time which varies greatly with the tool, work material and cutting conditions.[11] When turning nickel, steel and titanium alloys a strongly bonded interface is established very rapidly. However, in interrupted cutting operations, such as milling, continuous contact times are often much shorter than one second. Then the area of bonding may be considerably reduced or eliminated by the periodic exposure of the tool surface to the action of air and active lubricants. This is particularly true when extreme pressure lubricants are used with sulfur and chlorine additives.

Thus air itself acts to some extent as a cutting lubricant.[11] If cutting were carried out in oxygen-free outer space, problems of high cutting forces and extensive seizure would be encountered. The action of air modifies the flow of the chip at its outer edge when cutting steel. In this region, oxygen from the air can penetrate some distance from the chip edge and act to prevent seizure locally, i.e., at the position *H-E* in the diagram, *Figure 3.17*.

Figure 3.3 shows scanning electron micrographs of a steel chip cut at 49 m min $^{-1}$ (150 ft/min) - note the segmented character of the outer edge. Sections through this outer edge show a typical "slip-stick" action, *Figure 10.8*. The steel first sticks to the tool. However, the presence of oxygen in this region then restricts the bonding between tool and work material to small localized areas. Then, the feed force becomes strong enough to break the local bonds, and a segment of the chip slides away across the tool surface. The process is then repeated - successive segments first stick and then slide away to form the segmented outer edge of the chip. Oxygen is able to penetrate for only a short distance - e.g., 0.25 mm (0.010 in) - from the outer edge of the chip at this cutting speed. Further inside the chip body, seizure is continuous. *Figure 10.9* is a section through the same chip at a distance of 0.50 mm (0.020 in) from the edge and shows a flow pattern typical of seizure.

10.3.5 Importance of nitrogen in protecting the tool surfaces

In summary, oxygen plays a very important role in creating oxides on the tool surface and freshly cut work material surfaces.

These oxides "in the right places on the rake face" reduce friction and tool wear. However, "in the wrong places" oxides accelerate notch wear.

Deep grooves are often worn in the tool at the positions where the chip edge moves over the tool with this slip-stick action (*Figures 6.23* and *7.26*). This "grooving" or "notching" wear is associated with chemical interaction between the tool and work material surfaces and atmospheric oxygen. The rate of grooving wear is strongly influenced by jets of gas directed at this position. When cutting steel with high speed steel or cemented carbide tools, wear at this position is greatly accelerated by a jet of oxygen and retarded or eliminated by jets of nitrogen or argon. This same effect is observed when cutting with a CVD-coated tool where the coating is TiC. However, there is usually no grooving on tools coated with alumina, as long as this coating remains intact. Grooving is greatly reduced with TiN coated tools,[12] once again showing their benefits.

This illustrates an important issue. Namely that the nitrogen in air plays an important role in reducing oxidation of the tool when cutting steel and other metals at high speeds. This can be demonstrated by a simple experiment.[13] On those surfaces of the tool which exceed 400°C during cutting, colored oxide films are formed - the familiar "temper colors". If tools used for high speed cutting of steel are closely examined, two areas of the surface are seen to be completely free from the oxide films even though they were at high temperature during cutting. These regions, marked *X* in *Figure 10.10*, are on the clearance face just below the flank wear land, and on the rake face just beyond the worn area. The freedom from oxide films is not because these areas were cold during cutting - they were in fact closer to the heat source, and at a higher temperature than the adjacent oxidized surfaces.

40 μm

FIGURE 10.8 Section through steel chip near outer edge showing slip-stick action at interface in *sliding* wear region

FIGURE 10.9   Section through same steel chip as *Figure 10.8*, 0.5 mm from outer edge showing flow-zone characteristic of *seizure* at interface

The explanation is that, during cutting, these areas of the tool surface form are part of a very fine crevice. This is the region just below *G* in *Figure 3.17* on the clearance face, and just beyond *B-B'* on the rake face. The opposing face of this crevice is a freshly generated metal surface at high temperature. It is free from oxide or other contaminating layers and highly active chemically. It combines with and eliminates all the oxygen available in this narrow space, *leaving a pocket of nitrogen which acts to protect the tool surface from oxidation*. Further down the clearance face and at the edges of the chip more oxygen has access. Thus the tool is coated with oxide films, although the temperature is lower. This is shown diagrammatically in *Figure 10.10*. Also, *Figure 10.11* is a photomicrograph of the clearance face of a carbide tool used to cut steel at high speed. However, the protective action of atmospheric nitrogen is eliminated if a jet of oxygen is directed into the clearance crevice. The tool is then very heavily oxidized in the region which was previously free from oxide films (compare *Figures 10.11* and *10.12*).

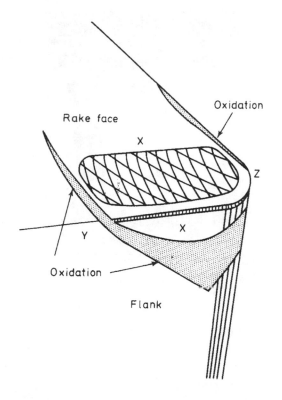

FIGURE 10.10 Occurrence of oxide films on tools used to cut steel at high speed[13]

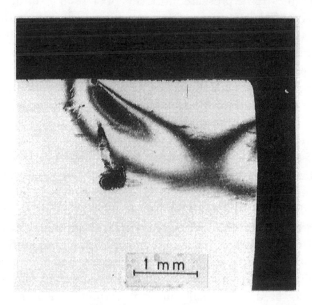

FIGURE 10.11 Clearance face of carbide tool used to cut steel at high speeds in air[13]

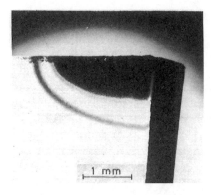

FIGURE 10.12   As *Figure 10.11,* but with jet of oxygen directed at clearance face[11]

10.3.6 Summarizing the role of air in cutting and implications for dry cutting

In summary, there are two counterbalancing effects of the 80% nitrogen and 20% oxygen in air. The oxygen in the air acts as a "lubricant", reducing the area of seizure. Simultaneously, the nitrogen in the air modifies this action and largely prevents a serious problem of oxidation of the tools. This is especially important when cutting high melting-point metals at high speed, since carbide tools are oxidized very rapidly when exposed to air at temperatures of the order of 900°C.

10.3.7 Extreme pressure lubricants

The active elements in extreme pressure lubricants operate in a similar manner to oxygen, but are much more effective. Chlorine forms chlorides by reaction with both tool and work materials. The chloride acts effectively only below the temperature at which it decomposes - about 350°C. Sulfides, formed by reaction at the interface, are effective up to about 750°C. Sulfur additives are thus more effective at high speeds and feed.

Although the cutting lubricants are applied as liquids, in most cases they must act in the gaseous state at the interface because of the high temperature in this region. A flood of lubricant is more effective than air because it allows penetration of the active elements into the interfacial crevice. This further restricts the area of seizure. In this sense, water acts as a lubricant as well as a coolant, penetrating between the tool and work surfaces and oxidizing them to restrict seizure.

One of the most effective cutting lubricants at medium to high speeds is carbon tetrachloride (CCl_4). The authors and colleagues have used it with care in their laboratories. However, because of its toxic effects, CCl_4 cannot be used in industrial cutting. *Figure 10.13* shows the influence of CCl_4 and water on the tool forces when machining steel over a range of cutting speeds.[14] Because of their action in reducing contact area, active lubricants effectively reduce the tool forces. They are most efficient at low cutting speeds and have a relatively small effect at speeds over 30 m min^{-1} (100 ft/min). By reducing forces, power consumption is reduced and temperatures may be lowered. As a final comment, carbon tetrachloride is *not considered as a lubricant at all under sliding* contact situations. With the transparent sapphire tools at very low speeds, CCl_4 caused extreme bonding at the rear of contact - termed as zone 2 adhesion by Doyle *et al.* and Wright.[11] This was most noticeable when cutting aluminum and lead.

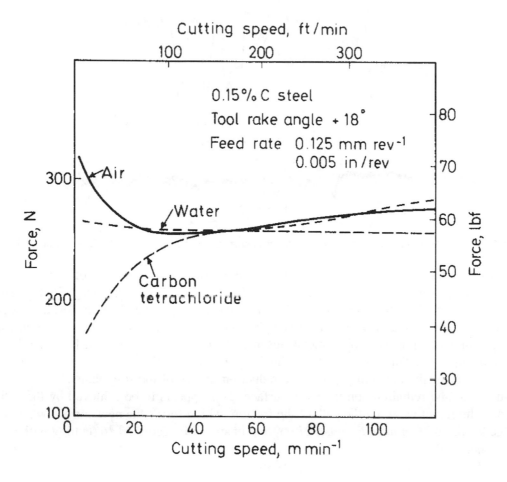

FIGURE 10.13 Influence of CCl_4 and water as lubricants on tool forces in relation to cutting speed. (After Rowe and Smart[10])

10.3.8 Surface finish

Improved surface finish is a major objective of cutting lubricants. In this respect they are particularly effective at rather low cutting speeds and feed rates in the presence of a built-up edge. As an example, *Figure 10.14* shows the surface finish traces made using a "surface profilometer". Experiments were on turned low carbon steel surfaces, produced by cutting at 8 m min^{-1} (25 ft/min) at a feed of 0.2 mm (0.008 in) per rev., both dry and using CCl_4 as a cutting fluid. The very great improvement in surface finish is caused by reduction in size of the built-up edge. Often, when cutting in air under such conditions, the built-up edge is very large compared with the feed, as shown diagrammatically in *Figure 10.15a* and in the photomicrograph *Figure 3.21*. Soluble oils act to reduce the built-up edge to a size commensurate with the feed, *Figure 10.15b*.

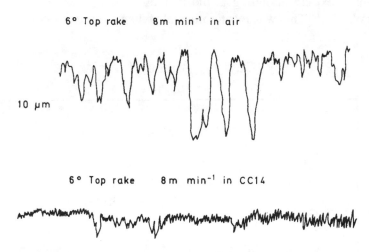

FIGURE 10.14   Surface profilometer traces of steel surfaces cut dry and with CCl$_4$ as cutting lubricant

Experiments have shown that the same result can be achieved by the use of distilled water or of a jet of oxygen gas. This suggests that the stability of a large built-up edge, which consists of many layers of work material (*Figure 3.21*), is at least partially dependent on a protective pocket of nitrogen when cutting in air. When air is replaced by oxygen, water vapor, or some chemically active gas they combine with fresh metal surfaces. Adhesion is then reduced between the layers of the built-up edge. This stable built-up edge is then much smaller.

The surface finish is also improved by a reduction in size of the fragments sheared from the built-up edge and remaining on the work surface. This appears to be achieved by the action of the active lubricant vapor on the path of the fracture which forms the new surface when a built-up edge is present. Compare *Figure 3.21* (cutting in air) with *Figure 10.16* (cutting with a chlorinated mineral oil).

10.3.9 Influence of cutting speed and tool material type

The influence of active lubricants on the rate of tool wear is very complex. In some conditions the rate of wear may be unaffected by the lubricating (as opposed to cooling) action of lubricants. In other cutting conditions, it may be greatly decreased or increased. There has been far too little study of the mechanisms by which the lubricants affect tool wear, and few general rules can be stated. Four examples are given to illustrate some effects of water-based (as opposed to oil-based) lubricants:

10.3.9.1   Cutting steels at low speed and feed using *carbide* tools

When cutting steels at low speed and feed using carbide tools, the active lubricants which reduce the size of the built-up edge, also increase the rate of wear and change its character.[13] The rate of flank wear is much increased and a shallow crater is formed close to the cutting edge, as shown diagrammatically in *Figure 10.15b*.

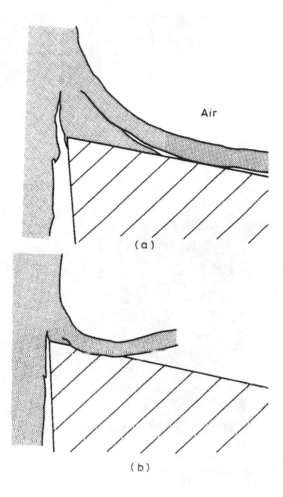

Air

(a)

(b)

FIGURE 10.15 (a) Built-up edge when cutting dry; (b) built-up edge when cutting with soluble oil lubricant [13]

Figure 10.17 shows the rake face wear on four tools after cutting *a)* in air, *b)* with a mineral oil, *c)* with distilled water, and d) with a jet of oxygen. The accelerated wear on the two latter tools is the result of interaction of the tool and work materials with an active gas or liquid environment to form. A small crater is formed where the chip contacts the tool after passing over the built-up edge (*Figure 10.15b*).

The mineral oil does not reduce the size of the built-up edge but does slightly reduce the small amount of wear on the rake face (*Figure 10.17a* and *b*). The craters formed with water (or soluble oil emulsion in water) are smaller and closer to the cutting edge. They have the same general shape as high-speed craters. They are not, however, formed by the same diffusion mechanism as the craters when cutting steel at high speed with WC-Co tools (*Figure 7.10*). These low-speed craters (*Figure 10.17c* and *d*) occur only at low speed and feed - e.g. 30 m min^{-1} and 0.1 mm/rev feed. Increasing cutting speed or feed changes the flow pattern around the cutting edge and eliminates crater wear.

FIGURE 10.16 Section through quick-stop showing built-up edge after cutting 0.15% C steel at low speed using chlorinated mineral oil lubricant

In summary, there are some unusual issues that occur with lubricants. The interactions between the lubricant and the tool are exceptionally dependent on cutting speed and tooling type. For example, at high speeds, the use of oil emulsion lubricant slightly *reduces* the rate of crater wear by diffusion on carbide tools. However, it causes another type of crater wear at low speed and feed. A large increase in wear rate also occurs when a soluble oil cutting fluid is used during cutting of cast iron with carbide tools at medium and low speeds. *This emphasizes the importance of understanding the mechanisms of tool wear for control of tool life.*

10.3.9.2   Cutting steels at low speed and feed with *high speed steel* tools

In continuous turning of steel with high speed steel tools the rate of flank wear may be greatly increased by use of water or a water-based cutting lubricant when compared with dry cutting. This has been observed by a number of research workers.[6,15,16] Under some test conditions, the rate of flank wear was accelerated by a factor of five or more[6] when lubricant was used, compared with dry cutting. This particularly occurred at low feed, and despite the fact that tool temperatures were lowered by the use of water-based fluid. Water-based emulsions accelerated the flank wear less than water alone. There is evidence to demonstrate that the water, by its action at the interface, modifies the flow close to the tool edge and changes the mechanism of tool wear to severe attrition.[15]

10.3.9.3   Cutting steels at *medium* speeds

König and Diederich[17] demonstrate that water-based cutting lubricants can greatly reduce the rate of cratering wear at medium cutting speeds. This is particularly true when cutting steel deoxidized with aluminium at high cutting speed using a steel-cutting grade of carbide. Penetration of the fluid or its vapors into the rear end of the contact area on the rake face promote the formation of relatively thick protective layers of oxide at the interface.

10.3.9.4   Cutting steel at *high* cutting speed (greater than 175 m min⁻¹)

Figure 10.18 compares the maximum rake face temperature near the rear of the crater when machining low carbon steel a) for dry cutting and b) for coolant supplied from above the cutting process (say in *Figure 2.1*). The application of a coolant is beneficial in low speed cutting operations but of doubtful value in high-speed turning of iron and steel.

Once again the inability of the coolant to significantly reduce tool temperatures at speeds greater than 175 m min⁻¹ is predominantly due to the conditions of intimate contact, or seizure, that occur over the major part of the chip-tool contact area. This part of the contact area is not accessible to the water-based lubricant. While the coolant can slightly penetrate to the friction region at the rear of the contact length, and also into the area created by the ragged edge of the chip, these effects do not greatly influence temperature, cutting forces or tool life at the higher speed.

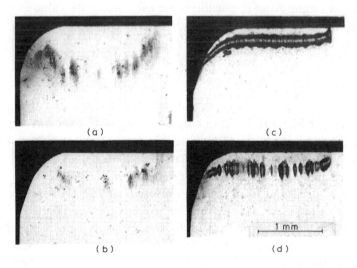

FIGURE 10.17   Wear on rake face of carbide tools when cutting steel at 30 m min⁻¹ (100 ft/min) at low speed.[13] (a) In air; (b) with mineral oil lubricant; (c) with water lubricant; (d) with jet of oxygen directed under chip in rake face

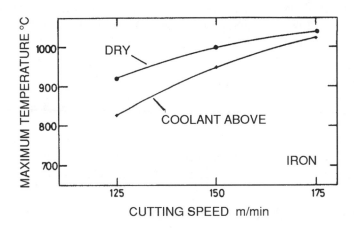

FIGURE 10.18 Reduced effect of a water-based coolant as speed is increased above 175 m min⁻¹

10.4 CONCLUSIONS ON COOLANTS AND LUBRICANTS

10.4.1 General observations

Understanding of the action of coolants and lubricants in metal cutting is still at a rather primitive level. The conditions are very complex, and it is still best to rely on experience and testing in the selection of cutting fluids. The results of such testing are shown in *Table 10.1*. The actions of lubricants in other engineering applications are of rather little value in guiding the selection and development of cutting and grinding fluids.[18] This is because of the conditions of seizure that act close to the tool edge.

10.4.2 Trends towards dry cutting

It is well known that the use of lubricants and coolants is under substantial criticism at the time of writing. It is feared by health professionals that long term exposure creates health hazards. These may range from minor yet unpleasant skin irritations, to respiratory problems, and to the more serious concerns about skin and other types of cancer.

Sheng[19] and colleagues have reviewed such data and related them to possible changes in the process plans for machining. It is possible to re-organize the process planning sequence and reduce the amount of fluid used for certain features on a machined part.

Another strategy is to *"target operations in the factory where cutting fluids may be of not much help or may even be making things worse"*. Some of the results in the last few sections may well have a significant impact on today's focus on "dry machining". Some examples include:

- In Sections 10.3.9.1 and 10.3.9.2, it was shown that carbide tools and high speed steel tools often behave poorly at low speeds in the presence of water-based lubricants. The evidence indicates that attrition wear takes place on the cooler tool surfaces. This is despite the presence of some lubrication effects. A cooler tool surface, attached to seized work material, promotes attrition and this outweighs any lubrication.

- *Figure 10.18* shows the reduced effectiveness of coolants in high-speed cutting (of some workmaterials). It is quite possible that very high speed cutting is best done dry for this reason in any case. Fluids might create no benefit and even leave ceramic tools vulnerable to craze-cracking. [20]
- Section 10.2 showed that a much smaller amount of coolant directed accurately at a "hot spot" can be effective. Although more expensive to engineer, this has great potential in reducing the volume of fluid used. A "mist coolant" spray similarly directed has been used commercially to solve difficult problems, as in the cutting of nickel based alloys.

Observations such as the above may eventually allow machine tool operators to first reduce, and then cease the use of coolants, without feeling that they have been unduly pressurized by government regulations to do so. They might even find beneficial cost reductions by some of the strategies suggested above.

To investigate dry machining, Malshe *et al.*,[20] machined aluminum A390 (aluminum-silicon alloy with 17% Si) using both polycrystalline diamond (PCD) and CVD diamond coated tools. They installed a particle counter over the lathe spindle to measure potentially injurious emissions from the tool surfaces. As might be expected, for both tool materials, the overall contribution to the mass of the overall emissions was low and was significantly lower than the proposed NIOSH standard for metalworking fluids. The PCD tool showed more emissions than the CVD-coated tool. This phenomena was attributed to the significant difference in the surface roughness of the cutting edge between the two tools. For the CVD tool the total emissions were $0.06 \ \mu g/m^3$ while the PCD tool had a rate of $0.1 \ \mu g/m^3$.

Tool wear in particular is greatly influenced by the atmosphere or environment. It is interesting to note that in dry cutting, air itself plays a subtle role. The oxygen in the air has a lubricating effect. This is because the oxides formed at the end of the chip-tool contact length, reduce seizure between the tool and the freshly cut metal. However, the oxygen accelerates notch wear at the ragged edge of the cut. The nitrogen in the air tends to reduce this notch wear. Trends towards dry cutting should build upon such research findings. In many situations, health-hazard lubricants can be replaced with less dangerous inert gas-jets.

Of course, at medium speeds, the water-based lubricants have a very beneficial effect. They reduce crater wear by diffusion or plastic deformation of the cutting edge. Given a gradual reduction in the use of cutting fluids in any given factory, these operations should be kept under coolant control for as many years as possible. Furthermore, difficult operations such as broaching and gun-drilling must rely on lubricants.

10.4.3 Recommendations

10.4.3.1 Creative solutions that avoid cutting fluids

The *first* practical objective should be to look more broadly at the "dry machining" challenge. Throughout Chapters 5, 9 and 10, several design issues and metallurgical effects have been described that offer creative solutions beyond the use of coolants/lubricants. These include:[19-23]

- The use of modified tooling designs that might reduce tool temperatures and thus reduce the need for coolant. *Figure 5.20* shows such an effect based on controlled contact tools. In other cases, careful selection of chip breaking grooves may reduce tool temperatures and also carry the chip away more easily.[21-23] This can often prevent the workpiece from overheating and distorting.

- The use of "free cutting" steels with the correct grade of carbide. An extremely important finding is described in Sections 9.7.3 and 9.7.4. *The MnS inclusions interact with the mixed crystal carbides to give "internal lubrication" effects.* By sharp contrast, the effects with the straight WC-Co grades are by no means as effective. A factory can make good use of this result by being meticulous in the choices of work material and tool material.
- The use of "calcium de-oxidized" steels where possible. Section 9.13.2 summarized the important effects of the silicates shown in *Figures 9.31 and 9.32*. Once again the formation of such layers is "very tool material dependent". These effects show some other options for eliminating coolant use when machining steel.
- Of course, if the new strategy does involve a switch from a standard steel to a "free cutting grade" it cannot be done without careful consultation with the original part designer. As difficult as "machinability" might be, no compromise can be made on the final component's fatigue life, fracture toughness or corrosion resistance. Integrated CAD/CAM systems that allow designers and fabricators to work cooperatively are therefore important software tools for such decision making. And as mentioned above, such systems can also reconsider the detailed process planning steps. For some components, where design features overlap, coolant can sometimes be reduced if the machining sequence is changed.[19]

10.4.3.2   Target operations that are, in any case, best done dry

The *second* practical objective should be to "*target*" those operations in a factory that are best done dry or at no loss without coolant. It has been shown above, and in the recent research[21] on "dry machining" that many aluminum alloys and cast irons can be machined dry with no loss of performance. Moreover, many operations with ceramic tools are best done dry.

10.4.3.3   Focus on tooling "hot spots" with pointed jets and mist coolant

The *third* practical step is to point the coolant nozzle only onto the "hot spots" of tools, or to use "mist coolant". This reduces the volume of fluid used. The results from such studies are shown in *Figures 10.1* through *10.5*.

10.4.3.4   Inert gases and cryogenic possibilities[22]

The *fourth* practical step is to consider jets of inert gases for some applications where notch wear of tools is a problem. Also, Hong and colleagues[22] have shown that liquid nitrogen can be directed through small channels in the chip-breaker pad and into the rear of the chip-tool contact area. This has a very beneficial effect when machining titanium alloys. It cools the chip-tool interface, curls the chip away more readily, improves surface finish on the bar and allows speed to be increased.

10.4.3.5   Reserve the regulated amounts of coolant/lubricant for the extreme operations

Hopefully once these "more obvious targets" have been addressed it will bring a particular factory's annual consumption underneath the regulated limits. As a *fifth* suggestion, this will then allow *some* lubrication - at a regulated limit - to be used in the more difficult operations. Such extreme processes include broaching and tapping operations, and those steel cutting operations at medium speeds where the use of cutting fluids still appears essential.

10.5 REFERENCES

1. Morton, I.S., *I.S.I. Special Report*, **94**, 185 (1967)
2. Cookson, J.O., *Tribology International*, 5th Feb. (1967)
3. Morton, I.S., *Industrial Lubrication and Tribology*, July/August, 163 (1972)
4. Smart, E.F. and Trent, E.M., *Proc. 15th Int. Conf M.T.D.R.*, p. 187 (1975)
5. Wright, P.K., *Trans. ASME, Journal of Engineering for Industry*, **100**, (2), 131 (1978)
6. Kurimoto, T. and Barrow, G., *Proc. 22nd Conf M.T.D.R.*, P. 237 (1981)
7. Kurimoto, T. and Barrow, G., *Annals C.I.R.P.* **31** (l), 19 (1982)
8. Hau-Bracamonte, J.L., *Metals Technol.* **8** (2), 447 (1981)
9. Mazurkiewicz, M. *et al.*, *Trans. A.S.M.E., J. Eng. for Ind.*, **111** (7), 7 (1989)
10. Rowe, G.W. and Smart, E.F., *Brit. J. App. Phys.*, **14**, 924 (1963)
11. Wright, P.K., *Metals Technol.*, **8** (4), 150 (1981)
12. Dearnley, P.A. and Trent, E.M., *Metals Technol.* **9**, (2), 60 (1982)
13. Trent, E.M., I.S.I. *Special Report*, **94**, 77 (1967)
14. Childs, T.H.C. and Rowe, G.W., *Reports on Progress in Physics* **36** (3), 225 (1973)
15. Childs, T.H.C. and Smith, A.B., *Metals Technol.*, **9** (7), 292 (1982)
16. Opitz, H. and König, W., *I.S.I. Publication*, **126**, 6 (1970)
17. KÖnig, W. and Diederich, N., *Annals of C.I.R.P.* **17**, 17 (1969)
18. Malkin, S., *Grinding Technology*, Society of Manufacturing Engineers, Dearborn, MI.
19. Bauer, D.J., Thurwachter, S., and Sheng, P.S., Trans. NAMRI/SME, 171 (1998)
20. Malshe, A.P., Taher, M.A., Muyshondt, A., Schmidt, W.F., Mohammed, Hafez, and Mohammed, Hazek, *Transactions of the NAMRI/SME*, **26**, 267 (1998)
21. Klocke, F., and Eisenblatter, G., *Dry Cutting - State of Research* in the Conference on High-Performance Tools, Dusseldorf, Germany, p. 159, November 1998
22. Hong, S. and Santosh, M., "Effects of liquid nitrogen as a cutting fluid in the machining of titanium" *International Journal of Machine Tools and Manufacture* (also see Hong's videotapes - Columbia University, New York, NY., 10027)
23. Jawahir, I.S., Dillon, O.W., Balaji, A.K., Redetzky, M., and Fang., *Proceedings of the CIRP International Workshop on Modeling of Machining Operations,* held in Atlanta, GA., Published by the University of Kentucky Lexington, KY, 40506-0108. p.161, (1998)

CHAPTER 11   HIGH SPEED MACHINING

11.1   INTRODUCTION TO HIGH SPEED MACHINING

Recently, there has been renewed interest in machining at very high cutting speeds. Some economic and materials science issues are thus reviewed in this chapter.

The first section outlines some of the reasons for the renewed interest in high speed machining. Twenty or even ten years ago, it would not have seemed reasonable to machine very large structures from a solid monolithic slab. However, modified component designs - shown in the lower photograph of *Figure 11.1*, eliminate costly and unpredictable joining operations in the factory.[1-3]

The behavior of workmaterials at high rates of strain is then reviewed. This is fundamental to an understanding of high speed cutting.[4-40]

Sections 11.6 to 11.8 describe some specific experiments on stainless steel, AISI 4340, aluminum aerospace alloys, and titanium alloys. These materials have different responses to high speed machining based on their material properties.

The last section discusses tool wear and tool life. A major question is whether or not adiabatic heating can lead to dynamic recovery or even localized melting in the shear zones of cutting. It was once hoped that such processes would occur and make a significant enough difference to the workmaterial properties that the local stresses on the tool would be reduced. So far, the evidence is that forces and stresses on the tool edge do not greatly reduce as speed is increased to very high values.

Beyond 200 m min^{-1} most materials show a "flattening out", or a very minor reduction, of the force *vs.* speed curve. In fact, there is far too little information on the type of tool wear that may be expected. If high speed machining is to be successful, then a study of tool wear mechanisms and data on tool life should be major considerations in future research work on this topic.

11.2 ECONOMICS OF HIGH SPEED MACHINING

New machining research programs at The Boeing Company provide an interesting motivation for high speed machining. The following example shows how design and manufacturing keep changing to suit a rather complex interaction between a) the availability of innovative manufacturing techniques and b) new economic conditions.

Inside the ceiling of a plane, structural members that resemble "giant coat-hangers" are spaced across the plane at intervals to give it torsional stability. Today, most are made from many conjoined pieces. This arrangement is shown in the upper photograph of *Figure 11.1*. By contrast, newer component designs - shown in the lower photograph of *Figure 11.1* - eliminate costly and unpredictable joining operations in the factory.

Thomas[1] has observed that such manufacturing innovations, *flowing-back* into the design phase, must be the new way of organizing the relationship between design and manufacturing. Similarly, Ayres[2], provides the succinct definition of computer integrated manufacturing (CIM) as *the confluence of the supply elements (such as new computer technologies) and the demand elements (the consumer requirements of flexibility, quality, and variety).*

Using Ayres' definition it can be observed that the new innovations, or the new *supply elements* include:

- Improved cutting tool technology and an understanding of how to control the accuracy of very high speed machining processes
- The availability of stiffer machine tools and very high speed spindles
- More homogeneous microstructures that give uniformity in large forging slabs
- The ability to carry out comprehensive testing, and show that these one-piece structures are at least if not more reliable than multiple-piece structures.

Meanwhile the new *demand elements* include:

- Escalating costs of joining and riveting operations which can only be semi-automated. Specifically these operations often require manual fixturing of the workpieces
- A preference for eliminating multiple fabrication steps which always demand more setup, fixturing, documentation and extra quality assurance
- General pressures on the whole of the airline industry, since deregulation, to cut costs and yet to improve the safety and the integrity of the aircraft.[3]

These trends introduce a great deal of complexity into the design and manufacturing process. The conclusion to be drawn is that *no single component or manufacturing process should be analyzed and optimized in isolation*. There will always be something that can be improved, simplified or made cheaper if design and manufacturing are viewed from a slightly wider system perspective.

How do work materials behave when the cutting speed is raised as high as 3,500 m min^{-1} (11,500 ft/min) for aluminum alloy wing panels?

What are the forces on the tool?

What is the effect on tool life?

These are some of the questions answered in this chapter.

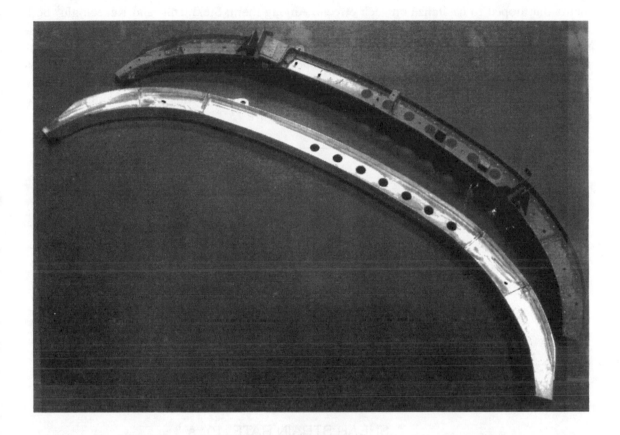

FIGURE 11.1 Integrated product and process design allows this aerospace component to be completely machined from the solid as shown in the lower photograph (Courtesy of Dr. Donald Sandstrom, The Boeing Company)

11.3 BRIEF HISTORICAL PERSPECTIVE

In the 1930s, Trent[4] and others provided clear metallographic evidence for adiabatic shear zones in high strain rate deformation processes. *Figure 5.6* shows such a shear band that occurred during a high strain rate impact experiment.

In 1957, Cottrell[5] addressed the *Conference on the Properties of Materials at High Strain Rate* with the following succinct summary: "If the rate of plastic flow is sufficiently high, there may not be time enough to conduct away the heat produced from the plastic working, and the temperature may rise sufficiently to soften the deforming material...intense rapid flow becomes concentrated in the first zones to become seriously weakened by this effect."

In 1964, Recht[6] developed a model for catastrophic shear instability in metals under dynamic loading conditions. He noted that in certain combinations of temperature and stress, the slope of the stress-strain curve could become zero or even negative. This would especially be the case if thermal softening dominated over strain hardening.

These general observations by materials scientists were the inspiration for machining experiments that hoped to capitalize on such effects. Among them, Siekmann[7] and Kececioglu[8] both reported their high speed cutting tests in 1958. Kececioglu's work is remarkable for its thoroughness on the shear strain rates and shear flow stresses in metal cutting. This paper is recommended for a wealth of experimental data on shear zone widths and primary shear zone stress values. Of particular note is that the primary shear zone average shear stress remains relatively constant over a rather large range of cutting speeds. To illustrate this point, the authors have replotted Kececioglu's results in the upper curve of *Figure 11.2*.

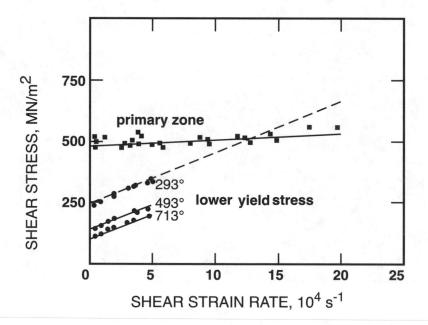

FIGURE 11.2   Comparison between Kececioglu's machining tests and Campbell and Furguson's high strain rate Hopkinson bar data for the yield stress of mild steel at various temperatures in degrees Kelvin (subtract 273 for degrees Centigrade)

During the late 1960s and throughout the 1970s, experiments with projectiles and ultra high speed cutting were conducted by several research teams including Recht's. The reader with a deep interest in high speed machining is especially referred to the work based on Arndt's[9] Ph.D. thesis at Monash University with R.H. Brown. The extensive citation list in Arndt's 1973 paper probably contains all the European, Japanese and U.S. work pertaining to high speed machining up until that time. The experiment at Monash used a military Bofer's gun to fire the workpiece passed the stationary strain-gaged tool. Cutting forces were measured at different speeds up to ballistic values. These experiments were further reinforced during the 1980s in a large, U.S. Air Force funded program led by General Electric, Inc. The results relating to material properties and chip formation are given by von Turkovich, Komanduri, Flom and colleagues.[10-15]

The high speed machining experiments up to that point held out a fascinating promise. The basic questions were: i) can the very high speeds lead to dynamic recovery or even local melting in the primary shear zone so that ii) the forces and stresses on the tool will be so much lower that tool wear will correspondingly be reduced.

These questions do not seem to have been adequately answered at the time of writing. To the authors' knowledge, no evidence experimental evidence has been obtained showing a dramatic "valley-like" drop in the cutting force *vs.* speed curve at some high speed. It is certainly the case that cutting forces gradually decrease in a smooth fashion as speed is increased (*Figures 4.12 and 4.13*). However, no dramatic drops have been demonstrated in the very high speed ranges that might be correlated with a sudden shift in material behavior, such as dynamic recovery or local melting in primary shear.

Perhaps even more relevant from a practical viewpoint is that none of the reported experiments have demonstrated a sudden reduction in tool wear at the very high cutting speeds. Indeed, researchers such as Arndt and Recht seemed to find that the tool life was extremely short in the ballistic experiments. The tools disintegrated after only an inch or two of cutting along the projectiles.

Such findings meant that interest in high speed machining dropped-off after the mid-1980s, and it is only recently that interest has been revived. Even now the interest is focussed on the higher production rates for aluminum and cast iron - not particularly the reduction of tool wear. Indeed, all the evidence points to the fact that most materials - even soft aluminum alloys - begin to machine with a *segmental or serrated chip* at some particular threshold value of cutting speed.

Albrecht[16] was the first to demonstrate that such chip forms accelerate attrition wear especially in carbide tools. This is because the segmental chip is associated with fluctuating fatigue type loading on the delicate tool edge.

An ultimate objective is thus to increase speed and yet minimize the degree of segmentation and reduce the fatigue type loading on the tool edge. Section 11.9 returns to this point with some specific practical recommendations.

The current research is thus aimed at understanding these segmental effects that are "triggered at some threshold speed" in most if not all materials. Before reviewing such segmental chip forms, some high strain rate properties of material are now considered.

11.4 MATERIAL PROPERTIES AT HIGH STRAIN RATES

The behavior of materials in the cutting operation is dependent on: a) the way in which initial yielding is affected by the high strain rates which arise, b) the strain hardening characteristics of the work material, and c) the time-dependent relationship between the flow stress and the heat generated during plastic deformation.[17]

During the 1960s and 1970s, Campbell's group at Oxford University developed an extensive laboratory for high strain rate deformations, especially focussing on the split Hopkinson bar technique.[18-20] In presenting such data it became customary to follow the notation of Rosenfield and Hahn[21] to describe the behavior in terms of four regions covering a wide range of strain rates and testing temperatures. Even though these papers are relatively old, the reader is encouraged to seek them out as one of the most reliable sources that exist for high strain rate data. Such experiments have not been carried out as much recently. This has been a result of their inherent expense and the fact that, since the 1980s, research funding in Europe, Japan and the U.S. has been more directed at robotics, automation and CAD/CAM.

The range of strain rates appropriate to machining begins at the boundary between "Regions II and IV" (a shear strain rate of ~10^4 per second), and extends well into "Region IV". In this last region, the rate of dislocation movement is influenced by phonon or electron damping mechanisms.[21] In this range the shear yield strength is observed to increase linearly with strain rate as described below and in equation 11.1.

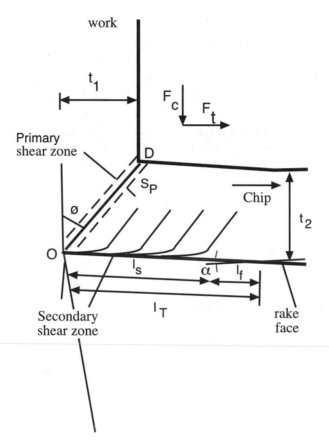

FIGURE 11.3 Simplified primary zone and secondary zone. Elements enter the primary zone and yield according to equation 11.1. Flow occurs through the primary zone according to equations 11.2 and 11.3. In the secondary zone, material starts out severely worked from the primary zone. Under conditions of seizure, more high strain deformation occurs along the tool face. Materials such as stainless steel, AISI 4340 and titanium maintain very high strength at high temperatures. Therefore they create very high shear stresses on the tool face.

A general analytical approach - taken in most of the machining research - is to compare the work material and tool material properties at the temperatures calculated with the equations in Chapter 5. However the workmaterial properties are also affected by the high strains and high strain rates that arise. Also, the time of heating the chip is very short whereas, in turning, the tool is stationary. All these factors must be considered when calculating stresses on the tool, or analyzing the segmental chip formation.

11.4.1   Initial yielding at the beginning of the simplified primary shear zone

As a first approximation, it can be assumed that all the plastic work of primary shear is done in a discrete, parallel-sided region of the type shown in *Figure 11.3*. In this case, an element, on entering the zone, rapidly attains its strain rate modified yield stress and is then further strained to the value γ_p given by equation 3.3. Information which relates only to the initial yielding behavior of the material can be drawn from the dynamic shear testing and punching of metals.[18-21] Effects such as strain hardening or softening, which occur after yielding, are then isolated; these are discussed in the following section.

In order to obtain high strain rate yield strength data, Dowling *et al.* employed punching methods to develop thin shear zones in flat strips of aluminum, copper, brass, and mild steel. In other work, Campbell and Ferguson carried out split Hopkinson bar experiments on mild steel. Their results for mild steel in the temperature range 293K< T < 713 K are incorporated in *Figure 11.2*. Owing to the dislocation damping mechanisms[18-21] the relationship between the shear yield stress and the shear strain rate (plotted on a linear scale) becomes linear in "Region IV".

This is most clearly seen in the Campbell and Ferguson data[18] for mild steel (*Figure 11.2*). Typical values of shear strain rate in the primary zone are between 2 and $20 \times 10^4 s^{-1}$ for commercial high-speed cutting of steel. Comparable strain rates have been obtained by Campbell and Ferguson in the high strain rate shear testing.

Such behavior allows a macroscopic viscosity coefficient μ to be defined. The value of μ may be obtained from the straight lines such as the shear yield stress curves in *Figure 11.2*, since

$$k_y = k_0(T) + \mu \dot{\gamma} \qquad (11.1)$$

where k_y is the shear yield stress and $k_0(T)$ is the stress at the beginning of linear behavior related to the test temperature.

11.4.2   Response of the material continuing through the primary shear zone

The primary and secondary shear zones are regarded as being separate from each other, and then each one is appraised in terms of the following constitutive equation.[22]

$$k(T, \gamma, \dot{\gamma}) = k_y(T_o, \gamma_y, \dot{\gamma}_y) + \left(\frac{\partial k}{\partial T} \Delta T + \frac{\partial k}{\partial \gamma} \Delta \gamma + \frac{\partial k}{\partial \dot{\gamma}} \Delta \dot{\gamma} \right) \qquad (11.2)$$

In the above equation, (k) is the flow stress at any temperature T, strain γ and strain rate $\dot{\gamma}$ in the zone.

As elements flow towards the shear zone, material first "hits" the entry boundary of the discrete parallel-sided primary shear zone and yields. It yields at (k_y) the shear yield stress, the yield strain γ_y, the yield strain rate $\dot{\gamma}_y$, and the starting temperature T_0. At this entry side, the material is still at the yield strain γ_y and is not significantly heated, remaining at T_0. This situation is represented by the first term in equation 11.2.

The equation is divided up in such a way as to separate the effects of initial yielding and subsequent flow. Subsequent flow is represented by the next three terms in equation 11.2. In the primary shear zone, the material is further deformed after yielding. Very high values of shear strain typically $\gamma = 2 - 5$ occur. Thus intense strain hardening is expected together with the dislocation damping mechanism already discussed. At the same time the plastic work done on the material causes a considerable heat input. These factors offset each other to some extent but, as Backofen[23] has pointed out, softening is time dependent. Thus equation 11.2 should contain a further term accounting for

$$(\partial k)/(\partial t) = -K\exp(-Q/RT) \qquad \text{(11.3)}$$

where Q is the molar activation energy. The workmaterial velocity normal to the shear zone may be calculated using the simple hodograph for machining and using the shear zone widths presented in Wright and Robinson's[22] or Kececiouglu's[8] work. An element typically spends only 200 μs in the primary shear zone. Although Bailey and Bhanvadia[24] have successfully applied the usually constitutive equations for hot-working to the primary zone behavior when machining aluminum, the temperatures obtained for steel are usually very modest. *Figure 11.4* shows that even in Siekmann's[7] ultra-high speed tests, the calculated *mean* primary zone temperatures are well below those normally regarded to be representative of hot-working for steel.

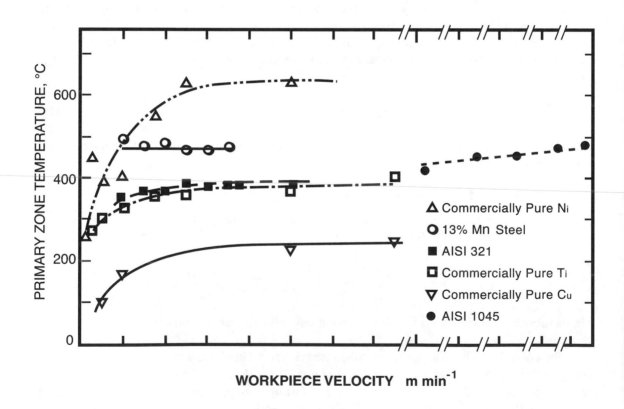

FIGURE 11.4   Temperatures calculated in the primary zone

Thus the indications are that material does not spend enough time in the deformation zone to be significantly softened. This probably accounts for Kececioglu's high and constant value of "dynamic shear stress" over all cutting speed ranges, up to and including the high speeds.

11.4.3 Response of the material passing along the secondary shear zone

In *Figure 11.3*, the secondary shear zone extends from point (O) at the tool edge to the end of the seized contact length. An element of material passing through this zone thus experiences many effects because the zone is at least 1-2 millimeters long depending on the work material. (By contrast the primary zone is only 0.1 - 0.2 millimeters wide). Thus in the secondary zone an element is reworked all the way along the interface. At the very cutting edge the element has just gone through the primary zone and then it turns to enter the front of the zone at (O) in *Figures 11.3* and *3.19*.

The details of the flow pattern in the secondary zone were described at length in Chapters 3 to 5. Note that the calculations indicate very high strain rates (e.g. $1 - 3 \times 10^4$ per second) and exceptionally high strains (e.g. 10 - 50). *Table 5.1* presents typical data. The temperatures in this area may be inferred from the many temperature profiles that are given in Chapters 5 and 9. When machining steels at 200 m min^{-1}, typical values are 600°C at the cutting edge, rising to 1,000°C further along the interface.

When machining stainless steels, alloy steels and aerospace alloys the temperatures are higher. However, recall from equation 11.3, there is a time effect and high temperatures may not mean softening. It is not possible to infer too much from the average stresses that can be calculated by dividing the measured shear force by the contact area. At the beginning of the contact region, near (O) stresses might be relatively lower than the average as described in *Figures 4.23* to *25*. However they rise quickly to an average value before reducing again over the rear fric tion region. As might be expected the average shear stresses increase with material strength and alloying.

The evidence from "quick-stop" specimens does indicate that the "difficult-to-machine" alloys create secondary shear zones with a great deal of strain hardening in the early part of the shear zone (say between O and *c'* in *Figure 5.5*). The rear of the shear zone may see some softening from warm working. However, the net result is that a very high shear stress is created when machining most alloys at high speeds.

The faster the cutting speed, the more the material in the early part of the secondary shear zone is influenced by "Region IV" dislocation damping and strain hardening. Furthermore, these effects are more pronounced in work materials such as stainless steel, alloy steels, titanium and nickel alloys. The overall result is that the chip's progress along the tool face is impeded by this growing shear force as cutting speed is raised.

Figure 11.5 shows a "quick stop" section that emphasizes the intense strain and seizure occurring just behind the cutting edge when machining "difficult to machine alloys" at comparatively high speeds. In summary this intense seizure at the rake face is equivalent to a very high friction stress, which grows even higher with increasing speed. With certain alloys it will be shown below that this is the "root cause" of the onset of the segmented chip form.

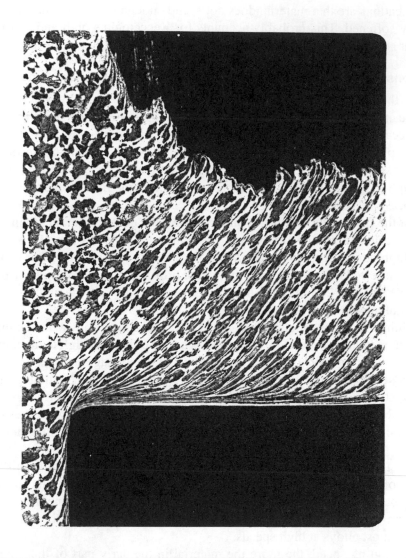

FIGURE 11.5 Intense strains and strain rates at the beginning of the secondary shear zone

11.5 INFLUENCE OF INCREASING SPEED ON CHIP FORMATION

A brief review is important at this point. In 1938, Ernst[25] classified chip types into a) discontinuous, b) over a built-up-edge, and c) continuous. For many years these were regarded as the "classical" possibilities. However, now the evidence shows that most materials enter a fourth possibility at high speeds; namely the serrated or segmented type of chip. The terms serrated, segmented and shear localized seem to be used synonomously in the literature. *Figure 11.6* shows the four chip types from the authors' laboratory work. These are for different work materials but were chosen from some of the "quick-stops" that best revealed the flow patterns.

a) Discontinuous

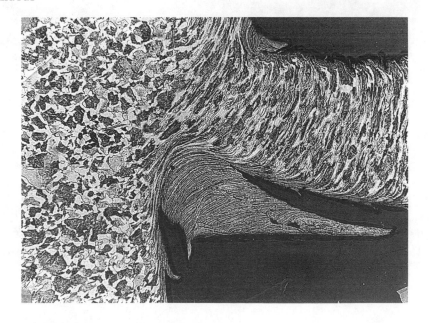

b) Over a built-up-edge

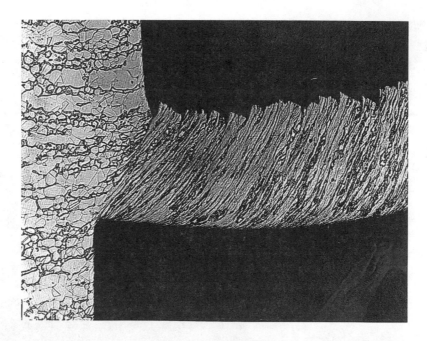

c) Continuous

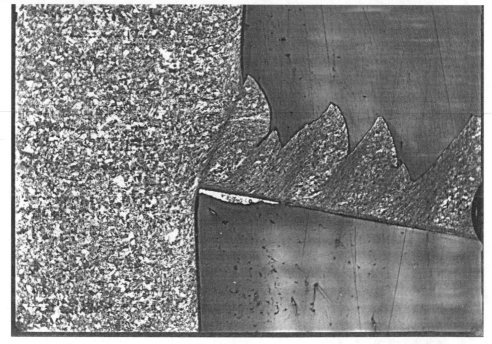

d) Segmented

FIGURE 11.6   Four types of chip: - discontinuous, over a built-up-edge, continuous, segmented

Ernst's [25] three classical descriptions are described first:

- Discontinuous chips form at very low speeds. Low temperatures cause the first material to strain harden and "stick" on the rake face. The incoming material is impeded by this stuck material and homogeneous strain occurs in large chunks of material ahead of the tool. There is often a "bulge" on the free surface. Eventually "something has to give" and a crack is initialed on a long shear plane. A discontinuous chip segment then pops free and "slips" off the tool. The "stick-slip" cycle then repeats.
- Built-up-edge effects form at modest speeds up to 30m min^{-1} for medium carbon steel. Temperatures are still low and a severely strain hardened "cap" remains at the cutting edge. A continuous chip flows over this cap (see *Figure 3.21*).
- Continuous chips form at higher speeds and hence higher temperatures. The temperatures eliminate the built-up-edge and a flow zone replaces it on the rake face. In the primary zone, "adiabatic shear lamellae" create a distinct shear zone, and chip flow is very smooth.
- The fourth type of chip may be termed segmented, serrated or shear-localized. The comprehensive research of Komanduri and colleagues has established much of the data in the two paragraphs below, on when particular materials experience this transition.[11-15]

There is a transition from a continuous chip to a segmented chip with increase in cutting speed for most materials. This type of chip persists with further increase in speed. No additional transitions or reversal to a continuous chip have been observed at least up to 30,488 m min^{-1} (or 100,000 ft./min). The deformation of the chip is inhomogeneous on a gross level. In a narrow band between the segments, deformation is very high. By contrast, inside the segments the deformation is relatively very low. Recht also observed that exceptionally narrow shear bands (as low as 0.002 mm) are separated by much larger regions of little strain. Physically, such segmental chips resemble a discontinuous chip.

The transition speed at which the chip form changes from a continuous to a segmented or shear-localized chip is found to be different for different work materials.[11-15] For example, it is only a few m min^{-1} or less in the case of titanium alloys, about 61 m min^{-1} in the case of nickel-base superalloys, and above 61 m min^{-1} and complete at about 244 m min^{-1} in the case of AISI 4340 steel (325 BHN). The speed at which catastrophic shear develops and the cutting speed at which individual segments are completely isolated are found to decrease with increase in the hardness of an AISI 4340 steel. Similar results are obtained with titanium alloys and nickel-base superalloys.

Sullivan Wright and Smith[26] showed that the transition speed from continuous to segmented for austenitic stainless steel is 35 m min^{-1}. They observed such chip types up to a high speed of 300 m min^{-1}, where tool life with carbide tools became impracticably low. These experiments are described in detail in the next main section.

As a final note in this overview section, it has been verified in all the research so far, that the segmented chip formation is not triggered by machine-tool vibration but is related to the *inherent metallurgical features* of the workmaterial for the machining conditions used. Of course, once these metallurgical instabilities are triggered, they drive regenerative chatter in the material, the toolholder and the machine tool.

Thus, any thing that can be done to reduce the intensity of the segmentation is always beneficial to the overall machining performance. And if such "difficult-to-machine" alloys can be processed on the stiffest available machine tools in a particular factory, then at least a challenging situation will be held under the best control.

11.6 STAINLESS STEEL

11.6.1 Overview

The detailed character of the segmental chip form is first discussed with reference to results from machining austenitic stainless steel.[26] The sequence of events in a typical cycle of segmentation formation has been established from a series of photomicrographs of "quick-stop" specimens. Associated dynamic cutting forces have been recorded.

When machining stainless steel, the essential feature of the chip formation is a varying shear strength of the work material at the chip/tool interface: a phenomenon that is similar to the "stick-slip conditions" described in Ernst's[25] work on the discontinuous chip formation. In a typical cycle, compressive stresses build up ahead of the tool as material sticks on to the rake face; selected shear then occurs on a primary shear plane of decreasing length as the chip moves away with greater velocity.

11.6.2 Photomicrographs of the chip forms

Figures 11.7 to *11.10* are representative of the various stages of the segmentation cycle. These photomicrographs cover the range of speed 50 - 150 m min^{-1}, but the mechanism of serration formation was the same for all conditions up to 300 m min^{-1}.

Figures 11.11 a-f summarize the descriptions given below, and above each schematic figure is the continuous chip formation.

Figures 11.7 and *11.11a* show that part of the cycle at which a shear instability has initiated along the long shear plane *AB*. The single slip band evident at position *B* resembles the form of the outer edge of the chip in *Figure 11.6c*, but it is clear that further shear does not proceed with a further slip band of equal length to *AB* as is the case with the continuous chip. Instead, the length of the shear plane reduces to give the length *CD* shown further along the chip in a previous cycle.

Strong evidence for this process is shown in *Figure 11.9*, where the quick-stop was obtained near the end of the selected shear part of the cycle. Shear at this stage has just occurred along a plane *CD*, and the length *AB* equivalent to *Figure 11.7* is also shown. The free edge of the chip, along regions such as *BD,* shows that the shortening of the shear plane is a gradual process, as summarized in *Figure 11.11a-c*.

A line drawn from the cutting edge to the free surface, shows that the instantaneous angle that the shear plane *AB* makes with the cutting direction is $\phi_{AB} \approx 18\,°$. By contrast, the instantaneous shear-plane angle of *CD* is $\phi_{CD} \approx 30\,°$.

These results show that the shear plane angle changes during the shear part of the cycle, as would be expected from the usual method of calculating ϕ. The shear-plane angle is given by the equations in Chapter 3. Dynamic variations in t_2 naturally lead to geometric variations in ϕ.

FIGURE 11.7 Chip formation showing shear initiating at 18° on the long shear plane AB

FIGURE 11.8 Chip formation showing shear continuing from the long shear plane AB

FIGURE 11.9 Chip formation showing shear finishing up at 30° on the short shear plane CD

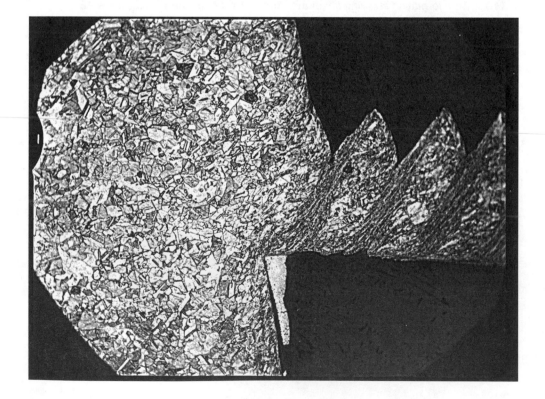

FIGURE 11.10 Chip formation showing shear temporarily arrested and a compressive bulge

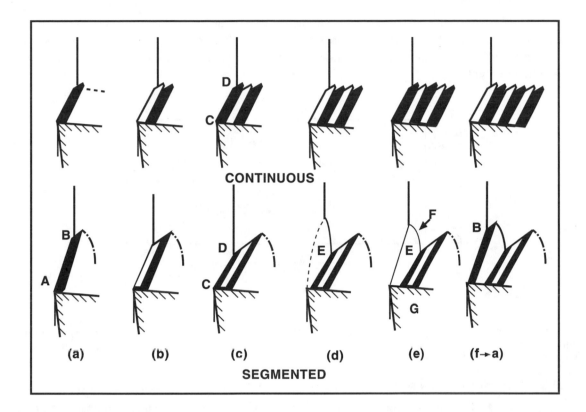

FIGURE 11.11 Schematic diagram of continuous chip formation occurring below 35 m min^{-1} and serrated chip formation occurring above 35 m min^{-1}. Lengths CD in the two chip forms are approximately equal. Also, ϕ_{cd} approx = 30°, whereas ϕ_{ab} approx = 18°.

The variation in shear-plane angle has been calculated - both using the equation in chapter 3, and using the mean minimum and mean maximum chip thickness measured from the serration dimensions in the photomicographs. These values are shown in *Figure 11.12* and are incorporated with the results from the measurement of in ϕ the continuous chip range up to 35 m min^{-1}. The latter have been obtained using the standard weighing and measuring technique that determines chip thickness and shear-plane angle. The results are of particular interest because it is the "high shear angle-thin chip-lower energy" situation that does not vary markedly with increasing speed. Speaking colloquially, it is as if "this is where the chip formation prefers to operate".

By contrast, it is the situation which is not initially favored at low speeds but which is triggered at 35 m min^{-1}. The observation that the shear plane is intermittently *longer* (and hence ϕ is smaller) at speeds above 35 m min^{-1} is a significant result. *Figure 11.12* is thus very relevant to the mechanism of segmented chip formation. It should be noted that *Figure 11.11* has been drawn to comply with this result, with the length *CD* of the segmented chip being of equal length to the shear plane in the continuous chip.

Once the selected shear part of the cycle has reached the shorter plane of length *CD*, the *shear process is arrested temporarily*. The evidence indicates that compressive stresses build up ahead of the sheared region (*ABDC, Figure 11.10*) to give a region of uniform low strain at *E* in

Figure 11.11. There is also prepile-up of the uncut material at F and the continuation of the secondary shear flow lines at *G*. The temporary arresting of shear and the compressive stresses are finally relieved by shear on the long plane *AB* again, so that *Figures 11.11e* and *f* mark the beginning of a new cycle.

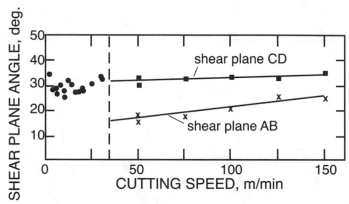

FIGURE 11.12 Shear plane angle measurements showing the "high shear angle-thin chip-lower energy" situation before the "threshold speed" and the "low shear angle-thick chip-higher energy" making its segmented contribution above 35 m min⁻¹.

11.6.3 Conditions in the secondary shear zone when machining stainless steel

In an analysis of the segmented chip formation in austenitic stainless steel, the observations to be explained are:

(i) the intermittent lengthening of the primary shear plane at speeds above 35 m min⁻¹ which continued up to the highest speed that was practically possible. This was 300 m min⁻¹ and was a short duration test with a carbide tool.

(ii) the fact that shear does not continue on this plane, so that *AB* reduces to *CD* in *Figure 11.7* through *11.10*. This process precedes the temporary arresting of shear and the build-up of compressive stresses ahead of the tool.

The above observations indicate that for the shear plane to become intermittently lower at higher cutting speeds, a sticking phenomenon of a periodic nature must be re-introduced in the secondary shear region. It is proposed, therefore, that a type of stick-slip mechanism develops at the rake face at high cutting speeds. This means that during the "arrested shear/compression" part of the cycle, the chip speed is significantly lower than the velocity during the "selected slip" part of the cycle.

The primary zone was discounted as the source of the large-scale heterogeneity because there appeared to be no reason for a sudden and intermittent change in the high strain-rate, warm-working conditions that were established in the continuous chip range up to 35 m min⁻¹, at which speed the temperature in the primary zone was calculated (using the equations in Chapter 5) to be 520°C. The possibility of a large-scale adiabatic shear instability emanating from the small-scale ones already occurring in the continuous chip type was eliminated by the sudden increase in length of the shear band at 35 m min⁻¹. Futhermore, since the chips remained non-magnetic, the possibility of a stress-induced transformation affecting the austenitic material in the primary zone was discounted. Indeed, a stress-induced transformation would be more likely in the high strain/low temperature conditions that would occur at speeds much lower than 35 m min^{-1}.

11.6.4 Proposed mechanism for segmented chips in stainless steel

In the secondary shear region the material properties vary dramatically along the contact length. Material is severely strain hardened in the first part of the contact length but, owing to the heat generated, temperatures at the very end of the zone can be in excess of 1000°C.

As already discussed, above a strain rate of $\dot{\gamma} \approx 10^4 s^{-1}$ the shear stress is found to increase significantly with strain rate because of the dislocation damping mechanisms which arise; this is usually termed "Region IV" behavior.[18-21] The approximate secondary shear zone strain rates have been calculated from the chip speed and zone thickness to give:

$$\dot{\gamma}_s = 1.5 \text{ x } 10^4 \text{ s}^{-1} \text{ at 30 m min}^{-1} \text{ and } 3.4 \text{ x } 10^4 \text{ s}^{-1} \text{ at 50 m min}^{-1}.$$

These values correspond to the beginning and then well into "Region IV" behavior for mild steel. The authors are not aware of any mechanical test data for austenitic stainless steel at such high strain rates; but it seems reasonable to assume dislocation damping will be occurring in a material stronger than mild steel and that the above values are in "Region IV" behavior for stainless steel.

11.6.4.1 Initiating a "stick" at the threshold speed because "Region IV" dislocation damping begins

As cutting speed is increased, the value of k_s jumps substantially, at some critical threshold speed, as this strain-rate effect "kicks-in". In the schematic cycle shown in *Figure 11.11*, the secondary-zone strain rate is highest when the chip speed is greatest (*Figure 11.11c*). A consequently high value of k_s "triggers" the "sticking" part of the cycle (*Figure 11.11d*).

11.6.4.2 Resuming the "slip" motion after the chip slows down, the strain rate is lessened, and there is some heat softening in the secondary shear zone

From the above, the material is both strain hardened and strain rate hardened in the initial part of the contact length, and the associated high value of k_s "triggers" the sticking part of the cycle. However, this hardened material is then softened over a short period of time by the heat retained in the tool and the bulk of the chip. Also, with the slowing down of the chip (*Figure 11.11e*), $\dot{\gamma}_s$ and k_s reduce again to allow the primary zone selected shear to restart at *AB*. The reduction in k_s allows the chip to "slip" away more easily, thus allowing the primary shear to proceed on a plane of decreasing length (i.e. increasing shear plane angle).

11.6.5 Summary of all the above effects

Figure 11.13 is a schematic of the Campbell *et al* and the Rosenfield and Hahn data on high strain rate properties.[18-21] Over "Region II" the variation of shear strength with strain rate is only modest. However, at "Region IV", dislocation damping triggers a rapid rise in shear stress with strain rate. *Figure 11.2* emphasizes that the primary shear zone of the metal cutting operation is certainly in "Region IV". The essential steps are:

Step 1: At some "threshold speed" the material at the front part of the secondary shear zone also enters "Region IV" behavior. This triggers more sticking friction on the rake face as k_s suddenly jumps to position (*ST*) in *Figure 11.13*.

Step 2: Chip flow is thus temporarily arrested and a "pile up" or "bulge" forms on the free surface.

Step 3: However, almost instantaneously, because of local heating and since the chip has slowed down, the conditions are about to return to slipping friction (*SL*) in *Figure 11.13*.

Step 4: When something "gives" in the system, the accumulated energy during pile-up means that the "shear plane has now been pushed down" to only 18° for stainless steel.

Step 5: The shear plane "can then return to where it wanted to be", quickly change in value, and resume at 30° with much less energy.

Step 6: However, this situation is only temporary: the increased chip speed takes the secondary shear zone material back to (*ST*) in *Figure 11.13* and the cycle keeps repeating on this basis.

This strain-rate dependent mechanism explains why the continuous chips obtained when machining medium-carbon steel at conventional speeds, also become segmented at ultra-high speeds. In fact, all materials are expected to enter "Region IV" at some deformation speed, implying that all materials will eventually machine with segmented chips if the speed is raised sufficiently high.

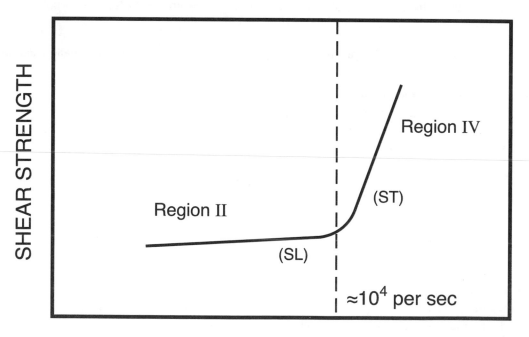

SHEAR STRAIN RATE

FIGURE 11.13 Schematic of strain rate data in "Region II" and "Region IV".[18-21] At the threshold cutting speed, the secondary shear zone material enters "Region IV" at (ST). The higher material shear strength causes "sticking" on the rake face. This impedes chip flow and causes the shear plane angle to segment to 18°. An instant later, the secondary shear zone material is weakened by thermal effects and the lower strain rate. It returns to (SL) and the chip accelerates on the rake face, allowing shear on a shorter plane at 30°.

11.7 AISI 4340

11.7.1 Overview

From the 1980's U.S. Air Force program, Komanduri and colleagues point out that segmented chips are more likely to form in the machining of materials with a limited slip system (e.g. hcp crystalline structure), poor thermal properties, and high hardness, such as alloy steels, titanium alloys, and nickel-based super-alloys. In contrast, continuous chips more are likely to form in the machining of materials with extensive slip system (e.g. fcc/bcc crystalline structure), good thermal properties, and low hardness, such as conventional aluminum alloys (e.g. A1 6061-T6) and low carbon steels (e.g. AISI 1018 steel).

11.7.2 Conditions in the primary zone when machining AISI 4340 steel

More recently, Hou and Komanduri[27] modeled the segmented chip forms that they observed from metallurgical sections to be triggered at 61 m min^{-1} in hardened AISI 4340 steel. They base their analysis on thermomechanical shear instabilities in the primary shear zone. Hou and Komanduri's model for how the segments are triggered at a threshold speed is therefore somewhat different from the Sullivan, Wright and Smith model more focussed on the "stick-slip" effects in the secondary zone. The two models are compared and contrasted in the last section 11.9 of this chapter along with the ideas of other investigators.

Hou and Komanduri note that there are several localized heat sources during chip formation. And when they are combined, they serve to *preheat* a large area ahead of the tool. Therefore, Hou and Komanduri first calculated this temperature in the whole of this "primary shear zone preheated area". Second, they calculated the temperature in a "specific shear band between the segments". They also calculated the strain in each area. They then applied the following equation to the two data sets in the two regions.

$$\sigma = (432.6572 - 0.3533T)\varepsilon^{(0.1213 + 6.4435\times10^{-5}T)} \tag{11.4}$$

This equation is a best fit from data in the *1954 ASM Metals Handbook*.[28] Note that it does not, unfortunately contain a strain rate term. Nevertheless, it allows an interesting strategy to be applied. First, a value of σ is calculated for the *bulk* of the preheated material in the overall general region of the primary shear zone. Second, a value of σ' is calculated for the strength in a discrete localized shear band.

Figure 11.14 shows Hou and Komanduri's [27] results for the variation of shear stress in the shear band, σ', at the shear band temperature and shear strength of the bulk material, σ, at the preheating temperature, with increasing cutting speed. The propositions from the theoretical analysis are as follows:

- When σ' is greater than or equal to σ, no shear localization takes place, since strain hardening predominates in the shear band. This is the case up to a predicted cutting speed of 52 m min^{-1}.
- Above this speed, thermal softening in the shear band predominates over strain hardening, i.e. $\sigma' < \sigma$. Hence, shear localization is imminent and segments soon form. This is the case for predicted values of cutting speed higher than 52 m min^{-1}.

- The intersection of these two curves gives the cutting speed for the onset of shear localization.

Experimental results of Komanduri *et al.*[11-14] showed the onset of shear instability in AISI 4340 at 60 m min^{-1}, being complete at 122 m min^{-1} - this compares reasonably well with the theoretical result that segmentation will be triggered at 52 m min^{-1}. Also, in *Figure 11.14*, the curves continue monotonically up to high speeds. In Hou and Komanduri's [27] analysis, this means that once the transition from a continuous to a shear-localized chip has occurred, no further change in chip type is then predicted for any increases of speed up to 1000 m min^{-1}. In this regard, the analytical results well support the experimental observations.

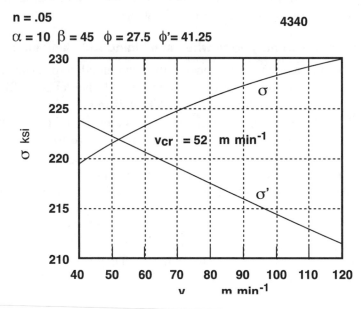

FIGURE 11.14 Hou and Komanduri's [27] analysis of onset of segmented chips 52 m min^{-1} when the strength of an individual shear band is less than the bulk

11.8 AEROSPACE ALUMINUM AND TITANIUM

11.8.1 Overview

Sandstrom and Hodowany[29] have used the finite element analysis (FEA) computer modeling techniques described in the next chapter to demonstrate the segmented chips in the high speed machining of 7050-T7451 aluminum and Ti-6A1-4V mill-annealed ELI titanium. In both cases their modeling speed was $3,060$ m min^{-1} ($10,000$ ft./min).

Segmented chips were predicted by their modeling work and were observed in the metallurgical chip sections from high speed "gas-gun" machining - an experiment similar to Arndt's[9] and Recht's.[6]

The "Mach2d" FEA code of *Third Waves Systems, Inc.*, was used. This is a Lagrangian code that employs dynamic effects, coupled transient heat transfer analysis, and adaptive FEA mesh-refinement and re-meshing. The code uses a conventional power strain-hardening law (similar to the equation in chapter 5) and a power series expansion in temperature for thermal softening. Very importantly, the code also includes a *power viscosity law* for strain rate sensitivity. Note that this viscosity term for strain rate has already been introduced in equation 11.1 as a fundamental material constant at high strain rates in "Region IV".

The importance of including the strain rate effects cannot be emphasized enough. For the aluminum alloy, Sandstrom and Hodowany did not obtain satisfactory agreement with experiment when "standard handbook" data were used. Disagreement was observed both in cutting force trends *vs.* cutting speed, and in observed chip morphology. However, these issues were resolved when they employed experimental shock physics techniques to derive material properties at high strain rates and elevated temperatures.

11.8.2   Chip morphology of 7050-T7451 aluminum and Ti-6A1-4V titanium

11.8.2.1   7050-T7451 aluminum

The computed effective plastic strain field for high-speed machining of the aluminum alloy is shown in *Figure 11.15*. It compares well with an image of an actual machining chip obtained under corresponding conditions. Once Sandstrom and Hodowany used the high strain rate data for the aluminum alloy, good agreement was observed both in absolute magnitude of cutting forces, and in cutting force trends *vs.* cutting speed.

From the *primary zone*, the average temperature rise in the bulk of the chip was found to be about 105°C. Similar temperature fluctuations have been experimentally measured during Kolsky bar tests of similar materials.[30] Individual values between 120 and 190°C were isolated in the shear localized thinner bands between the segments.

By contrast, in the *secondary shear zone*, the temperature and effective plastic strain rose dramatically. At the rear of secondary shear, temperatures approached 660°C, approximately to the melting point, and effective plastic strain exceeded 2,000 percent. The gradient of strain across the chip is greater than that of temperature because there has been time for heat to diffuse across the chip and into the tool. In contrast, shear is effectively localized in the secondary shear zone, resulting in a sharp strain profile.

11.8.2.2   Ti-6A1-4V mill-annealed ELI titanium

A similar result for titanium is shown in *Figure 11.16*. Frame C-1 shows the configuration before cutting begins. In frame C-2, a shear band is well developed and is being forced up the tool rake face; the tangential cutting force measured by Sandstrom and Hodowany was on its way to its peak value. At frame C-3, the material has just begun a new shear instability in the primary shear zone, relieving stress there; tangential force fell to nearly its minimum value. In frame C-4 the shear plane of C-3 has become localized and has begun to be forced up along the rake face; tangential force was again rising.

At frame C-5 shear is arrested but just about to begin again in the primary shear zone. The system has not yet relieved stresses there - thus tangential force was still at its maximum and was about to reduce as this shear plane developed. Note that, at an instant after C5, this will be a "long shear plane of low angle" similar to the Sullivan, Wright and Smith model. Once this shear band has been initiated, the shear plane gets shorter and the angle is greater. It is as if the system "prefers" this higher-shear-angle solution but cannot remain there because of the stick-slip phenomenon.

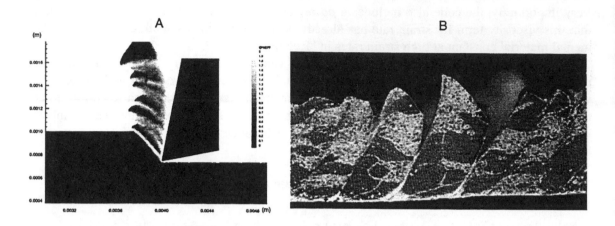

FIGURE 11.15 (A) Computed effective plastic strain field for 7050-T7451 aluminum alloy. Cutting speed: 3,060 m min^{-1} (10,000 SFM); carbide cutter-rake and relief angles: 11 degrees; uncut chip thickness: 254 μm (0.010 inch): orthogonal machining geometry. (B) Metallographic image of machining chip. Machining conditions same as A

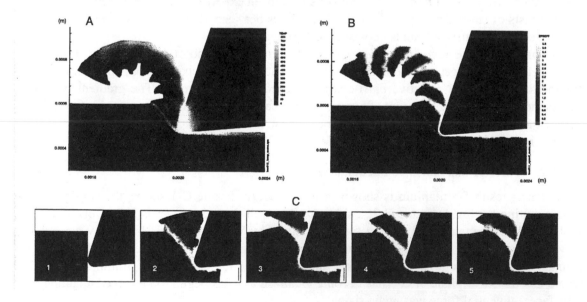

FIGURE 11.16 Modeling results for machining of Ti-6Al-4V mill-annealed ELI. Cutting speed; 10.16 m-sec^{-1} (2,000 SFM); undeformed chip thickness; 127 μm (0.005 inch). Cutting tool rake angle: 15 degrees; relief angle: 5 degrees; tip radius: 25.4 μm (0.001 inch). Orthogonal cutting geometry. (A) Field plot of temperature at end point of model run, corresponding to a cutting time 61.8 μsec. Temperature scale in degrees Celsius. (B) Field plot of effective plastic strain-same time as A. (C) Field plots of effective plastic strain at successive cutting times: (1) 0.0 μsec; (2) 3.6 μsec; (3) 7.7 μsec; (4) 11.6 μsec; and (5) 15.3 μsec

11.9 CONCLUSIONS AND RECOMMENDATIONS

11.9.1 Discussion of segmented chip forms in high speed machining

The research community is still at a relatively early stage of development in explaining all the issues associated with the segmented chips On the one hand, the overall cycle of events does seem to be well established today, and the general summary in *Figures 11.11* and *11.16* appears to be the agreed upon sequence by all investigators. On the other hand, the precise physical event that "triggers" the onset of segmented chips at a critical "threshold speed" is not as certain, and the authors remain "open-minded" on this issue.

Hou and Komanduri have developed the "thermomechanical primary shear zone instability model" for the AISI 4340 steel. This is schematically summarized in *Figure 11.14*. They compare the strength σ' of the material in one of the selected-shear bands with the bulk strength σ of a larger area of material in the general vicinity of the primary shear zone. They propose that when σ' is less than σ, segmentation will take over from continuous chips. This model is perhaps similar also to the general ideas proposed by Semiatin and Rao.[31] For completeness it should also be mentioned that Walker and Shaw[32] and Komanduri and Brown[33] have considered that inherent material weakness such as microcracks in the primary zone might catastrophically reduce the value of σ'.

By contrast, our own experiments on austenitic stainless steel have led to the "strain rate in secondary shear driven model". This is schematically summarized in *Figure 11.13*. At the "threshold speed", "Region IV" dislocation damping[18-21] begins to occur at (ST) in the front part of the secondary shear zone. When this material "sticks", the shear plane is pushed forward to 30°. But instantaneously, the heat softens this area the material drops in strength to (SL) and the chips "slip" for an instant until the cycle repeats. Sandstrom and Hodowany's correlations between chip motion and cutting forces in *Figure 11.16* seem to support this explanation.

Several observations thus lean the authors in favor of the "strain rate in secondary shear driven model". First, Sandstrom and Hodowany did not get good agreement between experiment and modeling until they used the high strain rate data for aluminum. Material property information at high strain rates is key to the analysis of segmented chips and it is unfortunate that Hou and Komanduri hinge their analysis on 1954 data for AISI 4340 that includes no strain rate term.

Second, adiabatic shear[6,34] is already taking place in the primary zone in the continuous chip cutting speed range. Black[35] provides a detailed explanation of the events, based on earlier, well-known machining experiments he carried out inside a scanning electron microscope. He emphasizes that within the primary zone itself, several adiabatic shear lamellae will operate more or less concurrently. And because they operate adiabatically the temperatures in them will be locally higher than the surroundings. Consequently, lower stresses, σ', will always be operating in such lamellae even in the continuous chip range. In fact, this seems to be fundamental to continuous chip formation in the first place. Given that these adiabatic shear lamellae are already operating in the continuous chips, there does not seem, to the authors, to be any reason for one of them to suddenly become unstable and trigger segmentation. Besides which, when segmented chips are triggered the instantaneous shear plane angle gets suddenly much longer (AB in *Figure 11.7*, and as seen in *Figure 11.12* for the shear plane angle results).

All this points to a "stick-slip" mechanism on the rake face as the "root-cause" of segmentation. Another analysis that concludes with this proposal is that of Manyindo and Oxley.[36] Their 1986 paper also provides a detailed analysis of movie films and photomicrographs. Slip line fields for the various steps in segmentation are also given. They also conclude that the segmentation occurs because of a "connection between the catastrophic shear part of the cycle and the conditions at the chip tool interface".[36]

Figure 11.17 is a schematic diagram which attempts to summarize the different chip types. It considers, as ordinate y, the *strain in the primary shear zone* versus, as abscissa x, increasing cutting speed and hence *increasing strain rate in the secondary shear zone*.

The diagram has then been arbitrarily divided into four ranges, similar to Ernst's[25] classifications, though not all materials exhibit all four ranges:

Speed Range I. At low speeds there is no work-softening of material on the rake face; gross seizure occurs. Material then undergoes pre-pile-up ahead of the tool. When an instability is initiated in the primary zone, the fracture strain of the material is less than the imposed geometric strain so that discontinuous chips are formed.

Speed Range II. Speeds are still relatively low, and cold-welding occurs on the rake face to give a built-up-edge. The stress gradients are such that the chip and built-up-edge are separate regions. An effective increase in the rake angle reduces the primary zone strain so that the chip is continuous.

Speed Range III. As speed is increased, the elimination of the cold-welding conditions at the rake face allows continuous chips to form, and the primary zone strain is significantly less than the fracture strain. But there is still sticking friction at the rake face, and the value of k_s determines ϕ (Chapter 3).

Speed Range IV. The effect of secondary-zone strain-rate becomes important and can reintroduce a stick-slip mechanism at the rake face. This causes the primary zone strain to vary and lead to segmented chips.

The response of different materials to these changes in strain, strain-rate and temperature accounts for the many types of chip that can be formed. Discontinuous chips will always form in inherently brittle materials such as cast iron. Highly ductile materials will always machine with continuous chips up to very high speeds. The chip formation in alloy steels, stainless steels, titanium alloys and other aerospace materials can also fall into Speed Ranges I, II or III, but in addition a stick-slip mechanism at high speeds can bring them into Speed Range IV.

11.9.2 Tool forces

Especially when machining aerospace alloys that have very thin wall-sections, the hope is that the forces between tool and work will be lessened at high speed - chiefly because of the reduced strength of the workmaterial resulting from the elevated temperatures that occur. If such forces can be reduced, then the thin walls will be less deflected

The conclusions with aluminum indicate that the *average* tool forces against the part being machined, do reduce by about 10 -15% as the speed is increased to high values. *Figures 4.12* and *4.13* are typical of most materials. Above 75 m min^{-1}, built-up-edge effects have mostly disappeared with steel and most materials, and then there is a gradual decrease in force. Actually, the largest fall in forces occurs in the low-to-medium speed range from about 30 m min^{-1} to 200 m min^{-1}. For all materials it seems, over the range from about 200 m min^{-1} to 3,000 m min^{-1}, the force curves "flatten out" and decrease by only 10 - 15%.

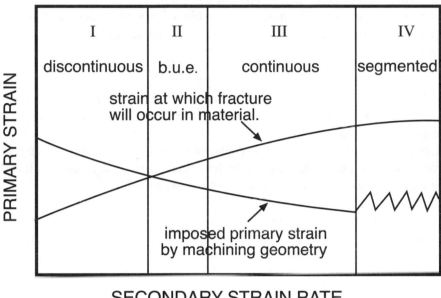

SECONDARY STRAIN RATE

FIGURE 11.17 Schematic variation of primary shear strain with increasing speed and hence secondary shear zone strain rate

Realistically, this small decrease in the *mean forces* is a minor advantage; it is vastly out-weighed by the tool wear difficulties that arise from the imposed *force fluctuations* that arise from the segmental chips.

One clear conclusion from the experiments thus far, is that there is no "*magical*" sudden drop in the forces at some very high speed. This possibility {reminiscent of an alchemist trying to turn "lead into gold"} was a hope of ultra-high speed machining in the early years and Arndt[8] provides a very careful and reasonable analysis of how this might occur. So far though, the experiments have yet to find the combination of events that can make it happen!

11.9.3 Tool wear

From a commercial viewpoint an understanding of such stick-slip conditions is important in tool wear. Albrecht was the first to show that the segmented chip type is associated with fatigue cracks on the flank face of worn tools, and marked attrition wear.[16] The fluctuating stresses at the cutting edge clearly encourage such wear. Thus, it is expected that if the segmented chip forms can be at least minimized in their intensity tool wear will be reduced. A dampening of the machine-tool vibrations associated with the segmented chip forms will also be advantageous.

The major conclusion on tool wear and life is that, high speed cutting will be feasible only if the cutting edge can sustain the applied normal stress. *Figure 11.18* is extremely important in this respect. It shows the hot compressive strength of a number of tool materials against temperature. The data has been taken from earlier chapters and clearly shows the superiority of TiC containing carbides over WC and in turn over high-speed steel. Of interest is that the slopes of these lines are extremely steep indicating the reason why cutting edges fail catastrophically once some limiting value of speed and hence temperature is reached.

Superimposed on these curves is a series of broken lines for workmaterials used in the authors' laboratories. These lines are the *mean* values of normal stress, calculated from the following equation:

$$\sigma_N = (F_c \cos\alpha - F_t \sin\alpha)/(Lw) \qquad (11.5)$$

In this speed range, the normal stress, σ_N, on the tool edge did not vary a great deal for a particular workmaterial thus the broken lines may be regarded as the steady-state *mean* normal stress level at medium to high speed.

Of particular interest is that these stress levels rank the workmaterials directly in terms of machinability. In the tests a high chromium fully hardened steel ($65Rc$) was the most difficult to machine with rapid tool wear by edge collapse and an austenitic 13% manganese steel was the next most difficult to machine.

The temperatures calculated in *Figure 11.4* do give some indication of the collapse speed. At one extreme, the primary zone temperatures for copper are so low that it is unlikely that the cross-over point in *Figure 11.18* can be reached. Indeed copper may be machined at exceedingly high rates even with steel tools. By contrast the calculated primary zone and hence cutting edge temperatures for both nickel and manganese steel are very high (*Figure 11.4*). So too are the normal stresses (*Figure 11.18*). These combined factors explain the relatively low speed collapse of tools used to cut these materials.

Finally it is re-emphasized that the occurrence of catastrophic, very thin shear zones in the primary region will not lead to a large and significant reduction in the stress, σ_N normal to the tool. This is because such intense adiabatic shear zones are always separated by relatively undeformed chip segments and, during the production of these, normal stresses on the delicate cutting edge will be high.

11.9.3.1   Economics versus chip morphology

As an overall conclusion, the economic benefits from high speed machining will be demonstrated in situations where new or modified part designs mean that large monolithic structures need to be fabricated by metal cutting (*Figure 11.1*). As with the facing of disc brakes and clutches with ceramic tools (in Chapter 8), the basic goal is to remove material quickly and achieve high part-throughput.[1-3, 37-39]

Cast iron machines with a discontinuous chip and tool stresses are low. Once the correct grade of tool has been chosen and a machine is carefully set up, high speed facing of cast iron is relatively easy. However, all the materials reviewed here in Chapter 11 develop a segmental chip at a rather low "threshold speed". These chip types are not so easy to deal with - they inherently create vibrations in the machining system and they cause attrition wear on the cutting tool.

Given that the high sticking effects at the rake face are believed to be the "root cause" of the segmented chip, the following are the obvious solutions to try in the factory for improve machinability. The common theme is to reduce sticking and hence prevent the strain rate in the secondary shear zone from entering "Region IV"[18-21] for as long as possible.

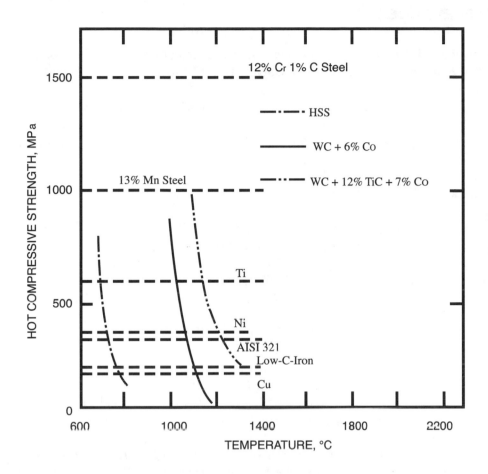

FIGURE 11.18 Normal stresses on the tool and hot compressive strength of tool materials

1. Application of low friction coatings to the tool. For most work materials, TiN is probably the best choice. When machining titanium, Chapter 9 indicated, based on Kramer's work[40] that other inert layers might be preferable.

2. Use of reduced contact length and/or chip-breaker type grooved tools that will reduce sticking on the tool.

3. Use of the nitrogen jets, as described at the end of Chapter 10, to reduce the chip-tool contact length. These are directed into the rear of the contact length and serve to curl the chip away more easily.

4. Where feasible, the use of "free cutting" additions in materials like stainless steel. Typically the "free cutting" additions are drawn out in the secondary shear zone and minimize the sticking. For other steels, even AISI 4340, new calcium deoxidation practices may also be fruitful areas to pursue. It was shown in Chapter 9 that, in the right circumstances, these provide inert silicate layers between the chip and the tool. These reduce sticking and hence the secondary zone strain rate.

11.10 REFERENCES

1. Thomas, R.J., *"What Machines Can't Do"*, See in particular Chapter 7, The Politics and Aesthetics of Manufacturing, pp. 246-258, University of California Press, Berkeley, Los Angeles, London, (1994)
2. Ayres, R.U. and Miller, S.M., *Robotics: Applications and Social Implications,* Ballinger Press, Cambridge MA, (1983)
3. The Economist Magazine, *Hubris at Airbus, Boeing Rebuilds*, **349**, (8096), 64 November 28th (1998)
4. Trent, E.M., *J.I.S.I.,* **1**, 401 (1941)
5. Cottrell, A.H., *Conf. on Props. of Materials at High Rates of Strain,* p.3, Inst. Mech. Eng., London (1957)
6. Recht, R. F., *Journal of Appl. Mech. Trans ASME*, **31**, 189 (1964)
7. Siekmann, H. J., *A.S.T.E.*, **58**, Paper Number 82 (1958)
8. Kececioglu, D., *Trans. ASME*, **80**, 158 (1958)
9. Arndt, G., *Proc. Inst. Mech. Engrs.,* **187**, 625 (1973)
10. von Turkovich, B. F., *Proceedings of the VIIth NAMRC Conference*, (University of Michigan) 241 (1979)
11. Komanduri, R. and von Turkovich, B. F., *Wear*, **69**, 179 (1981)
12. Komanduri, R., *Wear*, **76**, 15 (1982)
13. Komanduri, R. and Schroeder, T.A., *Transactions of American Society of Mechanical Engineers, Journal of Engineering for Industry*, **108**, 93 (1986)
14. Komanduri, R., Schroeder, T.A., Hazra, J., von Turkovich, B.F. and Flom, D.G., *Transactions of American Society of Mechanical Engineers, Journal of Engineering for Industry*, **104**, 121 (1982)
15. Flom, D.G., Komanduri, R. and Lee, M., *Annual Review of Material Science*, **14**, 231 (1984)
16. Albrecht, P., *Transactions of American Society of Mechanical Engineers, Journal of Engineering for Industry*, **84**, 405 (1962)
17. Stevenson, M.G., and Oxley, P. L., *Proc., Inst. Mech. Eng., London,* **185** 741 (1970)
18. J. D. Campbell and W. G. Ferguson, *Philosophical Magazine*, **21**, (169), 63 (1970)
19. Campbell, J.D., *Mater. Sci. Eng.,* **12,** 3 (1973)
20. Duffy, J., Campbell, J.D., and Hawley, R.H., *J. Appl. Mech.,* **38**, 83 (1971)
21. Rosenfield, A. L., and Hahn, G.T., *Trans. Am. Soc. Met.,* **69**, 962 (1966)
22. Wright, P.K and Robinson, J.L., *Journal of Metals Technology*, **4**, 240 (1977)
23. Backofen, W. A., *Deformation Processing*, Reading, Mass., Addison-Wesley pg. 271 (1972)
24. Bailey, J. A. and Bhanvadia, D. G., *Journal of Eng. Mat. and Tech.*, **95**, 94 (1973)
25. Ernst, H., The *Machining of Metals,* The American Society of Metals, Cleveland, Ohio, 1-34 (1938)
26. Sullivan. K.F., Wright, P.K. and Smith. P.D., *Journal of Metals Technology*, **5**, (6), 181 (1978)
27. Hou, Z.B., and Komanduri, R., *Int. J. Mech. Sci.*, **39**, (11), 1273 (1997)
28. Hoyt, S. L., *ASME Handbook, ed.*, New York, (1954) (6)

29. Sandstrom D.R, and Hodowany, J.N., *Proceedings of the CIRP International Workshop on Modeling of Machining Operations,* held in Atlanta, GA., Published by the University of Kentucky Lexington, KY, 40506-0108. p.217, (1998)

30. Hodowany, J. N., *On the conversation of plastic work into heat*, Ph.D Thesis, California Institute of Technology, Pasadena, CA (1997).

31. Semiatin, S.L., and Rao, S.B., *Journal of Materials Science and Engineering,* 61, 185 (1983)

32. Walker, T.J. and Shaw, M. C., *Proc. 10th M.T.D.R. Conference,* U.K., 241 (1969)

33. Komanduri, R. and Brown, R.H., *Transactions of American Society of Mechanical Engineers Journal of Engineering for Industry,* **103** (1), 33 (1981).

34. Rogers, H. C., *Ann. Rev. Mater. Sci.*, **9**, 283 (1979)

35. Black, J.T., *Journal of Engineering for Industry,* **101,** 403 (1979)

36. Manyindo, B.M., and Oxley, *P.L.B., Proc. Inst. Mech. Engrs.*, **200**, (C5), 349 (1986)

37. Tlusty, J., *Annals of the CIRP*, **42**, (2), 733 (1993)

38. Altan, T., Fallbohmer, P., Rodriguez, C.A., and Ozel, T., *High Speed Cutting of Cast Iron and Steel Alloys - State of Research* in the Conference on High-Performance Tools, Dusseldorf, p. 309, (1998)

39. Hock, S., *High Speed Cutting in Die and Mold Manufacturing - Practical Experience* in the Conference on High-Performance Tools, Dusseldorf, p. 333, (1998)

40. Kramer, B.M., Viens, D., and Chin, S., *Ann. of the CIRP*, **42**, (1) 111 (1993)

CHAPTER 12   MODELING OF METAL CUTTING

12.1 INTRODUCTION TO MODELING

Modeling methods are now discussed in five generic categories:
- Empirical modeling typified by Taylor's equation[1]
- Closed-form analytical modeling typified by Merchant's shear plane solution[2]
- Mechanistic modeling typified by DeVor et al.'s analysis of forces vs. chip thickness[3]
- FEA modeling typified by Sandstrom's high speed machining study (Chapter 11)[4]
- Artificial intelligence and other modeling methods that combine many of the above[5]

The goals of any kind of model are to predict physical behavior from known *a priori* conditions. Essentially, {known inputs + an accurate model = desired outputs}.

Colloquially speaking, it is desirable to predict the weather tomorrow, starting from today's known inputs of temperature, humidity, etc., and a model of how inputs are related to outputs. The metal cutting practitioner would like to know the tool life tomorrow, starting from today's input of the work material being purchased; the cutting inserts available; the features that have to be machined in the new "part-drawings" that have just arrived from the CAD/CAM sub-contractor; and how quickly the original client needs the part. The metal cutting theorist and model-developer would like to help with this question but, along the way, might also like to predict shear plane angle, cutting forces, and temperatures, as well as estimating the likely tool life at any given speed.

All the chapters in the book provide experimental data on why "machining forecasting" is about as reliable as "weather forecasting". Nevertheless, from an economic viewpoint, both are highly valuable to society. Thus, despite the approximations, frustrations and the all-too-frequent erroneous predictions, this work must carry on with the hope that ideas will build upon each other and create better models for the practitioner.

The key parameters that the day-to-day practitioner finds valuable are:
1. prediction of tool life
2. prediction of the accuracy of the component being machined
3. prediction of the surface finish on the component being machined
4. prediction of chip control
5. prediction of the loads on the tool, and/or workpiece, and/or fixtures

From the authors' experiences in industry, the five parameters above are more-or-less arranged in their order of importance. However, metal cutting is so diverse an activity that individual circumstances will juggle the order. In high precision machining of semiconductor components or mirrors, the accuracy is so critical that "#1 and #2" above will be reversed: the tool will be changed as often as desired accuracy dictates. In another circumstance, the machining of pure copper is "easy" from a tool-life point of view but "hard" from a surface finish viewpoint - here, "#1 and #3" might switch places.

Returning to the challenges of "machining forecasting", metal cutting does indeed pose great obstacles in comparison with other metal-processing operations. The upper diagram in *Figure 12.1* represents the metal extrusion process. The diameter of the undeformed material on the left is known, as is the diameter of the deformed material on the right, because it is all constrained to flow through a die of prescribed diameter. However, in the lower sketch of *Figure 12.1*, the free edge of the chip is unconstrained. The thickness of the undeformed material on the left is known, but the thickness of the chip is not known because there is nothing like a die to restrain the free, outer edge of the chip. As a result, the angle ϕ is not known *a priori* because it is not constrained, unlike processes such as extrusion. Thus, even if the models for machining and extrusion were as simple as the two equations on the right, in the lower case the unknown value of t_2 is a great obstacle to making any progress with predictions.

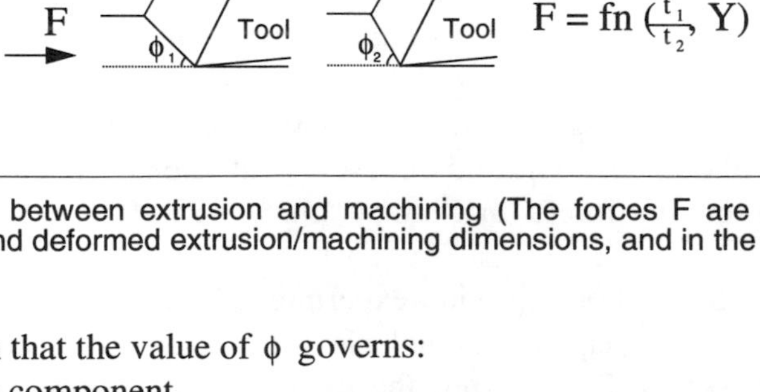

$$F = fn\left(\tfrac{t_1}{t_2}, Y\right)$$

$$F = fn\left(\tfrac{t_1}{t_2}, Y\right)$$

FIGURE 12.1 Comparisons between extrusion and machining (The forces F are a function of yield strength, Y, the undeformed and deformed extrusion/machining dimensions, and in the case of machining, the shear plane angle ϕ)

It has already been shown that the value of ϕ governs:
- the surface finish on the component.
- the stress σ on tool face, and forces on the fixtures and on the machine tool.
- the temperature T of the tool edge, the tool rake face, and the component.
- the power that needs to be exerted by the machine tool.

12.2 EMPIRICAL MODELS

12.2.1 Introduction

Given the interaction between ϕ and the four parameters in the above list, it has not been surprising that the research community has continued to persevere with their efforts and have approached the overall modeling issue in two basic ways:

1. Spending great amounts of time in trying to find a theoretical method that might predict ϕ, and consequently the other parameters

2. Abandoning the search for ϕ-predictions and developing empirical or mechanistic models that do not depend on first finding a value of ϕ. This was certainly Taylor's style in the early 1900s. He knew that tool life *vs.* cutting speed was crucial to operating his factories efficiently. Therefore he set his staff to machine enormous quantities of steel with a wide variety of tools and eventually arrive at the famous *Taylor Equation*. It is a totally empirical model that relates the cutting speed, V, and tool life, T, with the exponent n and constant C. These are particular to each tool-work combination.

$$VT^n = C \qquad \text{(12. 1)}$$

Plotted on log-log axes, a straight line is obtained for most tool-work combinations:

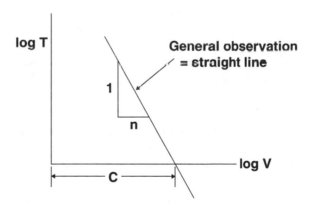

FIGURE 12.2 Double log plot of tool life *vs.* cutting speed

The *American Society of Metals Handbook* (Volume 3 - Machining), and *Metcut Research Associates' Machining Data Handbook*, include tabulated data for a wide range of potential work materials and their values of n and C. In addition, standard reference tables on recommended speeds and feeds for a wide variety of operations are given. All such data is empirical.[6,7]

Armarego and colleagues have probably created one of the most comprehensive data concerning other parameters. These are based on practical machining operations from which average forces and torque trends have been curve fitted using multivariable regression analysis. These provide *empirical-type* equations involving all the relevant operation variables. It has been found that these equations can greatly simplify the average force and torque predictions for a wide variety of milling, drilling and oblique cutting operations.[8-11]

Armarego has used these empirical-type equations in the development of constrained optimization analyses and software for selecting machining conditions (e.g. feed and speed) for optimum economic performance. Furthermore the equations have been used to select the tool geometry for improved (lower) forces and power requirements.

Examples of these equations are given in recent review papers.[12] The processes considered include: a) high speed steel, point-thinned general purpose drills, b) high speed steel, end-milling cutters and c) flat-faced turning tools used to machine S1214 free machining steel. The chip flow angle η_c with respect to the straight major cutting edge for the lathe tools has also been included in the empirical equations.[10] Similar equations have been established for face milling with TiN coated carbide inserts.[11]

In summary, these empirical equations, or models, allow a wide range of problems to be addressed. Once the experimental work has been put in place to establish the exponents and constants in the equations, they can be used by production personnel to set up and operate their machines.

From a pure research viewpoint, the empirical approach is sometimes criticized for relying on the original laboratory testing data. It is pointed out that the data are obtained under specific conditions which may not translate to the equally specific but slightly different conditions on a factory machine tool. As shown in Chapter 9, only minor changes to a calcium deoxidation procedure can make a big difference to tool life. On the other hand *all* modeling methodologies are prone to this criticism - whether they be analytical models, mechanistic models or FEA models, no matter how good the supposed model is. First, if the chosen inputs to the model do not match the conditions in practice, the "forecast" will be uncertain. Second, if the "internals" of the model do not embody the correct material constitutive equation, or have no way of accounting for the frictional variations through the secondary shear flow zone, then again the "forecast" will be uncertain.

Given this broader view, the empirical models in the *Handbooks* and in work such as Armarego's are extremely useful to the practitioner. This is all the more true, if the "forecasts" from the empirical testing are seen as "starting points" for operation. In the Handbooks in particular the recommended values of "*n* and *C*" are deliberately conservative. The personnel setting up the machine can be confident that no major damage will occur to tools or components. And then, "once things are stable on the machine" the cutting speed and feed can be prudently increased, if so desired, to increase productivity.

12.3 REVIEW of ANALYTICAL MODELS

From Chapters 4 and 15, three closed-form analytical models are summarized below:

12.3.1 The Merchant model

When the minimum energy principle is applied, the shear plane angle is predicted to be:

$$\phi = \pi/4 + \alpha/2 - \lambda/2 \qquad (12.2)$$

where ϕ = shear angle; α = rake angle; λ = friction angle

12.3.2 The Oxley model

The angle θ that the resultant cutting force makes with the shear plane is:

$$\theta = \phi + \lambda - \alpha \qquad (12.3)$$

12.3.3 The Rowe-Spick model

When the minimum energy principle is applied, the shear plane angle is predicted to be:

$$\cos\alpha \cos(2\phi - \alpha) - \beta\chi \sin^2\phi = 0. \qquad (12.4)$$

β = a constant between zero and one (not a friction angle). The value of χ is related to

$$L = \chi(OP)$$

$$L = \chi \frac{t_1}{\cos\alpha}$$

In the above, L is the chip-tool contact length. The family of curves in *Figure 4.10* shows that, as the contact length on the rake face of the tool increases, the minimum energy occurs at lower values of the shear plane angle and the rate of work done increases greatly. Friction, and more specifically seizure, play a great role in metal cutting and all models should consider it in detail.

Even though these closed-form analytical solutions were developed several years ago, the new student in the field is encouraged to review them in detail. When considered in the context of *Figure 4.10* they give great insight into the physics of cutting - i.e. the effect of rake angle and friction. Sandstrom's Java applet is highly recommended here.[13]

12.4 MECHANISTIC MODELS

12.4.1 Introduction

Following the early work by Koenigsberger and Sabberwal,[14] DeVor and colleagues have employed mechanistic methods over the past twenty years to model three-dimensional cutting processes such as milling and drilling. *The underlying assumption behind their mechanistic methods is that the cutting forces are proportional to the uncut chip area.* The constant of proportionality depends on the cutting conditions, cutting geometry and material properties. This section aims to give a general overview of the model-building methodology for these models. The review is based on a recent thesis by Stori, and his paper with King and Wright.[15,16]

12.4.2 Component accuracy - EMSIM

The "end-milling simulation" - EMSIM - uses a mechanistic model of the end-milling process for the prediction of cutting forces and tool deflections. This "indented-ski-slope" deflection is schematically shown in *Figure 2.6*. For the analysis, the flutes of the end-mill are decomposed into thin axial slices (in a class situation it is useful to pile-up a column of pennies, or coins, representing the sliced up end-mill). Force elements for each slice are then summed to predict the instantaneous cutting force. This mechanistic force model was developed by Kline, DeVor et al.[17-21] The elemental force components in the tangential and radial directions are shown in *Figure 12.3*, and can be expressed as:

$$\delta F_t(i) = K'_t D_z t_c \tag{12.5}$$

$$\delta F_R(i) = K_R \delta F_t \tag{12.6}$$

K_t' and K_R are empirical coefficients for the work material and tool material. D_z is the thickness of the axial disk. Also, t_c is the instantaneous chip thickness going around the arc-shaped slice. Examination of *Figure 12.3* shows that as the tool edge first "bites into" the surface labeled "**B**", the instantaneous value of t_c will be the same as the feed rate (f). But as the tool cuts around the arc to "**A**", it encounters a thinner and thinner instantaneous value of t_c.

In the ideal case, the instantaneous value of chip thickness t_c is approximated[15] by ($t_c = f\sin \alpha$), where *f* is the feed per tooth, and α is the angular position of the tooth in the cut, shown in *Figure 12.3*. However, the presence of *runout* greatly complicates the estimation of the instantaneous chip thickness. Runout occurs if the cutting tool axis is not perfectly aligned with the tool-holder and machine tool spindle. When the combined system spins around, the very tip of the tool might thus create two errors. It might vibrate back-and-forth {the case of parallel axis offset runout (PAOR)}, or it could scribe out a cutting circle on an axis that tilts relative to the tool's axis {the case of axis tilt runout (ATR)}.

For the basic case *without runout*, the angular position, α, of a tooth in the cut can be computed for each axial disk element, i, and flute, k. The i value will be an integer value depending on the number of axial slices used in the EMSIM run. A typical end-mill will have two or four flutes, thus k will be 1, 2, 3, or 4. In the equation below, the j value can be regarded as an integer value that "counts around" the arc during the computer simulation. So, for the simple case of the first slice and first flute, i = 1, and k = 1. Then, the first term on the right side of the equation below is one where $\theta(j)$ "counts around" the arc in *Figure 12.3* by (say) 100 integer steps. The second term in the {braces} on the right side of the equation is the vertical height of the engagement point above the bottom of the disc being analyzed. Again, to show a simple case, the first disc is i = 1 so the term in the square [brackets] becomes ($D_z/2$) which is just the mid-point of that first disc. This is multiplied by ($\tan(\lambda)/R$) to account for the helical twist. (Note, out of interest, if there were no twist, the mill would be a reamer with straight edges - and α would just track around in the same position as θ.

$$\alpha(i, j, k) = [\theta(j) + \gamma(k-1)] + \{[(i-1)D_z + D_z/2]\tan(\lambda)/R\} \tag{12.7}$$

In the above, γ is the angular spacing between flutes on the cutter, θ is the angular orientation measured in the same direction as α, and λ is the helix angle of the cutter. Another relationship that helps to understand the relative angles is ($t_c = R\theta/\tan\lambda$).

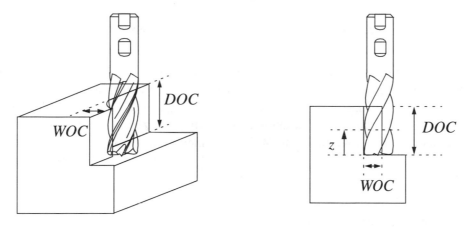

Cross-section of flute orientation

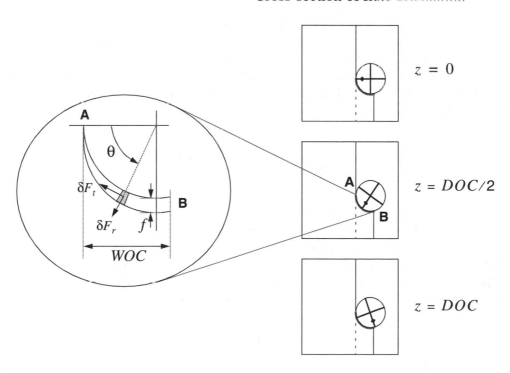

FIGURE 12.3 "Axial slices" through the "climb" end milling process.

For the prediction of the maximum surface form error, forces tangential to direction of feed (F_y) must be considered. In *Figure 12.3* the projection of the elemental force components F_t, and F_R onto the y-axis (acting away from the wall being machined) can be expressed as follows:

$$\delta F_y(i,j,k) = \delta F_t(i,j,k)\sin(\alpha(i,j,k)) + \delta F_R(i,j,k)\cos(\alpha(i,j,k))$$ (12. 8)

The tangential cutting force, F_y, is obtained by summing the elemental force contributions of all of the flutes engaged in a cut along the extent of the cut. The contributions for each of the flutes must be considered when $\alpha_{ex} \le \alpha(i,j,k) \le \alpha_{en}$, where α_{en} and α_{ex} are the entrance and exit angles, respectively, for the cut. The exit angle is approximately zero, $\alpha_{ex}=0$, and the entrance angle is a function of the radial width of cut (WOC) and the tool radius (R).

$$\alpha_{en} = \text{acos}(1 - WOC/R)$$ (12. 9)

Although in restricted cases the cutting forces may be obtained through a closed-form integration of equation 12.7, this becomes impossible as complications such as runout are introduced into the model. For this reason, EMSIM is implemented as a computer simulation based upon the elemental force contributions.

The prediction of the "ski-slope" effect is accomplished through a cantilever beam model for end-mill deflection. The instantaneous tangential cutting force is applied using a cantilever beam model at the force center of the distributed cutting forces. The surface error is taken to be the deflection of the tool at the point of contact between the cutter tooth and the finished surface. This is best visualized at the bottom of a pocket being cut. This point of contact is known as the surface generation point, SG. The form error at the surface generation point may be then expressed as follows:

$$\hat{\Delta}(SG) = \frac{F_y \overline{CF_y^2}(3\overline{CF_y} - L)}{6EI}$$ (12. 10)

L is the distance of the surface generation point from the fixed end of the cantilever beam. Usually this is where the milling tool clamps into the tool holder. E and I are the Young's modulus and moment of inertia of the end mill, respectively. The force center, $\overline{CF_y}$, is a virtual point of application. It represents a *point-force* load that would cause the equivalent bending moment about the tool holder as the actual *distributed-force* load which acts along the milling tool's side. The surface generation point is correlated with the angular position of the cutter, θ, by:

$$SG = R \cdot \theta / \tan(\lambda)$$ (12. 11)

Recent results are at http://madmax.me.berkeley.edu/webparam/demo2/summary2.html.

12.4.3 Component surface roughness - SURF

The "surface of the wall" generated by the peripheral cutting edges on an end mill has been modeled by Babin *et al.*[22,23], and an analytical solution has recently been developed by Melkote and Thangaraj.[24] This latter model is described below. Once again the reader might glance back, this time to the face *XY* in *Figure 2.7*, to identify the area under discussion.

Figure 12.4 shows the geometry of a typical end mill with a right hand helix. The end mill is modeled as a cylindrical surface of nominal radius R with N_t helical cutting edges on its periphery. The cutter is assumed to *rotate and translate* as shown in the figure. The final machined surface, assumed to be generated by the combined effect of the two motions, is described by the *trochoidal* path followed by the cutting edges.

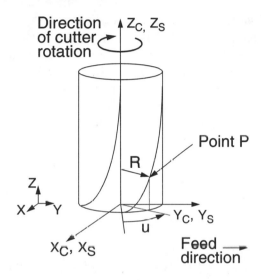

FIGURE 12.4 Milling geometry in ideal case with no runout. The tool rotates around Z_c perfectly aligned with the machine tool Z_s. Feed is then X and Y.

The position vector of a point P on the ith cutting edge is given by:

$$\vec{P} = \left\{ \frac{f|\theta_{rot}|}{2\pi} + R_i \cos(\theta_i + u) \right\} \hat{i} + \{ R_i \sin(\theta_i + u) \} \hat{j} + \left\{ \frac{R_i u}{\tan \lambda} \right\} \hat{k} \qquad \text{(12. 12)}$$

where f is the cutter feed, θ_{rot} is the rotation angle, λ is the helix angle, and

$$\theta_i = \theta_{rot} + (i - 1)\theta_p \qquad \text{(12. 13)}$$

The angle θ_p is the circular pitch of cutter teeth and the angle u represents the phase difference between the leading and trailing edges of the ith cutting edge due to the helix. The tooth path segments at a particular cutter axial location are determined by the intersections between the trajectories traced by successive teeth. The multiple intersections are found analytically from the multiple values of θ_{rot} at which the intersections occur:

$$\theta_{rot} = u + (i-1)\theta_p + \text{acos}\left[\frac{f(\theta_p + 2n\pi)}{4R\pi}\right] \tag{12.14}$$

where n = 0, 1, 2, 3... represents the number of cutter revolutions.

The ideal three-dimensional milled surface can be generated using the equations above and surface roughness parameters such as R_a, R_q, and R_{max} can then be computed from the profile data at any location along the cutter axis.

12.4.4   Roughness including runout

Eccentric motion of the cutter when mounted in the spindle can lead to runout - parallel axis offset runout (PAOR) and axis tilt runout (ATR). In *Figure 12.5*, the parameters *e* and β represent the magnitude and direction, respectively, of the cutter axis offset. In *Figure 12.6* axis tilt runout is represented by τ (magnitude) and ρ (direction). To generate a 3-D map of the modeled surface, the paths of discrete points, P_k, along each flute of the cutter must be determined. The cutter is divided into k discrete axial slices. For each slice, the position of the point P_k is found as the cutter is rotated through a complete revolution. This procedure is applied to each flute on the cutter varying k over the entire axial cut. Each point P_k is a position vector in an XYZ coordinate system:

$$P_k = x_k\mathbf{i} + y_k\mathbf{j} + z_k\mathbf{k} \tag{12.15}$$

where x_k, y_k, and z_k are determined as:

$$x_k = \frac{f\phi}{2\pi} + r_t\cos(\theta_i - \alpha) + \varepsilon\cos(\phi + \beta) + (l_t - z_k)\sin\tau\cos(\phi + \rho) \tag{12.16}$$

$$y_k = r_t\sin(\theta_i - \alpha) + \varepsilon\sin(\phi + \beta) + (l_t - z_k)\sin\tau\sin(\phi + \rho) \tag{12.17}$$

$$z_k = \frac{\sqrt{2r_t^2(1 - \cos\alpha)}}{\tan\lambda} \tag{12.18}$$

where θ_i in the above expressions defines the angle from the X axis to each flute, i, at the free end of the cutter:

$$\theta_i = \phi - i\theta_p \tag{12.19}$$

where i is varied from 0 to 3 for a cutter with four flutes. To determine the surface profile at each axial depth k, successive tooth path intersection points are found. The 2-D profile is defined by the points that connect the intersections. The 2-D profile is determined for each axial slice, then all slices are combined to form the 3-D surface map.

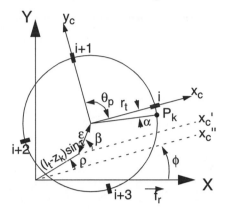

FIGURE 12.5 End milling with run out - top view

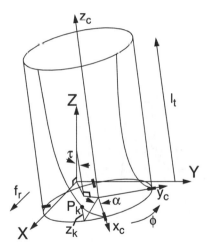

FIGURE 12.6 End milling with runout - side view (Figures 12.3 to 12.6 from Stori, King and Wright)

12.4.5 Case study: results from EMSIM

Figure 12.7a shows good agreement between EMSIM predictions and probing results for the form-error on the walls of a pocket machined in 6061 aluminum with a 0.5 inch high speed steel cutter.[25] Other results in the same set of experiments showed good agreement on forces (*Figure 12.7b*). Over time, these models are improving in their accuracy and scope, and can now be used by practitioners to obtain higher accuracy in processes such as end milling. Knowing the predicted deflection in *Figure 12.7a*, allows the "roughing and finishing" sequence of the CNC program to be closer to the "desired line" on the component being cut. However, the shape of the contour will not change unless *additional* finishing passes are used to "trim out" the inside corner of the pocket, where material might remain from a first finishing pass. This has to be done with care to ensure that the pocket is not "overcut" at the top surface.

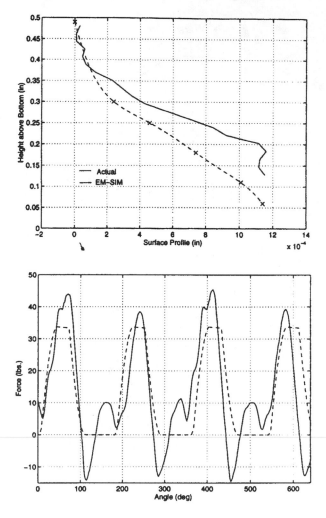

FIGURE 12.7 a) Contour plots showing the form error from equation 12.10. b) Comparisons between experimental forces (solid lines) and predicted forces (dashed lines from EMSIM). Results of Mueller, DeVor and Wright. Also see - http://madmax.me.berkeley.edu/webparam/demo2/summary2.html

12.5 FINITE ELEMENT ANALYSIS BASED MODELS

12.5.1 General introduction

For illustration purposes, *Figure 5.7* is repeated below next to a finite element mesh (*Figure 12.8*). On the left, the classical equations developed in texts such as Carslaw and Jaeger[26] create a *continuum* situation. By contrast, on the right, the finite element mesh, first used in this form for metal cutting by Tay *et al*,[27] creates a *discretized* situation. Here, many small, interconnected triangles, rectangles and similar shapes - called subregions or elements - give a piecewise approximation to the continuum equations. Finite element analysis reduces a continuum problem with infinite variations to a discretized problem with very large number but a finite number of unknowns. The roots of the FEA method are often traced to the Courant Institute in New York.[28]

The solution region is first divided into elements. Then the unknown field variable (in this particular case T) is expressed in terms of an assumed approximating function within each element. The approximating function - called an interpolation function - is defined in terms of the values of the field variable at specified corners of the mesh elements - these are called the nodes or nodal points. The nodal values of the field variable, and the interpolation functions for the elements, completely define the behavior of the field variable within the elements. For the finite element representation of any problem, the nodal values of the field variable become the new unknowns. Once these unknowns are found, the interpolation function defines the field variable throughout the assembly of elements.

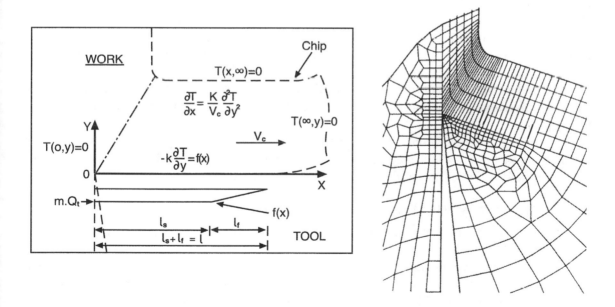

FIGURE 12.8 Continuum approach on the left *vs.* discretized FEA approach on the right

12.5.2 Developing the analysis

(a) **Discretize the continuum**: As reviewed by Black[29] the first step is to divide the solution region into the elements shown on the right of *Figure 12.8*. A variety of element shapes are used in the same solution region. The number and type of elements to be used in any given problem are matters of judgement, experience and trial and error. For example, the authors and colleagues have studied other metal-processing problems that involve high friction at a tool/work interface. It was critical to use a very fine mesh at that interface to obtain results in which theory and experiment matched each other.

(b) **Select the interpolation function(s)**: The next step is to assign the corner nodes to each element and then choose the type of interpolation function that represent the variations of the field variable over the elements. Often, although not always, *polynomials are selected as interpolation functions* for the field variable because they are easy to integrate and differentiate. The degree of the polynomial chosen, depends on the number of nodes assigned to the element, the nature and number of unknowns at each node, and certain continuity requirements.

(c) **Express the element properties**: The next step is to determine the matrix equations that express the properties of the individual elements. Four possible schemes are reviewed below: i) the direct approach, ii) the variational approach, iii) the weighted residual approach, iv) the energy balance approach.

(d) **Assemble the element properties to obtain the system equations:** To find the properties of the overall system, the matrix equation for each element must be integrated with all other matrix equations in order to finally express the behavior of the solution region or system. The assembly procedure relies on the fact that, at any node where elements are interconnected, the value of the field variable must be the same for each element sharing that node. Finally, before the system equations are ready for solution on the computer, they must incorporate the physics of the boundary conditions of the problem. For metal cutting they will include friction at the interface and the temperature conditions at the boundaries of the workpiece, toolholder and free edge of the chip.

(e) **Solve the system equations:** The assembly process of the preceding step gives a very large set of simultaneous equations that can be solved on a computer to obtain the unknown nodal values of the field variable - in this discussion the temperature T.

12.5.3 Approaches to the finite element analysis (FEA)

The four different approaches to formulating and calculating the properties of individual elements include:

(a) **Direct approach:** Structural analysis in civil engineering is the origin of the direct approach (equation 12.20). This direct approach is based on a direct stiffness method of the structure.

(b) **Variational approach:** Element properties can also be determined by the more versatile variational approach. The variational approach relies on the *calculus of variations* and involves extremizing the functions. In solid mechanics, the functional turns out to be potential energy, the complementary potential energy, or a derivative of these.

(c) **Weighted residuals approach:** A more versatile approach to deriving element properties is the weighted residuals approach. It begins with the governing equations of the problem and proceeds without relying on a functional or a variational statement. This allows a finite element analysis to be used on problems where no functional is available.

(d) Energy balance approach: This approach relies on the balance of thermal and/or mechanical energy of a system. The energy balance approach requires no variational statement and hence further broadens the possible applications.

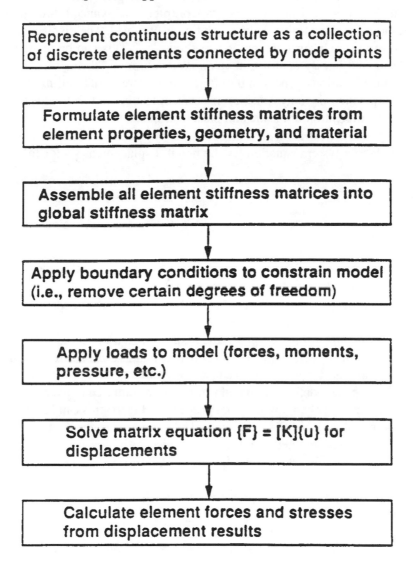

Represent continuous structure as a collection of discrete elements connected by node points

Formulate element stiffness matrices from element properties, geometry, and material

Assemble all element stiffness matrices into global stiffness matrix

Apply boundary conditions to constrain model (i.e., remove certain degrees of freedom)

Apply loads to model (forces, moments, pressure, etc.)

Solve matrix equation {F} = [K]{u} for displacements

Calculate element forces and stresses from displacement results

FIGURE 12.9   Overview of finite element solution procedure (Courtesy J.T. Black)

12.5.4   Application of FEA to the specific problem of machining

How do the matrix equations shown near the bottom of *Figure 12.9* get solved in a situation where the tool and work material are in relative motion? As summarized by Athavale and Strenkowski,[30] the two basic approaches are the Lagrangian and the Eulerian formulations, together with an arbitrary Lagrangian-Eulerian formulation.[31]

12.5.4.1 Lagrangian formulation

In a Lagrangian formulation, the mesh is "attached to the workpiece". In a virtual sense it resembles the experiments in which researchers etch a photogrid onto the workpiece and then study its deformation during cutting. The tool or workpiece is advanced through predefined displacement increments, and the finite element solution is obtained. The displacement increment is a function of the time step in *explicit* solution methods.[32,33] This can be related to the material removal rate during cutting. By contrast, in an *implicit* solution,[34-37] the time step has no physical significance. In the various papers in the literature, different material models are used: they include i) elastic-plastic, ii) only plastic, iii) viscoplastic, and iv) additive[33] as well as multiplicative[32] decomposition of elastic and plastic strains.

From a civil engineering (structural) viewpoint, the application of finite element analysis for stress analysis leads to a system equation for the nodal displacement vector {u} of the system as follows:

$$[K]\{u\} = \{R\} \tag{12.20}$$

where

[K] = the global stiffness matrix and
{R} = the load vector which include all its applied loads.

The solution of the system of the equation yields the nodal displacement vector {u} from which the element strains and hence stresses can be calculated.

By contrast from a metal cutting viewpoint, the assumptions used in the above equation 12.20, namely small displacement linear analysis and constant material properties, are no longer valid. In the updated Lagrangian formulation, because the elements move with the workpiece, they experience both large plastic deformation and rigid body motion. Under such circumstances, the larger deformations, and the changing material properties due to stress and strain in the material, need to be considered. The global stiffness matrix [K] is no longer constant. The stiffness matrix is dependent on the current geometry and the stress history imposed by the loading sequence. Since the governing equations are now nonlinear, the system must be solved iteratively on the computer.

The advantage of the updated Lagrangian formulation is that the tool can be simulated from the start of cutting to a steady-state. One of the disadvantages, however, is that the model requires large computational times to reach steady-state conditions. *In addition, a material failure and "parting-line or separation" mechanism has to be provided to allow the chip to separate from the workpiece.*

In the Lagrangian formulation, this "parting-line" or "separation" mechanism of the chip from the parent workpiece material ahead of the tool tip remains a significant and controversial issue for researchers. Nodes on this line are separated, and the line *"unzipped"*, when the tool tip is sufficiently close, or when a certain level of plastic strain is attained. Various schemes have been employed as criteria to part the chip: they include effective strain,[34] geometric distortion,[38] and crack generation and propagation.[39] To implement the chip parting line, several algorithms have been used: the slide-line,[34] node/element birth and death,[40] and remeshing.[32]

Obviously, this "unzipping" approach makes assumptions both on the location and stress level of the "failure", namely the cutting of the material. Sekhon and Chenot[41] and Marusich and Ortiz,[32] by contrast, have investigated mesh adaptivity and remeshing to allow for an arbitrary surface of separation. This approach is much more desirable because it more accurately models the cutting process. However, it is a time-consuming task to numerically perform the remeshing and mesh adaptivity and it is still dependent on the material model to correctly describe the "parting-line failure" in the workpiece just ahead of, or at the very tool tip.

A related problem in a Lagrangian formulation is the computational instability due to the large distortion of some elements. This problem may eventually cause unrealistic results or premature termination of the analysis. Severe element distortion also results in a degradation of the accuracy. To address this issue it is useful to redefine the mesh system periodically - a process called *rezoning*. The rezoning involves the assignment of a new mesh through interpolation. The capability of rezoning has been included in many finite element codes for large displacement, large strain, plastic deformations.[42]

12.5.4.2 Eulerian formulation

In the Eulerian formulation, the workpiece material is assumed to flow through a meshed control volume - the cutting zone. This Eulerian viewpoint focuses attention on a particular point in space and then examines the phenomena occurring there.

The cutting process is treated as a large deformation process involving a viscoplastic material. It is assumed that elastic effects are negligible. Therefore, the constitutive law becomes a viscous flow equation relating flow stress to the instantaneous strain rates.

$$\{\sigma\} = [C\{(\mu)\}]\{\dot{\varepsilon}\} \tag{12.21}$$

where

μ = the viscosity which is dependent on the strain rate

The global matrix for all elements in the chip and workpiece is found by summing the contribution for each element. This results in the following equation:

$$[K]\{U\} = \{P\} \tag{12.22}$$

where
[K] = the global stiffness matrix
{U} = the nodal velocities
{P} = the vector of all applied nodal loads

There are *three iterative cycles* that are performed during this simulation:

(1) The first involves solving the viscoplastic equation for velocity and strain-rate distribution in the chip and workpiece. An estimate of the strain-rate is initially used to calculate the viscosity at ambient temperature conditions. After each iteration, the strain-rate is compared to the initial value. The iterations are continued until the initial and calculated strain-rate converge.

(2) Once the velocities have been determined, the temperature is calculated. Because, in general, the thermal properties of both the workpiece and tool will depend on the temperature, an iterative solution is again required until the temperatures converge.

(3) The final step consists of updating the chip geometry to ensure that the computed velocities on the chip surfaces are parallel to the free surfaces. As Athavale and Strenkowski[30] point out, one disadvantage of the Eulerian formulation is that the boundaries of the chip free surface i) need to be known in advance and ii) also need to be adjusted iteratively during the simulation.

This issue can be addressed in an updated Eulerian approach as follows. At the end of each iteration step the calculated velocity field is integrated. Then, the displacement obtained can be used to create a new geometry, which defines the next Eulerian step. During such calculations, the free surface of the chip, and the chip shape, must be determined explicitly. Also, the workpiece velocities must be determined explicitly. The incoming bulk workpiece velocity must obviously be tangential to the element sides that represent the external physical boundaries of the free surface of the workpiece and newly forming chip. Similarly, the two boundaries of the formed chip - one along the rake face and the other opposite on the chip's free surface - must be repositioned iteratively. The repositioning of these two chip boundaries may be independent[43] or dependent.[37] Also note that for the Eulerian formulation, the strains have to be computed from the strain-rates by integrating along stream lines. Thus the basic method is not applicable to a discontinuous workpiece that the tool would cut in an interrupted fashion. Space-time Eulerian formulations[44] are being developed for discontinuous workpieces, discontinuous chips and segmented chips and these will see further uses in the future.

12.5.4.3   Summary

Despite these challenges, the main advantage of the Eulerian formation is that it eliminates the need for an explicit material failure or separation criterion. For Eulerian models, the chip parting criteria occurs at the stagnation point at the tool tip. And this is obtained from the velocity distribution. Further, physical separation of the chip and workpiece nodes are not required, since the mesh is not attached to the workpiece surface.

12.5.5   Brief historical perspective of FEA use in metal cutting research

The reader is first referred to a comprehensive review by Chen and Black.[45] This contains most of the significant citations up to the early 1990s. Second, the collection of papers edited by Jawahir, Balaji and Stevenson[46] contains contributions from many of today's investigators in the FEA area.

The earliest papers appeared in 1973. Klamecki[47] analyzed the stress state transition from plane stress at the workpiece surface to plane strain in the central region of the shear zone, and Tay *et al* [27, 48-50] presented a two-dimensional model of the temperature distribution generated during orthogonal machining. It used the following two-dimensional energy equation,[27] for the heat conduction and convection:

$$k_t \left(\frac{\partial^2 T}{\partial x^2} + \frac{\partial^2 T}{\partial y^2} \right) - \rho C_p \left(U_x \frac{\partial T}{\partial x} + U_y \frac{\partial T}{\partial y} \right) + \dot{q} = 0 \qquad \text{(12. 23)}$$

where k_t is the thermal conductivity, ρ is the density, C_p is the specific heat, T is the temperature and $\dot{q}$ is the volumetric heat generation rate.

The following boundary conditions apply:

$T = T_s$ on a surface S_T where T is defined,

$$-k_t\frac{\partial T}{\partial n} = q \text{ on a surface } S_q \text{ where q is defined.}$$

$$-k_t\frac{\partial T}{\partial n} = h \ (T - T\infty) \text{ on a surface } S_h \text{ where is h is defined.}$$

By the late 1970s, Tay, Stevenson, de Vahl Davis and Oxley[49,50] had successfully adapted the analysis to avoid the need for a flow field as input, partly based on Stevenson and Oxley's earlier work.[49] Also, a basic mesh for a given tool geometry was automatically adjusted for shear angle and contact length over wide range.

Stevenson, Wright and Chow[51] also used the tool etching method described in Chapter 5 to verify the FEA predictions. *Figure 12.10* shows very good agreement, with a discrepancy between experiment and calculation of only about 50°C, although in this case the calculation gives the higher temperatures. Also, the isothermals of the calculation in the second case are displaced along the rake face from those of the experiment, suggesting an error in the measurement of the contact length used as input to the calculation.

Such an error could also have arisen from the formation of the crater. The contact length scar which was measured on the tool may have resulted from the early period of the cut before the crater formed. Formation of the crater would promote chip curl and gradually reduce contact length as the cut continued.

Also in the late 1970s, Muraka, Barrow and Hindjuka[52] published a finite element analysis for calculating the temperature distributions in orthogonal machining. They used the same governing equation and boundary conditions as Tay *et al.*, and obtained the solution by using the Galerkin method.

The effects of coolant on temperature distribution in metal cutting were then studied both experimentally and theoretically by Childs, Maekawa and Maulik,[53] using an analysis similar to that of Tay *et al.* An interesting conclusion reached, compatible with the findings in Chapter 10, was that the effect of coolant was least on the rake face, greater on the flank face and greatest on the tool and tool holder.

In related work, Maekawa, Nakano and Kitagawa[54] found that the use of a high-thermal-conductivity tool such as diamond, was more effective in reducing the rake temperature than the use of a coolant. This tendency became more obvious when a low-thermal-conductivity material such as a titanium alloy was machined.

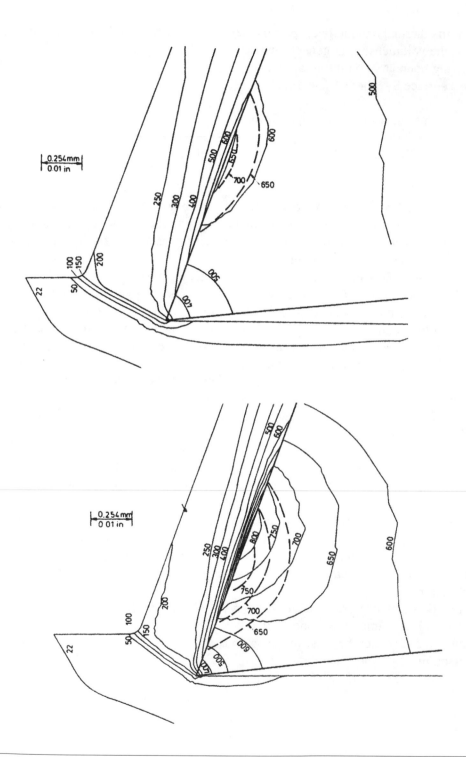

FIGURE 12.10 Experimental results from the tool etching method (dashed lines) in Chapter 5 and the corresponding FEA predictions Cutting speeds:61 m min^{-1} and 106 m min^{-1}

For example, *Figure 12.11a* shows the steady cutting temperature when machining the titanium alloy in air using a) ceramic, b) carbide and c) diamond tools. *Figure 12.11b* shows a similar result for machining medium-carbon steel. The intervals of isotherms in both figures are 100°C. The cutting temperature for the titanium alloy is higher for every tool type, even though a low cutting speed and feed was set as compared to the steel machining. An interesting result is that the edge temperature for the ceramic tool, whose thermal conductivity is the lowest, markedly increases to as high as 1300°C.

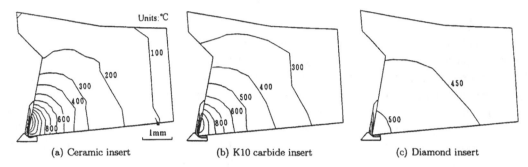

(a) Ceramic insert	(b) K10 carbide insert	(c) Diamond insert

Isotherms near tool tip in machining titanium alloy

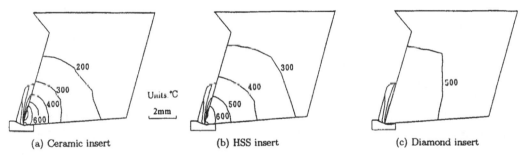

(a) Ceramic insert	(b) HSS insert	(c) Diamond insert

Isotherms near tool tip in machining carbon steel

FIGURE 12.11 Maekawa et al.'s calculated isotherms near tool tip a) above: when machining titanium alloy b) below when machining medium-carbon steel

12.5.5.1 Lagrangian examples

Following the early FEA-machining papers of the 1970s, both Lagrangian and Eulerian approaches have been developed over the last twenty years.

In 1991, for example, Komvopoulos and Erpenbeck[55] used the general purpose code ABAQUS and an updated Lagrangian formulation. A *distance criterion* was used as the chip separation criterion. They investigated two possible constitutive deformation laws: i) elastic-perfectly plastic (EPP) and ii) elastic-plastic with isotropic strain hardening and strain rate sensitivity (EPSHR). Unworn and worn (cratered) tools with strongly adherent built-up edges were used for comparison. A total of five finite element simulations with different material models, different friction coefficient values and different tool geometries were studied to find their effects on the cutting process. Coulomb's law was used for the modeling of the tool-chip interface friction - the reviews of the seizure conditions in Chapters 4 through 9 indicate that their results could still be re-run with different boundary conditions and some new insights will emerge.

Lin and Lin[56] also developed a thermo-elastic-plastic finite element model based on the updated Lagrangian formation and the Prandtl-Reuss flow rule. A *strain energy density* was used as the chip separation criterion. This is a material constant representing the energy absorption capability, and was determined from a uniaxial tensile test.

12.5.5.2   Eulerian examples

In 1990, Strenkowski and Moon[57] employed an improved Eulerian model. In this modified version, a separate procedure was used to predict the chip geometry and chip/tool contact length. The chip geometry was predicted by requiring that material velocity normal to a free surface be zero. This method was originally proposed by Zienkiewicz *et al.*[58] Once the chip geometry was determined, the contact length was predicted by checking the normal stress at every node along the tool-chip interface. A positive normal stress at any node along the interface meant that this particular node had been released from the tool/chip interface. The predicted contact length was then determined and used to modify the thermal conduction path between the chip and the tool.

Also, Strenkowski and Hiatt [59] coupled the *stresses* resulting from the improved Eulerian model with a fracture mechanics model proposed by Broek.[60] It allowed the prediction of the ductile regime in a single point diamond turning operation on a brittle material. In further work, Strenkowski, Larson and Chern[61] coupled the *temperatures* resulting from the improved Eulerian model with a tool-wear by diffusion model. In this work, *thermal infinite elements* were used for the flow zone in secondary shear. The infinite element[62] differs from the standard finite element because one of its edges is made to extend to infinity - such an element geometry is useful for unbounded field problems. Given the nature of the flow zone (e.g. a thin zone where shear strains are between 20 and 50 - *see* Chapters 4 and 5) the use of an element which is very long and thin has great intuitive appeal. Once Strenkowski *et al.* had calculated the temperature distributions in such a way, the depth of the diffusion wear along the tool rake face was then determined from the diffusion coefficients.

12.5.6   Case study I: modeling chip-breaker grooves

Jawahir and colleagues developed an Eulerian formulation that followed the Hu-Washizu variational principle.[63] The global stiffness matrix was based on equation 12.22 in the earlier section, and similarly the constitutive equation was:

$$\sigma = \mu \dot{\varepsilon} \tag{12.24}$$

where σ is the effective stress, μ is the strain-rate and temperature dependent viscosity $\mu = \mu(\bar{\sigma}, \dot{\bar{\varepsilon}}, T, \varepsilon)$ and $\dot{\varepsilon}$ is the strain-rate. The thermal effects were developed from the same 2-D equation as Tay *et al*'s.

The model also included the chip-groove on the tool rakeface. The boundary conditions used for the deformation analysis and the thermal analysis are shown in *Figure 12.12*.

First, the normal component of the velocity at the tool-chip interface (U_x) is set to be zero. Second, a frictional, tangential stress is prescribed at the tool-chip interface and the groove backwall. In the "controlled contact" or land region - *see Figure 4.11* - it is assumed that the chip is initially in entire contact. The normal stress at the nodes is calculated and it is assumed that the tool and the chip are in contact as long as the normal stress is compressive - similar to Stren-kowski and Moon's work. A similar procedure is followed for the groove backwall. Thus, the tool-chip contact length is determined after the computational iterations. The friction on the "controlled-contact" land is divided into regions of sticking and sliding friction. Coulomb friction is assumed at the groove backwall.

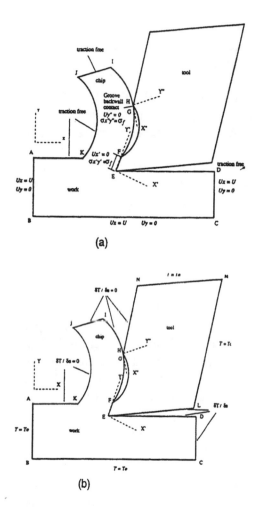

FIGURE 12.12 Boundary conditions for (a) deformation analysis; and (b) thermal analysis [63]

12.5.7 Case study II: finite element analysis of stresses in the tool

Kistler and colleagues carried out an investigation to calculate the stresses in a cutting tool and then compared the results with the experiments of Bagchi and Wright described at the end of Chapter 4.

The problem was essentially two dimensional, with nodes being restrained in the "depth" direction. The geometry and mesh for the figure below were created using MSC/PATRAN [64,65] before an ABAQUS version 5.4 input file was created. The mesh was oriented in the direction of the movement between tool and workpiece. The analysis deck created by MSC/PATRAN for ABAQUS was edited to modify material parameters for use with a "damage/plasticity model"[65] and to set the problem up to use a conjugate gradient analysis technique. Post-processing was performed by ABAQUS.

The "damage/plasticity model"[65] was a material model which accounted for the deviatoric deformation resulting from the presence of dislocations and dilatational deformation and ensuing failure from the growth of voids. This model allowed the calculation of accumulated damage in a finite element. The capability of the ABAQUS finite element code was used to turn finite elements "off" when a user-defined parameter level was reached in each element. This eliminated elements which otherwise would have had large distortions (causing convergence problems in the analysis) but which physically had failed or had lost their load carrying capability. This adequately represented the cutting/failure/chip formation mechanisms without requiring remeshing or mesh adapting. As in all the experiments reported in Chapter 4, the highest normal stresses were localized at the cutting tip. The analytical stresses in the cutting direction compared well with the corresponding experimental stresses *Figure 12.13*. The most notable difference between two stress profiles was that the spatial gradient of the stress magnitude decreased more quickly with distance in the finite element analysis. Kistler attributed this difference to the effects in the chip tool contact area. The FEA assumed frictionless conditions for convenience, and calculated a contact length of 0.58mm. In the experiments, the contact length was 0.85mm. These different values of contact length almost certainly account for the differences in the normal stress profile.

12.5.8   Case study III: cutting temperature of ceramic tools

El-Wardany, Mohammed and Elbestawi carried out investigations into the different factors which influence the temperature distribution in Al_2O_3-TiC ceramic tool rake face during machining of difficult-to-cut materials, such as case hardened AISI 1552 steel (60-65 Rc) and nickel-based superalloys (e.g. Inconel 718). A commercially available finite element program (ALGOR) was used to calculate the temperature distributions.[66] Temperature measurements on the tool rake face were performed using a thermocouple based technique and the results compared with the finite element analysis. Experiments were then performed to study the effect of cutting parameters, different tool geometries, tool conditions, and workpiece materials on the cutting edge temperatures.

Of particular interest, it was found that there is an optimum value of rake angle where the cutting edge temperature was minimum. Details are given in *Figure 12.14*. Initially the edge temperature reduced with a "more negative rake angle" from -6° to -20°; but beyond -20°, to even more negative values, the temperatures increased again. The same trend was obtained even for a higher feed of 0.15 mm/rev. El-Wardany *et al* provide the following explanation: At -6° rake angle the "air-gap" between the tool's clearance face and the cut surface is relatively small. Moving from -6° to -20° rake angle, the tool gets "rotated/lifted up" away from the surface and there is a greater chance of cooling by convection. But eventually, the shearing of the material in the secondary shear zone gets to be so intense with more aggressively negative rake angles, that the overall heat generated is larger and temperatures rise again.

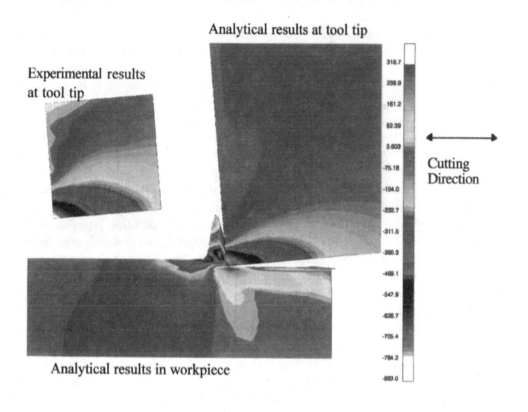

FIGURE 12.13 Comparison of cutting direction stress (MPa) with Bagchi and Wright's experiment

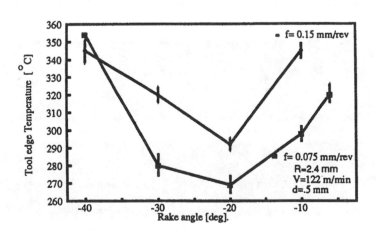

FIGURE 12.14 The effect of (negative) rake angles on cutting edge temperatures during turning of hardened steel at $V = 122$ m min^{-1}, $d = 0.5$ mm, $R = 3.2$ mm and $f = 0.075$ mm/rev.

12.5.9 Case study IV: finite element analysis of burr formation in 2-D orthogonal cutting

Park and Dornfeld[67] developed a finite element model of burr formation in 2-D orthogonal cutting using plane strain assumptions and investigated the influences of various process parameters. A general purpose FEM software package, ABAQUS, was used to simulate the chip and burr formation processes, especially the transition from steady-state cutting to burr formation on tool exit. An adiabatic heating model was adopted to simulate the heat generation effects due to plastic work of the workpiece and chip. Also, based on a ductile failure model in ABAQUS, the metal cutting simulation procedure was developed to separate the chip from the workpiece and to give a final burr/breakout configuration. The burr formation mechanism is divided into four stages: initiation, initial development, pivoting point, and final development, *Figure 12.15*. The initiation stage represents the point where the plastically deformed region appears on the edge of the workpiece. In the initial development stage, significant deflection of the workpiece edge occurs, and a bending mechanism initiates burr formation. In the pivoting point stage, material instability occurs at the edge of the workpiece. A burr is further developed under the influence of a negative deformation zone ahead of the tool. Hence, plastic bending and shearing are the dominant mechanisms in this stage. However, if the material cannot sustain highly localized strain in front of the tool edge, then fracture is initiated and leads to the edge breakout phenomenon.

The influence of exit angles of the workpiece, tool rake angles, and back-up materials on burr formation processes in 304L stainless steel were investigated. The burr formation mechanisms with respect to five different exit angles were found, and the duration of the burr formation process increases with an increase of exit angle, resulting in different burr/breakout configurations. Also, with fixed cutting conditions and workpiece exit geometry, the influence of the rake angle was found to be closely related to the rate of plastic work in steady-state cutting because the larger the rate of plastic work in steady-state cutting, the earlier the burr initiation commences. In order to effectively minimize the burr size, three cases of back-up material influence on burr formation processes were also examined. It was found that the burr size could be effectively minimized when back-up material supports the workpiece to just below the pre-defined machined surface.

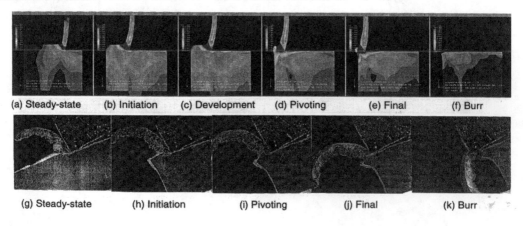

(a) Steady-state   (b) Initiation   (c) Development   (d) Pivoting   (e) Final   (f) Burr

(g) Steady-state   (h) Initiation   (i) Pivoting   (j) Final   (k) Burr

FIGURE 12.15 Comparison between finite element simulation (top) and SEM (bottom) pictures of burr formation mechanism in 2-D orthogonal cutting

12.5.10 Summary

The first Case Study based on Jawahir *et al.*'s FEA work shows that it will be possible to model different styles of chip-breaker and hopefully optimize their geometry for minimum temperature and prolonged tool life.

Similarly, Kistler and colleagues confirm that the FEA methods allow a detailed study of the normal stresses on the tool. A promising area for future work, is to re-run the analyses on geometries such as those shown in *Figure 8.2* for ceramic tools. The FEA methods allow a wide variety of such cases to be investigated once they have been checked against experiment - such as the transparent tool work for stress and the tool etching method for temperature.

The third Case Study indicates that there may well exist an "optimum" negative rake angle for minimum temperature when turning hardened steel.

The fourth case study allows burr height and depth to be related to cutting conditions.

12.6 ARTIFICIAL INTELLIGENCE BASED MODELING

12.6.1 Overview of basic concepts in artificial intelligence

The last two main sections highlight the mathematical sophistication of the mechanistic models and the FEA-based models. Furthermore, the case studies show their great potential for the future design of cutting tool geometries. Nevertheless it is clear that they still rely on empirical testing for frictional boundary inputs and material properties at high strain rates. Also, the models are complex in nature and the best of the FEA models - the Eulerian models with remeshing are very time consuming to run on today's computers.

It is therefore not surprising that many other modeling approaches have been attempted. These include neural network approaches, genetic algorithm approaches and artificial intelligence (AI) based approaches. The last of these, AI-based modeling is considered in this section.

Artificial Intelligence is an area of computer science that attempts to re-create human intelligence "inside the computer". Thus for metal cutting, an AI-based model attempts to re-create the skills and expertise of a manufacturing craftsperson. The desired outcome is that a computer controlled CNC machine tool can be set up and then operate itself without human intervention. In essence, the goal is to create an "intelligent, self-monitoring and self-diagnosing" CNC machine tool.

Knowledge Engineering is a sub-field of AI in which computer scientists interview skilled craftspeople and codify rules that express the craft-oriented knowledge. At the highest level, the goals are to capture the general procedures, methods, and generic skills of the field. At the lowest level, highly detailed "If...then" rules are captured and encoded in a computer base.

Expert Systems contain the rules that are gathered during knowledge engineering. They contain many hundreds of rules, many of which can be highly quantitative. By contrast, some important rules can only be expressed in qualitative terms. For example, a machinist might say to a knowledge engineer, "I will keep increasing the cutting speed in 10% increments if the chips are silver colored and watch them until they are light-to-dark-blue". In this case there is no quantitative data on the specific temperatures involved. But this "rule of thumb", formally called a heuristic, contains very valuable information on the trade-off between productivity and tool life. Namely, if the chips are silver in color, the cutting speed is probably "too safe" and the machine is not making parts profitably. But if the chips are beyond dark-blue and then purple, the speed is probably "too high", tool life will be short, and again profitability will suffer. Vague as these color labels are, if the expert system contains an "If...then" table that captures this information it can be used to run the machine tool within "safe but productive" limits. One criticism might be that different machinists, based on their personal experiences, will give slightly different information in the above scenario. This generally means that careful knowledge engineering garners many different people's view, tries to average them as best possible, and then compares the predictions from the expert system with the predictions from the best craftspeople available. The research of Bourne, Hayes and the authors is mentioned here for further citations in the application of artificial intelligence to manufacturing. [5,68]

12.6.2 The role of the expert machinist

Given this brief introduction, the AI-based modeling approach asks this simple question: *How does a skilled craftsperson set up a machine tool, choose the cutting conditions, and then monitor the operation for a successful outcome?* Once the question has been answered and the detailed steps codified the model can be put to use. The approach is so deceptively simple, that it is often criticized as "just being a great deal of common sense" based on simple observations of what people do on a day-to-day basis in a factory. But given the complexity of all the other modeling methods in this chapter, it is an interesting and thought-provoking procedure to review.

In today's factory situation the general way of setting the cutting conditions is as follows:

12.6.2.1 Step 1: carry out general planning

Consult the CAD/CAM files to identify the following main considerations: i) the precise material specifications (including hardness and heat-treatment), ii) the desired part tolerances and surface finishes, iii), any "unusual features" on the part that might require special fixturing and iv) the batch size.

12.6.2.2 Step 2: select tooling type in relation to batch size

Select the cutting tools that will be used based on batch size. Recall from the summary in Chapter 8, that in a "job-shop" setting, high speed steels are likely to be the most convenient choice. By contrast, in a high volume automobile manufacturing situation, carbide and ceramics are the desirable choice.

12.6.2.3 Step 3: select specific process and the tools that go with them

Identify the specific cutting processes that will be used to obtain the specific part features needed. Ideally, the part can be made with straight-forward milling, turning and drilling operations so that unusual fixturing, tooling and set-ups will not be needed. But in some cases, deep features will make end-mills vibrate and require a much reduced cutting speed. Or a slot in a part will require the stock to be "hanging-out" of a vise, again prompting a reduced cutting speed.

12.6.2.4   Step 4: pick the first round of speeds and feeds in the "Machinability Data Handbook"

Consult Metcut's *Machinability Data Handbook*, for the recommended cutting speeds and feeds for this particular material's composition and heat-treatment. The *Handbook*, specializes in tables for the different cutting tool types and the different cutting processes (e.g. end milling *vs.* face milling *vs.* turning) described in Steps 2 and 3 above.

12.6.2.5   Step 5: modify the chosen speeds and feeds based on part tolerances

Review the recommended conditions based on the tolerances needed. In today's economically competitive environments, focusing on high quality, it is quite likely that tight tolerances and very good surface finish are being specified. This means that both "roughing" and "finishing" passes will be selected. Each of these will demand different speeds and feeds.

12.6.2.6   Step 6: modify the chosen speeds and feeds based on surface finish

The surface finish is a direct function of the tool's nose-radius and the feedrate. It is thus likely that for the finishing pass the feed rate will be reduced to obtain the desired surface finish and then the speed will be increased to maintain productivity.

12.6.2.7   Step 7: begin the roughing and finishing passes on the machine tool

Cutting can begin once the fixturing and tools have been assembled into the machine tool. It is clear from all the steps above that many iterations have already been made but that the conditions may still be too conservative. As mentioned earlier in the chapter, the *Machinability Data Handbook* usually sets conservative bounds on the problem so that "no surprises" will arise when cutting actually begins. Thus the CNC programmers or the machinist will watch the first few parts coming off the machine and make adjustments where desirable. Recall at the end of Chapter 3, *Figure 3.30*, the machinist will be watching and listening for many trends and signals from the machining operation. During roughing, the desire is to remove material fast - chip color is often the best guide. During finishing, the desire is to obtain a smooth finish - the visual condition of the surface is monitored in real-time and then checked with a profileometer.

12.6.3   Information in the AI-based model and expert system

Clearly, a big part of this approach relies on the quantitative information in the *Machinability Data Handbook*. For the other information, the AI-based model simply documents the above steps and as much as possible "fills-out" a supplementary expert system with "If...then" rules. Some rules might still be quite quantitative. Boothroyd,[69] and others, have derived expressions relating the feed per tooth (f), and the tool radius (R), to the center-line average roughness, R_a. From purely geometric considerations, the ideal surface roughness is given by:

$$R_a = 0.0321(f^2/R) \qquad \text{(12. 25)}$$

Here, a "look-up-table" can be created containing this information with some supplementary heuristics that might relate, say, to coated *vs.* uncoated tool materials where the former might give a further improvement on smoothness. Other information on what percentage the speed should be reduced for a "part/fixture overhang" or a "deep-pocket/long-tool situation" will be much more heuristically based.

12.6.4   Case Study: AI-based modeling of "best speed and feed"

When a skilled machinist begins the first roughing pass, the hope is that the *Machinability Data Handbook* values of speed and feed are "safe but productive". Analogically, the machinist is searching for the "cruising speed" for the cut where there is a nice balance between high rate part output but not too much tool changing and re-setup.

At first, the machinist will make sure that the correct grade of cutting tool has been chosen. There are always situations where the chosen tool is too brittle and a catastrophic failure occurs when the first part is cut. There is no choice but to substitute a new tool grade from the charts in Chapter 7, *Tables 7.3*.

Eventually the machining will "settle down" and then the machinist will carry out fine-tuning adjustments. The goal is to "push the chosen tool into the safe but productive region" as described in the following AI-based model. A machinist knows that the rate of metal removal for any of the particular tooling types is set by the "collapse of the cutting edge". This is an instantaneous version of the wear by plastic deformation of the tool edge shown in *Figure 6.24, diagram #2*. It occurs if the local stress and temperature in this region exceed the hot compressive yield strength of the tool material in the vulnerable cutting edge region. A skilled machinist tries to judge this by chip color and vague intuitions about the stress on the edge from vibrations and tooling deflections. However, these are understandably very intuitive and not based on measured temperatures or stresses. The analysis below focuses on these measurements and their correlations with the tool condition.

12.6.5   A "fail-safe" diagram

To illustrate the AI-based model, the plastic deformation of the cutting edge is now analyzed in more detail. *Figure 12.16* presents laboratory compression tests on different tool materials. These plots show the relationship between the yield strength and temperature. However, *Figure 12.16* should be viewed as more than just the physical properties of the tool materials presented in a way commonly found in *Handbooks*. It also represents a "*fail-safe*" diagram for the edge of a cutting tool made from one of the materials. A point in the lower left space, below the curves, represents a safe condition where the tool edge will not yield and plastically deform. Conversely, a point in the top right area, above the curves, represents a condition where the materials will yield and deform. Ideally, the cutting speed and feed rate need to be adjusted so that the temperature and stress values of the tool's cutting edge are in the safe region. However, operating conditions cannot be "too safe." To machine and produce parts quickly, the speed and feed rate, and consequently temperature and stress, must be pushed into the "safe but productive" shaded region, just under the curves in *Figure 12.16*.

Whether or not the tool is operating in this "*safe but productive*" region is controlled by the cutting conditions that the machinist or manufacturing engineer selects. For a constant depth of cut d, the functional relationships between the *control input variables* speed V and feed rate f, and the *physical state variables* temperature T and stress σ may be summarized as:

$$T_{edge} = function(K\phi\alpha Vfk_p) = AT_{mean} = BT_{max} \tag{12.26}$$

$$\sigma_{edge} = \sigma_{max} = Cf = 1.75\sigma_{mean} \tag{12.27}$$

Where K represents the thermal properties of the tool material; ϕ is the shear plane angle; α is the tool rake angle; V and f are speed and feed; k_p is the shear strength of the workmaterial being cut; and A, B, and C are constants.

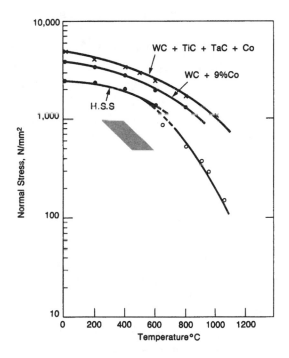

FIGURE 12.16 Elevated temperature yield strength of tool materials. The shaded area is the "*safe-but-productive*" region for temperature and stress, when machining with high-speed steel. The top two curves are for cemented carbide tools made from tungsten carbide (WC), titanium carbide (TiC), and tantalum carbide (TaC), cemented together with cobalt (Co). The lower curve is for high-speed steel (HSS.). From Wright and Bourne[5]

The AI-based modeling approach, requires a knowledge of the particular stress σ_{edge} and particular temperature T_{edge} in this region. The preceding equations were written to begin with these terms.

12.6.6   Real-time monitoring of temperature and stress

To implement the AI-based model, i) a remote thermocouple is needed to monitor *mean* temperature and ii) a dynamometer or a strain gaged toolholder is needed to monitor *mean* cutting stress. These are well-established, relatively inexpensive sensors. For the turning operation in

Figure 2.1, the remote thermocouple is placed between the cutting tool insert and the supporting shim, directly below the chip-tool contact area. The readings from this remote thermocouple can be correlated with the mean temperature of the rake face surface. The dynamometer can be used to support the tool-holder and to directly measure the principal cutting force F_c. This can be divided by the chip-tool contact area ($a=Lw$) to give the mean stress on the tool.

These two sensors provide the raw data shown as the first step in *Table 12.1*. The second step is to convert this data, in real time, to the mean temperature and stress values. Third, these mean values can be converted, again in real time, to the localized temperature and stress values at the very cutting edge of the tool.

TABLE 12.1   Strategy for estimating tool edge temperature and stresses in real-time

STEP	Temperature Monitoring	Stress Monitoring
1. Obtain raw data as shown in the next two columns >>	Measure temperature at remote thermocouple on bottom of tool insert.	Measure principal cutting force (F_c) with dynamometer.
2. Calculate mean stress and mean temperature, based on reverse solution and using controlled contact tools >>	Solve heat-diffusion equations to render T_{mean} on top, rake face in center of contact area.	Use a tool with controlled contact area (a) and calculate mean stress $\sigma_{mean} = F_c/a$.
3. Estimate localized stress and temperature of tool edge using heuristics. Also estimate the maximum temperature toward the rear of the contact area	Use generic-temperature profile to determine both the temperature of the tool edge T_{edge} and the maximum temperature T_{max}.	Use generic-stress profile to determine σ_{edge} (which is also σ_{max}).

12.6.7   Temperature monitoring

In *Figure 2.1*, the remote, chrome/alumel thermocouple is 6 mm away from the chip-tool contact area. This dimension reflects the typical thickness of a cemented carbide, cutting tool insert. The mean temperature T_{mean} on the top of the tool in the center of the contact area can be found by solving the three-dimensional heat diffusion equations for the complete cutting tool and its holder. This analysis treats the contact area a as a frictional, planar, heat source lying at the center of an ellipse. Heat diffuses from this source to the remote thermocouple.

A reverse solution allows T_{mean} to be calculated from the remote thermocouple reading.[70,71] When machining steels, at cutting speeds typical of commercial practice, the remote thermocouple readings are in the range 100 - 150°C. The thermal analysis shows that these *thermocouple readings should be multiplied by 6.25* to obtain T_{mean} on the chip/tool contact area. This gives an estimate of mean temperatures in the range 625 -950°C, depending on the chosen cutting speed. This agrees well with the temperatures obtained by Tay *et al.* using FEA and the experiments from the tool etching method.

Data from the many examples in Chapters 5 and 9 have then been averaged over a range of cutting conditions. For cutting steels at commercial speeds and feeds, the results show that a starting *heuristic* to use for steels is:

$$T_{edge} = 0.76 \ T_{mean} \qquad \text{(12. 28)}$$

$$T_{max} = 1.18 \ T_{mean} \qquad \text{(12. 29)}$$

These heuristic constants can thus be used to convert the remote thermocouple reading (*see its position in Figure 2.1*) directly to the edge or maximum temperature. The cutting-edge temperature can be used to control wear by plastic deformation, and the maximum temperature used to control wear by diffusion. It is re-emphasized that this is an AI-based approach. The starting values for the heuristics (values of 0.76 and 1.18) are deliberately approximate. However, over time, the idea is that these will be adjusted from the learning that goes on during normal operations of the machine.

12.6.7.1 Stress monitoring

Piezo-electric dynamometers and strain-gauged tool-holders are, today, the most practical devices for obtaining the raw, cutting force data. Although it is relatively easy to measure the forces, there is no direct relationship between them and the physics of the chip formation process (*see* Chapter 4). Actually, the simple tensile test is the best illustration that forces are meaningless unless they are related to the area under load. Thus, to exploit the forces, the chip-tool contact area a must be specified, and then the stress can be calculated ($\sigma_{mean} = F/a$). Actually, this is the same as in any other engineering design situation.

To correlate forces with mean stresses during real-time machining, it is possible to use controlled contact area tools. These predetermine the value of the area that bears the load. Of course, these tools are mainly designed to provide the practical advantages such as chip-shape control and the overall reduction in cutting temperatures. To implement the AI-based model, the mean stress can be converted to the maximum stress on the tool edge, to allow wear by plastic deformation to be controlled. The results in Chapter 4 show that for many cutting conditions and work materials, a conservative starting point for the *heuristic* is to use the lower bound of the experimental stress variations:

$$\sigma_{edge} = \sigma_{max} = 1.75\sigma_{mean} \qquad \text{(12. 30)}$$

The result has been used in the real-time determination of the stress acting on the edge of the tool. {Wear by fracture is also strongly influenced by the maximum stress acting on the tool edge.}

12.6.7.2 Summary

To reiterate, this AI-based control scheme establishes the "cruising speed" for cutting where rates of metal removal are "safe but productive". The model has been illustrated by analyzing one of the major wear mechanisms, plastic deformation of the cutting edge. Thermocouples and dynamometers can be used for real-time machining to ensure that operations remain safe, but productive in the shaded area of *Figure 12.16*. In this way, the inevitable approximations in the modeling and the unpredictable variations in actual machining can be accounted for by the sensor feedback (*see Figure 2.1*)

12.7 CONCLUSIONS

In this chapter, the division by section heading is more to create some structure and to identify certain "styles of approach". It pays not to "label" any of the methods because there is considerable overlap between all the modeling methods reviewed.

Actually, Taylor was a complete empiricist and did just focus on collecting data for later use. His famous equation came from replotting has experimental data on log-log paper. Armarego *et al*'s work is also empirical at first glance but they base their equations on the mechanics of chip flow and there is a great deal of thought behind the way the equations are set up.

The Merchant model remains the clearest physical picture of the metal cutting problem. But even this needs an empirically derived coefficient of friction. Oxley and Rowe-Spick had the goal of improving Merchant's model and worked to find closed-form predictive analyses for forces and temperatures. Again, the analyses provide new insights but both use empirical constants to calculate the final value of the shear plane angle.

The mechanistic models do not attempt (unlike Merchant/Oxley/Rowe-Spick) to predict the shear plane angle. In addition, they rely on empirical constants such as K_t' and K_R in equations 12.5 and 12.6. DeVor *et al*'s work seems mechanistic at first glance but it also includes the empirical constants $K_t{'}$ and K_R. Recent experiments by Stori and Wright are displayed on the following Website <http://madmax.me.berkeley.edu/webparam/demo2/summary2.html>. These show very good agreement between the predictions and values obtained from a coordinate measuring machine.[83]

The finite element models in metal cutting provide a more global view of the calculation and they now allow tool designs and cutting conditions to be evaluated as shown in the three Case Studies in section 12.5. Chen and Black's review paper provides an excellent summary of the early FEA methods.[47-52 and 72-82] This is reproduced below with some recent additions. The AI-based model allows for a different approach where approximate heuristics from the machining research are used to operate a machine tool in a "safe but productive" region. Finally, Molecular Dynamic simulation is mentioned here as another fruitful avenue to pursue. Komanduri and colleagues, and Ueda and colleagues, are cited as sources for such information. [84-86]

TABLE 12.2 FEA models in metal cutting (Courtesy of J.T. Black and Z.G.Chen).

YEAR	FIELD	AUTHORS	REF.	COMMENTS
1973	stress	Klamecki	[47]	3-D, incipient chip formation
1973	temperature	Tay	[27]	Steady state temperature
1974	temperature	Tay et al.	[48]	Steady state temperature
1979	temperature	Muraka et al.	[52]	Steady state temperature
1980	residual stress	Lajczok	[72]	Steady state temperature
1982	stress & temp	Usui & Shirakashi	[73]	Steady state temperature
1983	temperature	Stevenson et al.	[51]	Comparison with experiment
1984	stress	Iwata et al.	[74]	Temperature
1984	instant. stress	Liu et al.	[76]	Thermal & mechanical
1985	stress	Strenkowski and Carroll	[80]	Incipient stage to steady state
1986	stress	Strenkowski and Carroll	[81]	Eulcrian model
1988	stress	Carroll and Stren-kowski	[82]	Lagrangian & Eulerian models
1988	temperature	Childs et al.	[53]	Effects of coolant
1988	stress & temp	Klamecki & Kim	[77]	3-D, incipient chip formation
1989	stress & temp	Howerton & Stren-kowski	[78]	Eulerian model, BUE formation
1990	stress & temp	Strenkowski and Moon	[57]	Modified Eulerian model
1990	stress & temp	Strenkowski and Hiatt	[56]	Eulerian model, ductile/brittle regime
1991	stress & temp	Strenkowski, Lar-son and Chern	[61]	Eulerian model, tool wear
1991	stress & temp	Komvopoulos et al.	[55]	Lagrangian
1992	stress & temp	Lin & Lin	[56]	Lagrangian
1994	3D stresses	Strenkowski and Lin	[79]	3D tool forces and chip flow
1997	stress & temp	Athavale & Stren-kowski	[37]	Material damage model
1997	tool stresses	Kistler	[64]	Comparisons with experiment
1998	stress & temp	Jawahir et al	[46]	Collection of modeling papers including many FEAs

12.8 REFERENCES

1. Taylor, F.W., *Trans. A.S.M.E.*, **28**, 31 (1907)
2. Merchant, M.E., J. *Appl. Phys.*, **16,** (5), 267 (1945)
3. Devor, R.E., Kline, W.A., and Zdeblick, W.J., *Eight North American Manufacturing Research Conference Proceedings*, p. 297, (1980)
4. Sandstrom D.R, and Hodowany, J.N., *Proceedings of the CIRP International Workshop on Modeling of Machining Operations,* held in Atlanta, GA., Published by the University of Kentucky Lexington, KY, 40506-0108. p.217, (1998)
5. Wright, P.K. and Bourne, D.A., *Manufacturing Intelligence*, Addison Wesley, (1988)
6. *ASM Metals Handbook*, Volume 3 - Machining
7. Metcut Research Associates, *Machining Data Handbook*, 3rd Edn. (1980).
8. Armarego, E.J.A, Smith, A.R.J. and Gong, Z.J., *Annals of CIRP*, **39** (1) pp. 41-45 (1990
9. Armarego, E.J.A. and Zhao, H., *Annals of CIRP*, **45** (1) pp. 65-70 (1996).
10. Samaranayake, P. and Armarego, E.J.A., *Pacific Conference on Manufacturing* (PCM '98) (1998).
11. Armarego, E.J.A. and Wang, J. 4th *Int. Conf. on Advanced Manufacturing Systems and Technology*, Udine, Italy, pp. 97-105, (1996).
12. Armarego, E.J.A., *Proceedings of the CIRP International Workshop on Modeling of Machining Operations,* held in Atlanta, GA., Published by the University of Kentucky Lexington, KY, 40506-0108. p.95, (1998)
13. Sandstrom, D., <http://www.halcyon.com/sandstr/ApTool.html>
14. Koenigsberger, F. and Sabberwal, A.J.P., *Int. J. MTDR*, Vol. **1**, pp. 15-33, (1961).
15. Stori, J.A., *Machining Operation Planning based on Process Simulation and the Mechanics of Milling*, Ph.D., thesis University of California, Berkeley, See pages 40-54 in particular(1998)
16. J.A. Stori, C.King and P.K. Wright," *ASME Journal of Manufacturing Science and Engineering,* (1999)
17. S.G. Kapoor, DeVor, R.E., Zhu, R., Gajjela, R., Parakkai, G., and Smithey, D., *Proceedings of the CIRP International Workshop on Modeling of Machining Operations,* held in Atlanta, GA., Published by the University of Kentucky Lexington, KY, 40506-0108. p.109, (1998)
18. Kline, W.A., DeVor, R.E., and Lindberg, J.R.," *Int. J. Mach. Tool Des. Res.*, Vol. 22, No. 1. pp. 7-22 1982.
19. Kline, W.A., DeVor, R.E., and Shareef, I.A., *ASME Journal of Engineering for Industry*, Vol. 104, Aug. 1982, p. 272.
20. Sutherland, J.W., and DeVor, R.E., *ASME Journal of Engineering for Industry*, Vol. 108, Nov. 1986, p. 269.
21. Ehmann,K.F., Kapoor, S.G., DeVor, R.E., Lagoglu., I., *Journal of Manufacturing Science and Engineering*, 119, 655, (1997)
22. Babin, T.S., *The Modeling, Characterization, and Assessment of End Milled Surfaces*, Ph.D. Dissertation, University of Illinois at Urbana-Champaign, 1988.
23. Babin, T.S., Lee, J.M., Sutherland, J.W., and Kapoor, S.G., 1985, *Proceedings 13th North American Manufacturing Research Conference*, pp. 362-368.

24. Melkote, S.N. and Thangaraj, A.R., *ASME Journal of Engineering for Industry*, Vol. 116, May. 1994, p. 166.

25. M.E. Mueller, R.E. DeVor, and P.K.Wright, *Transactions of the 25th North American Manufacturing Research Institution*, **25**, 123, (1997)

26. Carslaw, H., S., and Jaeger, J.C., *Conduction of Heat in Solids*, 2nd. Edition, Oxford University Press, London.

27. Tay, A.O., Stevenson, M.G., de Vahl Davis, G., *Proc. Inst. of Mech. Engrs.*, **188**, 627, (1974)

28. Courant, R., *Bulletin of the American Mathematics Society*, **49**, (1943)

29. Black, J.T., Personal communications, Auburn University, (1998)

30. Athavale, S.M., and Strenkowski, J.S., *Proceedings of the CIRP International Workshop on Modeling of Machining Operations,* held in Atlanta, GA., Published by the University of Kentucky Lexington, KY, 40506-0108. p.203, (1998)

31. Rakotomalala, R., Joyot, P., and Touratier, M., *Communications in Numerical Methods Engineering*, **9**, 987, (1993)

32. Marusich, T.D. and Ortiz, M., *International Journal for Numerical Methods in Engineering*, **38**, 3675, (1995)

33. Lin, Z.C., and Lin, S.Y., *Journal of Engineering Materials and Technology*, **114**, 218 (1992)

34. Caroll J.T. III, and Strenkowski, J.S., *International Journal of Mechanical Sciences*, **30**, (12), 899 (1988)

35. Obikawa, T. and Usui, E., *Journal of Manufacturing Science and Engineering*, **118**, 208 (1996)

36. Maekawa, K. and Shirakashi, T., *Journal of Engineering Manufacture*, **120**, (B3), 233 (1996)

37. Strenkowski, J.S. and Athavale, S.M., *Journal of Manufacturing Science and Engineering*, **119**, 681 (1997)

38. Zhang, B., and Bagchi, A., *Proc. of ASME IMEC, Material Issues in Machining and Physics of Machining Processes, II*, Eds. Stephenson, D.A., and Stevenson, R., p. 157, (1994).

39. Obikawa, T., Sasahara, H., Shirakasi, T. and Usui, E., *Journal of Manufacturing Science and Engineering*, **119**, 667 (1997)

40. Lin, Z-.C., *International Journal of Machine Tools and Manufacture*, **34**, (6), 770 (1994)

41. Sekhon, G.S., and Chenot, J.L., *Engineering Computations*, **10**, 31 (1993)

42. Engelmann, B., *Mechanical Engineering,* **113**, (3), 48 (1991)

43. Yang, J. A., *A predictive Model of Chip Breaking for Groove-Type Tools in Orthogonal Machining of AISI 1020 Steel*, Ph.D. Dissertation, North Carolina State University, (1992)

44. Masud, A., *A Space-Time Finite Element Method for Fluid-Structure Interaction*, Ph.D. Dissertation, Stanford University, (1993).

45. Chen, G.Z., and Black, J.T., *Manufacturing Review*, **7**, (2), 120 (1994)

46. Jawahir, I.S., Balaji, A.K., and Stevenson, R., *Editors of the Proceedings of the CIRP International Workshop on Modeling of Machining Operations,* held in Atlanta, GA., Published by the University of Kentucky Lexington, KY, 40506-0108. (1998)

47. Klamecki B. E., *Incipient Chip Formation in Metal Cutting-A Three-Dimension Finite Element Analysis*, Ph.D. dissertation, University of Illinois at Urbana-Champaign, (1973)

48. Tay, A.O., Stevenson, M.G., and de Vahl Davis G., *Proceedings of the Institute of Mechanical Engineers,* **188**, 627 (1974)

49. Oxley, P.L.B., and Stevenson, M.G., *Journal of the Institute of Metals*, **95**, 308 (1967)

50. Tay, A.O., Stevenson, M.G., de Vahl Davis, G., and Oxley, P.L.B., *International Journal of Machine* Tool Design and Research, **16**, 335 (1976)

51. Stevenson, M.G., Wright, P.K., and Chow, J.G., *J. of Eng. for Industry* (ASME), **105**, 149 (1983)

52. Muraka, P.D., Barrow, G., and Hinduja S., *International Journal f Mechanical Science*, **21**, 445 (1979)

53. Childs, T.H.C., Maekawa, K., and Maulik, P., *Materials Science and Technology*, **4**, 1006 (1988)

54. Maekawa, K., Nakano Y., and Kitagawa, T., *JSME, International Journal*, Series C, **39**, (4) (1996)

55. Komvopoulos, K., and Erpenbeck, S.A., *Journal of Engineering for Industry*, **113**, 253 (1991)

56. Lin, Z.C., and Lin, S.Y., *Journal of Engineering Materials and Technology*, **114**, 218 (1992)

57. Strenkowski, J.S., and Moon K.-J., *Journal of Engineering for Industry*, **112**, 313, (1990)

58. Zienkiewicz, O.C., Jain, P.C., and Onate, E., *International Journal of Solids and Structures*, **14**, 15 (1978)

59. Strenkowski, J.S., and Hiatt, D., *Fundamental Issues in Machining (ASME)* PED Vol. **43**, 67 (1990)

60. Broek D., *Kluwer Academic Publishers,* Dordrecht, The Netherlands, (1988)

61. Strenkowski, J.S., Larson, and Chern, ASME Tribological Aspects in Manufacturing PED Vol. **54** and TRIB Vol. **2**, 279 (1991)

62. Bettess, P., *International Journal of Numerical Methods in Engineering*, **11**, 53 (1977)

63. Jawahir, I.S., Dillon, O.W., Balaji, A.K., Redetzky, M., and Fang, N., *Proceedings of the CIRP International Workshop on Modeling of Machining Operations,* held in Atlanta, GA., Published by the University of Kentucky Lexington, KY, 40506-0108. p.161, (1998)

64. Kistler, B.L. *Finite Element Analyses of Tool Stresses in Metal Cutting Processes* Sandia Report, SAND-97- 8224, UC-406 (1997)

65. Bammann, D.J., Chiesa, M.L., McDonald, M.L., Kawahara, W.A., Dike, J.J., and Revelli, V.D., *AMD*, **107**, 7 (1990)

66. El-Wardany, T.I., Mohammed, E., and Elbestawi, M.A., *International Journal of Machine Tools and Manufacture*, **36**, (5), 611 (1996)

67. Park, I.W. and D.A. Dornfeld, "A Study of Burr Formation Processes using the Finite Element Method - Part I and II", *Trans. ASME, J. Eng. Mats.*, to appear, 1999

68. C.C. Hayes, S. Desa and P.K. Wright, *American Society of Mechanical Engineers, Winter Annual Meeting, Special Bound Volume on Concurrent Product and Process Design*, December 1989, San Francisco, CA., DE-Vol. **21** and PED-Vol. **36**, 87 (1989)

69. Boothroyd, G., *Fundamentals of Metal Machining and Machine Tools*, Marcel Dekker (1990)

70. D.W. Yen and P.K. Wright, *Trans. ASME, Journal of Engineering for Industry*, **108**, 252 (1986)

71. J.G. Chow and P.K. Wright, *Trans. ASME, Journal of Engineering for Industry*, **110**, (1), 56 (1988)

72. Lajczok, M.R., *A Study of some Aspects of Metal Machining Using the Finite Element Method*, Ph.D. dissertation, North Carolina State University, Mechanical and Aerospace Engineering Dept., (1980)

73. Usui, E., and Shirakashi, T., *ASME Publication PED*, **7**, pp. 13 (1982)

74. Iwata, K., Osakada, K., and Terasaka, Y., *Journal of Engineering Materials and Technology*, **106**, 132 (1984)

75. Liu, C.R., Lin, Z.C., and Barash, M.M., *High Speed Machining,* 167 (1984)

76. Liu, C.R., Lin, Z.C., and Barash, M.M. *High Speed Machining*, 181 (1984)

77. Klamecki, B.E., and Kim, S., *Journal of Engineering for Industry*, **110**, 322 (1988)

78. Howerton D.H., and Strenkowski, J.S., *Transactions of the North American MAnufacturing Research Conference*, **17**, 95 (1989)

79. Strenkowski, J.S. and Lin, J.C., *Manufacturing Science and Engineering* MED Vol. **4**, 273, (1996)

80. Strenkowski, J.S. and Carroll, J.T., *Journal of Engineering for Industry (ASME),* **107**, 349 (1985)

81. Strenkowski, J.S. and Carroll, J.T., *Manufacturing Processes Machines and Systems* (Proceedings of the 13th NSF Conference on Production Research), 261 (1986)

82. Carroll, J.T., and Strenkowski, J.S., *International Journal of Mechanical Sciences*, **30**, (12) 899 (1988)

83. Stori, J.A., and Wright, P.K., http://madmax.me.berkeley.edu/webparam/demo2/summary2.html

84. Komanduri, R., and Hou, Z.B., "Thermal Modeling of the Metal Cutting process" - to be published in the *International Journal of Mechanical Sciences* (1999/2000)

85. Komanduri, R., and Raff, L.M.," Molecular Dynamics (MD) Simulation of Machining" - to be published in the *Proceedings of the Institution of Mechanical Engineers,* London (2000)

86. Ueda, K., Fu, H., and Manabe, K., *Machining Science and Technology*, **3**, 1, 61 (1999)

CHAPTER 13

MANAGEMENT OF TECHNOLOGY

13.1 RETROSPECTIVE AND PERSPECTIVE

In his book *Man the Tool Maker*, published by the British Museum, Kenneth P. Oakley states the proposition: "Human progress has gone step by step with the discovery of better materials of which to make *cutting* tools, and the history of man is therefore broadly divisible into the Stone Age, the Bronze Age, the Iron Age, and the Steel Age".

Certainly, in the last 100 years - since the invention of high speed steels by Taylor and White - the accelerated development of new materials for cutting tools has made the most important contribution to machining efficiency. Despite the considerable advances in the machine tools, historical events show that the discovery of a new "genus" of cutting tool is the "trigger" for advances. High speed steels were so revolutionary in their time that the machinery industry had to "scramble" to respond. Similarly, with the broader use of cemented carbides in the 1940s, and ceramics in the 1970s, the machine tool industry was, in each era, faced with the pressure to build stiffer and faster machines with greater spindle speeds. NC machines, invented in 1950, allowed efficiencies of another kind, but the higher speeds made possible by each era of tool material still remained, and will continue to remain, the primary driver for the overall design of lathes and milling machines.

The ability of any type of cutting tool to withstand the stresses and temperatures at its edge is the key factor limiting the speed of cutting and the forces and power which can be employed. *A primary research goal is therefore to obtain an increase in the high temperature strength of any tool material.* However, this must be balanced against secondary effects, particularly, impact strength and diffusion resistance.

How will Kenneth Oakley's quote apply in the next 100 years?

- From the point of view of the *work material*, the consumer demands of high speed travel will continue to extend the Steel Age into the Space Age. Titanium, nickel, and high-silicon aluminum alloys are now commonplace in the aerospace and automobile industries. Machining science and technology must respond to these alloys, no matter how difficult they are to machine.

*   From the point of view of the *tool material*, there seems to be an ironical return to the Stone Age, as diamond, ceramic and other non-metallic materials slowly gain acceptance. These have a very high strength at elevated temperatures. However, while these have been commercially available for some time, it is only recently that their *consistency* has improved, leading to a market demand in relation to the cemented carbides. In addition to consistent performance, there is also the question of cost *vs*. performance. In this Chapter 13 on Conclusions, it worth discussing this point in some more detail below.

13.2 CONCLUSIONS ON NEW TOOL MATERIALS

The commercialization of new tool materials, all too often, falls into a category described by Busch and Dismukus as the "*advanced technology trap*". "In these situations, the new high performance technology is developed with the intention of having it "*substitute*" for an incumbent, lower performance alternative and, while the performance of the new technology is measurably better, unfortunately it is also significantly more expensive. Undaunted, the commercialization activities proceed assuming that the market-place will recognize the performance advantages and be willing to pay for them. Too often, this is simply not the case".[1]

The developers of cutting tools are always faced with this *substitution* paradigm. The potential market users - the machine shops - are already engaged in a satisfactory business and they are not "in desperate need" of a new *invention*. Invention is the other paradigm of commercialization, where a new product could not exist without the invention of a new material. One obvious example includes the transistor: in 1947, Shockley, Brattain and Bardeen and the directors at Bell Labs., recognized the need for this invention to replace the troublesome vacuum tube. And once invented, the transistor spurred the integrated circuit, the microprocessor and all elements of today's computerized society.

Thus, when a material is commercialized through *invention*, it becomes, or is a critical part of, a new product or system that enables or provides unique and valued functionality. By contrast, in *substitution-based* commercialization, new materials technologies compete on the basis of their cost and performance relative to the existing products. The technology offering the best balance of cost and performance will be the winner in the market place. Thus ceramic or diamond-coated cutting tools may well continue to displace tungsten carbide. But only where and when the balance of cost and performance is considered favorable to that of the carbides.

What are the market forces that might displace cemented carbide in favor of ceramic or diamond? For the especially troublesome machining problems - with aerospace nickel, titanium or high-silicon aluminum - the cutting tool salesperson will be welcomed by most users to carry out some trials on their machinery. And in today's competitive environment, most machine shops will still be willing to try new designs, coatings (and coolants) hoping for improved performance. But how much are they willing to pay?

Experience has shown that the willingness to pay for the improved performance is consistently overestimated by those in charge of technology commercialization. In most commercial markets, a new technology will not replace an existing one, even if the price increase is only

50%, *no matter how good the performance is*. In many markets, especially large mature markets, the proponents of "better" technologies are frustrated in their substitution attempts, even if the price increase is only 10%.

There are many reasons for user reluctance. Occasionally, the purchasing department is out-of-touch with the factory floor and there is a fixed budget for tooling. But most often, the CNC programmers, set-up engineers and operators have been disappointed before by new substitution tooling aiming to replace the already reliable cemented carbides. In the past they may have found that the increased expense of a ceramic or diamond tool is not justified by the *overall* performance as measured by "part-throughput". And needless to say, if only one or two catastrophic failures have been experienced from a new tool material, the staff will be extremely wary. As described at the end of Chapter 8, this explains why the "rather old-fashioned" high speed steel is still a stalwart performer and still capturing over one billion dollars a year in U.S. sales - about half of the market share in the total of 2.5 billion dollars of tool materials (*see - Table 8.3*).

The conclusion to be drawn is that, as good as ceramics and CVD-diamond might be, to gain market share, they probably need to cost no more than 10% more than cemented carbide. Thus, to enjoy more widespread commercial success, the cost of CVD diamond must be reduced by several orders of magnitude. In their review, Busch and Dismukes[1] indicate that the DC Arc-Jet deposition method is likely to be the most effective CVD process if measured by deposition cost per square. Butler and Windischmann's review is an interesting survey of all CVD-diamond methods.[2]

13.2.1 Specific observations and recommendations: I

The conservative but understandable environment in the user-community prompts the conclusions at the end of Chapter 8. There is still much to be gained from research into coated high speed steel and coated carbide tools.

High speed steel tools coated with TiN reduce the seizure effect at the tool rake face. This is the primary driver for their improved performance. Secondary benefits then include: reduce temperatures in the secondary shear zone and hence in the tool; reduced diffusion wear during long cuts; smoother chip flow patterns and hence improved surface finish; a reduced depth of work hardening into the surface of the work piece. *Figure 6.29* shows the reduced temperatures and the possible increase in cutting speed that the coatings provide. However, while all the experimental evidence shows that cutting speeds can be increased, this may not be the best way to view the improvement. This is partly because "the coating still sits on a rather vulnerable substrate". As shown in *Figure 6.30*, if the coating breaks or if it delaminates, the substrate offers no more resistance than an uncoated tool. Thus, the conservative recommendation is to increase speed by a modest amount - say 10% - and then the coated tools can be counted on for much longer life and a much more consistent product quality (*see* Section 6.10). The research community should bear in mind that in today's economic climate, a consistent performance that allows a CNC machine to operate for many hours of uninterrupted service is often better than increases in speed.

Cemented carbide tools coated with multiple layers are the current trends and in Chapter 7, results for a wide variety of layering types are described. TiN, TiC, TiCN, and Al_2O_3 are the common basis for multiple layered structures, but many "sandwich" combinations are being investigated today as reported by Komanduri.[3] In triple-layer "sandwiches" the middle layer is often TiCN as shown earlier in *Figure 7.33*. Multiple layering allows the final thickness of the "sandwich" to be as much as 10-15 μ m. This is greater than can be achieved by trying to create the whole layer in one operation. When used in turning and milling operations, the successive layers of very thin coats, rather than one heavy coat, give a more stable structure. This stability has been shown to be a result of a lower crack propagation rate at the interface.

For example, work by Cho and Komvopoulos[4,5] compared WC-Co tools that were overcoated with i) a two-layer TiC/Al_2O_3 (Al_2O_3 on the outside) and ii) a three-layer $TiC/Al_2O_3/TiN$ (TiN on the outside). The three-layer tools performed better on AISI 4340 hot-rolled steel. Around the nose radius and on the flank face, the three-layer tools were considerably better. Cho and Komvopoulos proposed that the TiN outer layer on the three-layer tool is more effective (than the Al_2O_3 on the outside of the two-layer tool) in dissipating the external work of plastic shearing imposed on it from the stresses involved in chip formation.

This is an interesting conclusion. It shows that the Al_2O_3 might seem the more abrasion resistant outer layer coating but *what counts the most is the ability of the outer layer to dissipate the energy of the imposed shear stresses*. Viewed at the atomic level the TiN coatings seem, in these experiments, to have been more effective in allowing dislocation flow in the overall "sandwich" of the coating, thereby reducing the tendencies for microcrack growth and subsequent coating loss by delamination.

PVD nanolayers are expected to find a place in the market, where as described in Section 7.9.7, they create a variety of properties on the surface of the tough carbide core. The nanolayers in particular provide alternate layers of tough and hard materials. Somewhat similar to the above findings[6] the evidence indicates that such "nano-sandwiches" arrest microcracks at the interfaces. There is also the possibility of adding extra layers for oxidation resistance, and solid lubrication for example with nanolayers of molybdenum disulphide. [6,7] Further, as described by Kramer[8,9] there is much opportunity to investigate materials that have natural barrier against diffusion and solution into difficult-to-machine aerospace materials such as the titanium alloys.

Tool edge preparations are the other promising area to pursue. The overall confidence in new more brittle tool materials will grow if users can be 100% certain that no catastrophic failures will occur. It does seem that the appropriate edge preparation as shown in *Figure 8.2,* can redistribute the stress throughout the tool. The high normal stress at the edge "acts more inwards" and the possible tensile stresses further back along the rake face are reduced.

Experimental work with the birefringent sapphire and with the FEA analyses are recommended for these studies. It is interesting to note that the edge preparations do not need expensive material developments - rather careful control of the usual edge-preparations that are already being done by the manufacturers of inserts. At the time of writing, this may be one of the biggest opportunities for "advances at minimum investment".

13.3 CONCLUSIONS ON MACHINABILITY

The tool material advances can be complemented by subtle adjustments to work materials that create "free machining" varieties of standard alloys. Of course there is a limit as to how much substitution can be allowed. First and foremost, the design attributes of the component being manufactured must drive material selection. No compromises should be made on basic strength, fracture toughness or corrosion resistance, merely to obtain a few percent increases in "free machining" ability.

Having made this cautionary introduction, it is clear that "free machining" additions can play a major role in production. In Chapter 9, a variety of additions to aluminum, brasses, steel, stainless steel and nickel alloys were described. In general, the role of the additions is two-fold:

- In all cases the additives are elongated in the secondary shear zone. This acts as an "internal lubrication" reducing the shear stresses and temperatures on the rake face. Consequently, tool life is extended and chip flow is smoother. Since rake face forces are reduced, there is also the possibility that the sub-surface strain in the work material will be reduced. A study of this correlation is recommended for future research.
- Second, in some cases, the fracture strain of the material in primary shear is reduced to the extent that microcracks form there. This has the beneficial effect of curling and breaking the chips more easily - a great benefit on unattended automatic and CNC machines.

There is a stronger correlation than generally realized between the specific tool material being used and the "free-machining" additions in re-sulfurized steels. The reader is invited to refer back to Section 9.7 and 9.13, where the benefits of machining with the mixed-carbide grades are recommended over the WC-Co grades in this instance.

The retention of a sulfide layer at the interface - with the mixed crystal, steel-cutting grades of carbide tool - is most probably the result of a bond formed between the MnS and the cubic carbides of the tool material. The bond is strong enough to keep it anchored to the tool surface. It resists the flow stress imposed by the chip. To be effective however, enough sulfide must be present in the work material. This is needed to replace those parts of the sulfide that are, from time-to-time, carried away by the under surface of the chip. It was emphasized that the blanket term "machinability" is not a unique property of a work material.

The above results show an important point with MnS based "free cutting steels": It is only when using the steel-cutting carbide grades, or tools coated with TiC or TiN, that the formation of a large built-up edge is prevented and long tool lives are found.

Similar subtleties arise with calcium deoxidized steels (*see-* section 9.7.6.4 and *Figure 9.31*). In the best of situations, these create glassy silicates in the secondary shear zone, again reducing the stresses and temperatures on the tool with a consequent life improvement. But, as above, these benefits are related to the specific tool materials being used. The evidence thus far indicates that the most beneficial combination is the calcium deoxidized steel with the mixed crystal carbide. Further investigations into these combinations is recommended, together with an analysis of the role for coatings such as TiN and CVD-diamond.

13.3.1 Specific observations and recommendations: II

In the above extremely specific cases, the authors have emphasized that "machinability" is more dependent on tool material than work material.

This observation will perhaps be even more the case in terms of addressing the challenges that come with high speed machining in Chapter 11. That chapter ends with some specific recommendations on how to minimize the effects of the segmented chip formation and hence alleviate the fluctuating stresses on the tool. In addition the studies of Kramer[8,9] are worth referring to and expanding in future work.

Kramer proposes that different work/tool combinations will be the key to success in high speed machining: i) for machining the high-silicon aluminum, abrasion wear is the biggest problem, but tool stresses are low - the diamond coated tools are recommended, ii) for machining alloy steels and nickel-based super alloys, diffusion wear is the biggest problem at high speeds - ceramic and oxide tools are recommended for use because of their chemical stability but they need to be consistent in their performance. Also, further research to further improve the hot strength of cemented carbides at elevated temperature is recommended. Finally, in this area, further research into sialons and cubic boron nitride (CBN) is suggested - these have solubility problems at present, despite their admirable high strength at high temperatures. The goal of the research would be to see if there is some combination of speed and feed that would prompt a transition at some critical temperature to a lower diffusion wear regime, iii) for the machining of titanium alloys, all tool wear problems seem to be evident - while measures can be taken to choose chemically inert tool materials that also have a high-strength at elevated temperatures, more attention to lubrication and innovative cooling techniques[10] are suggested.

13.4 CONCLUSIONS ON MODELING

The ability to run large computer simulations of machining, promises to provide much more information on typical tool stresses and temperatures which can then be correlated with tool deflections, component surface finish and tool wear. It will also be possible to investigate different tool geometries and chip breakers as described in the Case Studies of Chapter 12. Prediction of segmented chip forms in high speed machining has also been possible - Chapter 11. It has been emphasized however that these mechanistic and FEA analyses hinge on two critical issues:

- Material properties at the high strain rates, high strains and elevated temperatures of machining
- Knowledge of the boundary conditions - namely seizure conditions - at the chip-tool interface

Despite the large expense of running the experiments, there is a need to revisit the classical work of Campbell[11] and others: to review the data; to re-run some of the experiments with modern sensors and data-gathering equipment; and to extend the range of work materials to the aerospace and alloy steels now being used commercially in much greater quantity. Without this data the great potential of the Eulerian FEA models will fall short in terms of accurate predictions for the practitioner.

13.4.1 Specific observations and recommendations: III

What does the practitioner value in this context? In the introduction to Chapter 12, it was emphasized that tool life, component accuracy, and surface finish are the key concerns in practice. The values of stress, force, and temperature are of little interest on a day-to-day basis. The table below by Athavale and Furness[12] shows that the mapping between incoming materials and designs with tool and condition selections are the keys to success. Modeling research *Table 13.2* should keep these high level objectives in mind as desirable long term solutions for the practitioner.

TABLE 13.1   Current practice in industry. The task of the Manufacturing Process Engineer is to map inputs on the left to specific machine instructions on the right. Modeling and analysis should address such issues for specific guidance to the practitioner. Based on information from Athavale and Furness at the Ford Motor Company.

Incoming data and information →	Selected parameters for operation
Approved vendors & Part drawings	Process plans
Floor space & Plant capacity	Stations
Part materials	Process conditions
Part complexity	Tool & Fixturing layouts
Cycle time	Cutting tool selections
Cost objectives	Tool life estimates and Run-offs

TABLE 13.2   Current practice in "academic research". The outcomes in the third column should ideally help the practitioner with parameter selections in the table above.

Modeling Approach	Main Equation & Formulation	General Prediction Goal
Taylor's empirical testing methods for tool life	$VT^n = C$	Gives the suggested value of V, for a desired life, T.
Merchant's force circle	$\phi = \pi/4 + \alpha/2 + \lambda/2$	Predict ϕ, for equations on stress & temperature
Mechanistic modeling, for example the EMSIM model for form error	$\hat{\Delta} = \dfrac{F_y \overline{CF_y}^2 (3\overline{CF_y} - L)}{6EI}$	Estimates of the error in vertical walls, or errors in surface finish (SURF)
Finite element analysis	Eulerian example: $\{\sigma\} = [C\{(\mu)\}]\{\varepsilon\}$	Prediction of stress and temperature fields
Artificial Intelligence method for tool life vs. metallurgical data	$T_{edge} = 0.76 \ T_{mean}$ $\sigma_{edge} = 1.75\sigma_{mean}$	Real-time monitoring of temperature & stress for tool-life control

13.5   MACHINING AND THE GLOBAL ECONOMY

The reader is first invited to glance back at *Figure 1.1*. While metal cutting is commonly associated with big industries that manufacture big products (automotive, aerospace, home appliance, etc.) the machining of metals and alloys plays a crucial role in a range of manufacturing activities, including the ultraprecision machining of extremely delicate components.

Figure 13.1 specifically focuses on machining precision. These data, based on the results of Taniguchi[13] and Dornfeld,[14] illustrate three main ranges of machining precision. These are also shown in Taniguchi's *Table 13.3* grouped under the headings of *m*) mechanical, *e*) electrical, and *o*) optical.

- *normal machining* delivers the precision needed for *m*) automobile manufacturing, *e*) switches and *o*) camera bodies
- *precision machining* delivers the precision needed for *m*) bearings and gears, *e*) electrical relays, and *o*) optical connectors
- *ultra-precision machining* delivers the precision needs for *m*) ultra-precision x-y tables, *e*) VLSI manufacturing support, *o*) lenses, diffraction gratings and video discs.

The data emphasize that the precision at any level, has been more easily achieved as the last few decades have gone by. In fact, lines have been drawn at the years 1980 and 2000 representing the years in which the first edition and this edition of *Metal Cutting* have been published. Considerable advances have occurred.

The greatest benefit has probably come from NC control, where the axes of factory floor machines have been driven by servomechanisms consisting of appropriate transducers, servo-motors, and amplifiers with increasing sophistication of control. This closed-loop control of the machinery motions has probably had the biggest impact on the improvements in precision and accuracy over the last 50 years. Important advances in machine tool stiffness have also occurred. Advances in this field have especially been the focus of the research work by Tlusty and colleagues.[15]

It is valuable to compare *Figure 13.1* with *13.2*. In semiconductor manufacturing, the *minimum* line widths in today's semiconductor logic devices are typically 0.25 to 0.35 microns wide. These line widths have been decreasing rapidly since the introduction of the integrated circuit around 1960. A large number of technological improvements in VLSI-design, lithography techniques, deposition methods, ion-beam etching and clean room practices have maintained the size reduction shown in *Figure 13.2* over time.

Up until now that is: the semiconductor industry is concerned that today's optical lithography techniques are not accurate enough to maintain the trend in the Figure. The natural limit of UV-lithography, semiconductor manufacturing today is generally cited to be line widths of 0.13 to 0.18 microns (*see* Madden and Moore[16]).

This has prompted major research programs in advanced lithography, sponsored by alliances of semiconductor manufacturing companies. One such project,[17] using extreme ultraviolet (EUV) lithography and magnetically levitated stages, has the goal of achieving line widths below 0.1 microns perhaps eventually reaching 0.03 microns by the year 2010. While such technologies are not expected to be commercially available soon, they represent examples of how the trends can continue to be satisfied.

The fabrication of such magnetically levitated stages, demands the most precise metal cutting technologies available. This observation is just one example of the interesting coupling that occurs in *Table 13.3*. The precision machining equipment allows the precision VLSI and optical equipment to be made, which in turn allows the machining equipment to be better controlled and then, even more precise. This is a spiral of increasing capability where all technologies drive each other to higher achievements.

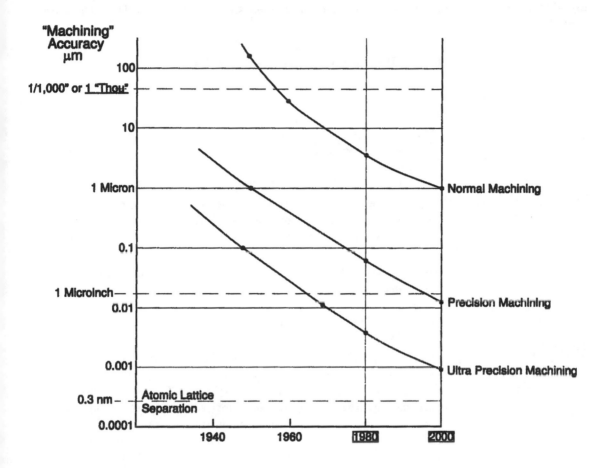

FIGURE 13.1 Variations over time in machining accuracy. These are grouped in three categories: normal, precision and ultra-precision. For reference on the y-axis 25 microns = "one thou". Based on the work of Taniguchi[13] and Dornfeld[14]

TABLE 13.3   **Products machined with different levels of precision (Courtesy of Norio Taniguchi and David Dornfeld)**

Examples of Precision Manufactured Products

	Tolerance Band	Mechanical	Electronic	Optical
Normal Machining	200 μm	Normal domestic appliances, automotive fittings etc.	General purpose electrical parts, e.g. switches, motors, and connectors	Camera, telescope, binocular bodies
	50 μm	General purpose mechanical parts for typewriters, engines etc.	Transistors, diodes, magnetic heads for tape recorders	Camera shutters, lens holders for cameras and microscopes
Precision Machining	5 μm	Mechanical watch parts, machine tool bearings, gears, ballscrews, rotary compressor parts	Electrical relays, condensers, silicon wafers, TV color masks	Lenses, prism, optical fibre and connectors (multi-mode)
	0.5 μm	Ball and roller bearings, precision drawn wire, hydraulic servo-valves, aerostatic gyro bearings	Magnetic scales, CCD, quartz oscillators, magnetic memory bubbles, magnetron, IC line width, thin film pressure transducers, thermal printer heads, thin film head discs	Precision lenses, optical scales, IC exposure masks (photo, X-ray), laser polygon mirrors, X-ray mirrors, elastic deflection mirrors, monomode optical fiber and connectors
Ultra-precision Machining	0.05 μm	Gauge blocks, diamond indentor top radius, microtome cutting edge radius, ultra-precision X-Y tables	IC memories, electronic video discs, LSI	Optical flats, precision Fresnel lenses, optical diffraction gratings, optical video discs
	0.005 μm		VLSI, super-lattice thin films	Ultra-precision diffraction gratings

Notes:
CCD charge couple device
IC integrated circuit
LSI large scale integration
VLSI very large scale integration)

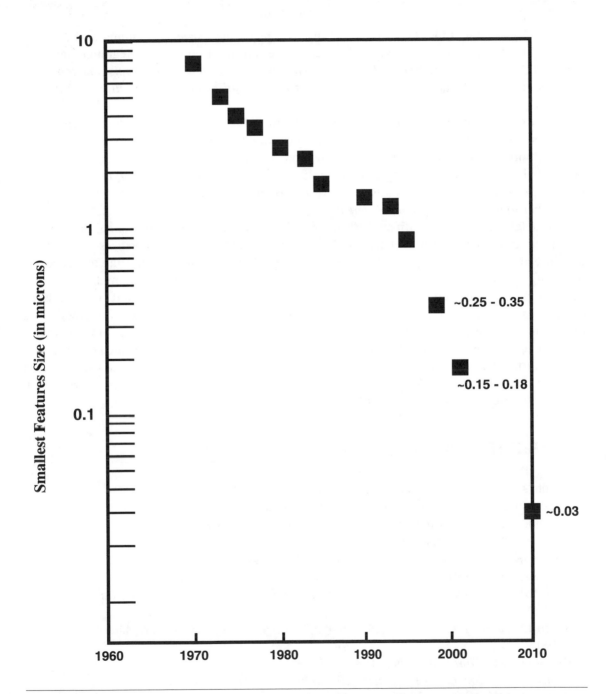

FIGURE 13.2 Trends in the precision of semiconductor transistor logic devices. Today typical values are 0.25 to 0.35 microns falling to 0.13 to 0.18 microns as the book goes to press. Research projects are aiming for below 0.1 microns and possibly 0.03 microns by the year 2010. More information on such research is given in the Semiconductor Association Roadmap (see - SIA Semiconductor Industry Association <http://www.semichips.org>)

13.5.1 Specific observations and recommendations: IV

The reader is invited to refer again to the discussions on high speed machining at the beginning of Chapter 11. Ayres[18] provides the succinct definition of computer integrated manufacturing (CIM) as *the confluence of the supply elements (such as new cutting tools and computer technologies) and the demand elements (the consumer requirements of flexibility, quality, and variety).*

Many examples of this confluence are shown in *Table 13.3.* In particular, the pressing demands of the semiconductor industry for narrower line widths spur all sorts of innovations in the machining of magnetically levitated tables, precision lenses, optical scales and diffraction gratings shown on the bottom right of *Table 13.3. This confluence of demand and supply creates a spiral of increasing capability, where all technologies are "co-dependent" and drive each other to higher levels of achievement.*

Thus metal cutting is a significant industry, even though at first glance, it is small in comparison to the customer industries it serves.[19,20,21] For over 200 years, the advances in all major industries have critically depended on the supporting advances from the machine tool industry including: shipbuilding, railroad, gun-making (*see* Preface), construction, automobile, aircraft, home appliance, and consumer electronics industries - all these have many thousands of employees engaged in activities that depend on machining as a core production method. And increasingly, particularly in last 20 years, the semiconductor industry and the semiconductor-equipment-making industry (which total 200 billion dollars a year) have depended on precision metal cutting. The economic conclusion is worth reiterating: *Machining and the machine tool industry are key building blocks for industrial society, since they provide the base upon which all other industries perform their production.*

13.6 REFERENCES

1. Busch, J.V. and Dismukes, J.P., *Diamond and related materials*, **3**, 295 (1994)
2. Butler, J.E., and Windischmann, H., *Materials Research Society Bulletin*, 23, (9), 22 (1998)
3. Komanduri, R., *Tool Materials*, Encyclopedia of Chemical Technology, Fourth Edition, Volume 24, 390 (1997) John Wiley and Sons Inc.
4. Cho, S.-S., and Komvopoulos, K., *Journal of Tribology (ASME)*, **119**, 8, (1997)
5. Cho, S.-S., and Komvopoulos, K., *Journal of Tribology (ASME)*, **120**, 75, (1998)
6. Shaw, M.C., Marshall, D.B., Dadkhah, M.S., and Evans, A.G., *Acta Metall. Mater.* **41**, (11), 3311 (1993)
7. Keem J.E., and Kramer, B.M., U.S. Patent 5,268,216 to Ovonic Synthetic Materials, (Dec. 7, 1993)
8. Kramer, B.M., *Thin Solid Films,* **108,** 117 (1983)
9. Kramer, B.K., *Journal of Engineering for Industry (ASME)*, 109, 87, (1987)
10. Hong, S. and Santosh, M., "Effects of liquid nitrogen as a cutting fluid in the machining of titanium" *International Journal of Machine Tools and Manufacture* (1999)
11. J. D. Campbell and W. G. Ferguson, *Philosophical Magazine*, **21**, (169), 63 (1970)

12. Furness, R., *Proceedings of the CIRP International Workshop on Modeling of Machining Operations,* held in Atlanta, GA., Published by the University of Kentucky Lexington, KY, 40506-0108. (1998)

13. Taniguchi, N., *Precision Engineering,* **16**, (1), 5 (1994)

14. Dornfeld, D.A., *Lectures in Precision Engineering* at the University of California Berkeley (1999)

15. Smith, S., and Tlusty, J., *Journal of Manufacturing Science and Engineering,* **119**, 664, (1997)

16. Madden, A.P, and Moore, G., *The Red Herring Magazine,* 64 (April 1998)

17. Peterson, *Science News,* **152**, 302 (November 8th., 1997)

18. Ayres, R.U. and Miller, S.M., *Robotics: Applications and Social Implications,* Ballinger Press, Cambridge MA, (1983)

19. *A Technology Roadmap for the Machine Tool Industry,* The Association for Manufacturing Technology. McLean, VA (1996).

20. *1997-98 Economic Handbook of the Machine Tool Industry,* The Association for Manufacturing Technology. McLean, VA (1997).

21. Komanduri, R., *Symposium on US Contributions to Machining and Grinding Research in the 20th Century.* Reprinted from Applied Mechanics Reviews Volume 46, Number 3, March 1993 and available from the ASME Book Number AMR 126. (1993).

CHAPTER 14

EXERCISES FOR STUDENTS

14.1 REVIEW QUESTIONS

14.1.1 Essential features of machining

1. Present in note form some practical and fundamental reasons why the machinability of an alloy is difficult to assess. Make comparisons with the extrusion process.

2. Is it reasonable to describe pure copper as "easily machined"? Give reasons.

3. Define or describe in short, 3-4 line paragraphs, with equations or a diagram:
- •i) Shear plane angle and its measurement.
- •ii) Merchant's force circle.
- •iii) The force needed to achieve minimum energy.
- •iv) The basic equation from the Rowe-Spick theory.

$$dw/d\phi = (kus)_{primary} + (\beta kus)_{secondary} = 0. \tag{14.1}$$

4. Why is the shear plane angle an important point of focus in machining?

5a. Draw Merchant's force circle and derive the theoretical expression for the shear plane angle: develop the geometrical equations to show

$$\phi = \pi/4 + \alpha/2 - \lambda/2 \tag{14.2}$$

where ϕ = shear angle; α = rake angle; λ = friction angle

5b. Then develop a relationship for the theoretical cutting force. In this case do not present the details of the differentiation method but explain the underlying principles and show the result.

5c. Comment on how well this analysis will fit experimental data.

6a. Draw the velocity triangle for the following single shear plane model of cutting.

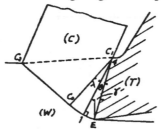

FIGURE 14.1 Single shear plane with fan-shaped secondary shear zone (controlled contact tool).

6b. Draw the velocity triangle for this fan-shaped primary and secondary zone combined.

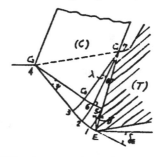

FIGURE 14.2 Fan-shaped zones in both primary and secondary.

6c. Draw the same field for an unrestricted tool face.

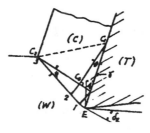

FIGURE 14.3 Unrestricted tool face.

7. Use the Rowe-Spick theory to calculate the shear plane angle for the following information on free-machining steel.
- Speed = 100 m min^{-1}

• Rake angle = 18°
• Feed rate = 0.285 mm/rev
• Contact length = 0.6 mm
• β is part-sticking/part-sliding = 0.8.

8. A commercially pure copper machined with a 6° positive rake angle tool and a "quick stop" section shows that full sticking friction occurs on the rake face. The undeformed chip thickness is 0.30 mm per rev and the shear plane angle is 18°. What chip tool contact length is expected from the Rowe Spick analysis?

9. The following data was recorded when cutting a low alloy steel.
• Cutting force = 630 N
• Thrust force = 410 N
• Rake angle = 6°
• Chip width = 1.72 mm
• Contact length = 1.05 mm
• Temperature of tool edge = 775°C
Assume that the normal stress on the rake face is distributed evenly over the contact area and then analyze the "ability to withstand wear by plastic deformation of the edge" for each of the tool materials.

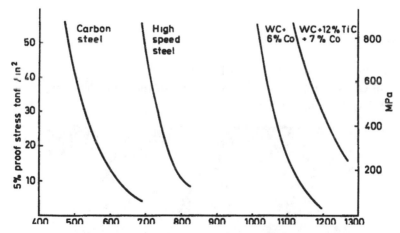

FIGURE 14.4 5% proof stress of tool materials

10. The figure below is a "quick stop" section taken from the machining of commercially pure copper. The following data has been recorded:

• Cutting speed = 210 m min^{-1}

• Feed rate = 0.2 mm. (undeformed chip thickness)

• Shear plane angle = 14.81°

• Width of primary shear zone = 0.093 mm

Calculate

• i) The primary shear strain.

• ii) The primary shear strain rate.

• iii) The bulk chip speed, and show this in a velocity triangle.

FIGURE 14.5 Chip section of commercially pure copper

11. Categorize the chip forms shown in the figure below of "quick stop" sections of AISI 1045 steel. What range of cutting speed could be estimated for the four photographs? (Feed = constant = 0.1 mm per rev, rake angle = + 6°).

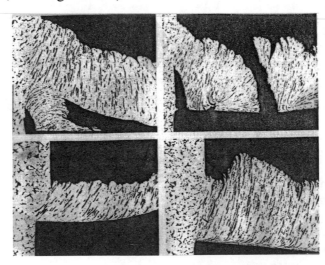

FIGURE 14.6 Four chips from the same medium carbon iron

12. Calculate the range of power requirement for the results shown below when machining austenitic 13% manganese steel at a feed rate of 0.3 mm per rev.

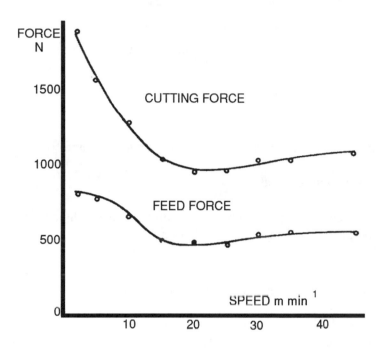

FIGURE 14.7 Forces when machining austenitic 13% manganese steel

13. The following machining data for a bar of commercially pure copper should be used to calculate:
a) The primary zone temperature
b) The temperature profile along the rake face to position 1.4mm

Undeformed Chip thickness	=	0.2mm
Undeformed Chip width	=	1.5mm
Rake angle	=	6 degrees
Shear plane angle	=	13 degrees
Main cutting force	=	640 N
Thrust force	=	315 N
Test bar initial temperature	=	17 degrees C
Length of seizure content	=	1.4 mm
Thickness of secondary zone	=	0.07 mm
Cutting Speed	=	100 m min^{-1}

Use typical room temperature values for the conductivity, density and specific heat of copper. With these values and *Figure 5.1*, it should be found that approximately 52% of the heat goes into the bar. Assume that plane strain conditions occur.

14.1.2 Supplementary questions on machining economics

14a. A batch of simple parts is made on a manual lathe. Verify that the *cost per component, C,* may be given in terms of the {operator's wage + overhead} = W, by:

$$C_{TOTAL} = WT_L + WT_M + WT_R \left[\frac{T_M}{T}\right] + y\left[\frac{T_M}{T}\right]$$

(14. 3)

- WT_L = "non-productive" costs, which vary depending on loading and fixturing

- WT_M = actual costs of cutting metal

- WT_R = the tool replacement cost shared by all the components machined. This cost is divided up among all the components because each one uses up T_M minutes of total tool life, T, and is allocated $\frac{T_M}{T}$ of WT_R.

- Using the same logic, all components use up their share $\frac{T_M}{T}$ of the tool cost, y.

14b. Which costs might be reduced by moving the job from a manual to a CNC machine? At the same time, which costs might be increased? Factors such as batch size and part complexity influence such decisions.

15. The right hand side of the above equation 14.3 can be re-formulated in terms of cutting speed. The values of T_L and T_R are constants but the other times are speed dependent:

$$VT^n = C$$

(14. 4)

is the Taylor equation that expresses the tool life T in terms of V.

$$T_m = (\pi dl)/1000fV$$

(14. 5)

is an expression for the time to machine a round bar in a lathe where the length of the bar is (*l*), its diameter (*d*), the feed rate is (*f*) and the cutting speed is *V.*

Units are peculiar to the standard industrial ways of expressing speed in meters per minute and feed in terms of millimeters. Length and diameter are also in millimeters. To make all the units compatible, the meters per minute are multiplied by 1000.

It is possible to calculate the optimum cost per component with respect to cutting speed. Essentially, the idea is to *differentiate equation 14.3 above with respect to V and find the minimum in the curve below - Figure 14.8.*

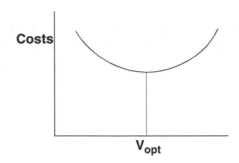

FIGURE 14.8 Optimal cutting speed to minimize costs

On the left side of the figure, costs are too high - this is because the machine tool is "being driven too slowly" and the machining time T_M per component is too long. On the other hand, on the right side, the costs are too high - this is because the machine tool is "being driven too fast", the tools wear out too fast and both T_R and tool costs increase. {In a teaching situation one can make the race-car analogy - too slow means too long around the track but too fast might mean too many visits to the pits}.

The first step is to maximize the feed rate, f. This is done by choosing the surface finish (R_a) as the main criterion that controls part quality. This is given by the equation shown near the end of Chapter 12, where (R) is the nose radius of the tool.

$$R_a = 0.0321(f^2/R) \qquad (14.6)$$

The second step is to perform the differentiation of equation 14.3 using the sub-equations 14.4 and 14.5 to isolate the parameter V.

The expressions are rather cumbersome. Detailed analyses are presented in other machining text books (such as Cook or Armarego and Brown - *see* Bibliography). Only the final equations - usable in the next three examples - are given below. The value of T appearing in the following equations is the value of tool life that will give minimum cost with variations in V. The cutting speed, V, at the minimum cost is also shown in these equations and in the above figure as V_{opt}. At this optimum set of values, all the variable parameters (f, V, and T, etc) are denoted with a superscripted star (*).

The third step is to make note of some minor additions to the procedure.

- First, the Taylor equations can also be prepared in terms of feed rate. However, increases in feed rate are "less damaging" to the tool life than increases in speed - thus a value of n_1 also appears which is the exponent that is obtained from tool life testing leading to plots of (T *vs.* f). The value of n_1 first appears in the third equation below - equation 14.9.
- Second, since the Taylor equations are now a function of both V and f, the constant (C) is replaced by the constant (K), which combines both the feed and speed constants. This is also shown in equation 14.9.

• Third, to account for the variables in the main equation 14.3 that are not directly related to change in speed, another constant (R) is formed that combines the tool cost, y, {operator + machine cost} = W, and the tool replacement time, T_R. All times are measured in minutes, and all costs in cents. As a specific example - in the first case study - (R = 3 + 70/11.67). This is because there are 4 edges in the insert (280/4 cents), and the operator + machine costs are 700/60 cents per minute.

$$R = T_R + (y/(W)) \tag{14. 7}$$

$$T^* = R\left(\frac{1}{n} - 1\right) \tag{14. 8}$$

$$T^* = K(V^*)^{-1/n} (f)^{* -\frac{1}{n_1}} \tag{14. 9}$$

$$V^* = \frac{K}{T^*(f^*)^{\frac{1}{n_1}}} \tag{14. 10}$$

$$\left(C^* = W\left(T_L + \frac{T_M^*}{1-n}\right) \text{ - or - } \left(C^* = W\left(T_L + T_M^*\left(1 + \frac{R}{T^*}\right)\right)\right)\right) \tag{14. 11}$$

In summary, the above equations relate the optimized tool life, T^*, the recommended cutting speed, V^*, and the recommended feed rate, f^*, to get the minimum in the parabolic graph shown earlier. Equation 14.5 gives T_M^*.

15a. The tool life relationship for a single pass turning operation on 18-8 stainless steel with a WC-TiC-Co tool is:

$$T_{(mins)} = \frac{7.5 \times 10^9}{V^5 f^{2.15}} \tag{14. 12}$$

where V is the cutting speed in m min^{-1} and f is the feed rate in mm per rev. The "throwaway" insert costs \$2.80 but offers 4 cutting edges; the machining plus overhead costs are \$7.00 per hour; the time taken to change a cutting edge is 3 mins., and the non-productive time per component is 30 seconds. The component diameter is 25 mm and its length is 100 mm.

If the feed rate must not exceed $f^* = 0.3$ mm per rev to achieve a satisfactory surface finish, calculate:

- i) the optimum tool life, T*
- ii) the optimum cutting speed, V*
- iii) the optimum cutting time per component, T_M*
- iv) the minimum cost per component, C*

15b. In a production turning shop, past records have shown that the tool life varies with speed and feed (and single pass - depth of cut, d = 1.5mm.) as shown below:

TABLE 14.1 Machining data

Speed m min^{-1}	Feed mm per rev.	Tool life (min.)
140	0.1	110
170	0.1	37
140	0.2	40

Derive an expression of the form, $T = KV^{-\frac{1}{n}}f^{-\frac{1}{n_1}}$, and estimate the tool life for this operation at a speed of 155 m min^{-1} and feed of 0.15 mm per rev. Outline the assumptions used to obtain this estimate.

15c. A piece of stainless steel is to be finish machined in a single pass operation. The dimensions of the bar are 150 mm in length and 50 mm in diameter. The *ASM Handbook* tables are consulted for the material and the tool life relation is given by

$$T = \frac{8.3 \times 10^5}{V^{4.34}f^{1.95}} \tag{14. 13}$$

The cost of the cutting edge of the tool is 40 cents and the {operator + machine} costs are 8 cents per min. It takes the operator 1 minute to change the tool and the largest feed rate for an acceptable surface finish is 0.3 mm per rev.
   Use the equations to calculate
- i) the optimum tool life
- ii) the optimum speed
- iii) the optimum cutting time per component
- iv) the minimum cost per component when $T_L = 20$ seconds.

14.2   INTERACTIVE FURTHER WORK ON THE SHEAR PLANE

Use Netscape with Java capability to access <http://www.halcyon.com/sandstr/ ApTool.html>. Dr. Sandstrom of The Boeing Company has built an interesting Java applet that investigates the variables in the Ernst and Merchant theory of metal-cutting.

Print out[1] the chip formation for the following cases. Also complete the table.

Rake Angle (degrees)	Friction Coefficient (0 to 1)	Friction Angle write in (degs)	Shear Angle write in (degs)
0	0		
+45	0		
+45	0.5		
+45	1.0		
+6	0		
+6	0.5		
+6	1.0		
-6	0		
-6	0.5		
-6	1.0		
-4	20		
-42	1.0		

---

1. Note: to print from a Webpage first highlight and copy the information into an already open "wordprocessing file" on the desktop and then print from there.

| CHAPTER 15 | # BIBLIOGRAPHY AND SELECTED WEB-SITES |

15.1 INTRODUCTION

1. Further reading is given under the following headings:
- Special texts/collections on metal cutting
- General texts on manufacturing that include chapters on metal cutting
- Texts that link to rapid prototyping
- Texts that link to forming
- Suggested general reading
- Suggested subscriptions

15.2 TEXTS ON THE METAL CUTTING PROCESS

2. Armarego, E.J.A., and Brown, R.H., "The Machining of Metals", Prentice Hall, Englewood Cliffs, New Jersey (1969)
3. Boothroyd, G., "Fundamentals of Machining and Machine Tools", Marcel Dekker, New York (1989)
4. Cook, N.H., "Manufacturing Analysis", Addison Wesley, Reading MA (1966)
5. Hoffman, E.G., "Jig and Fixture Design", Delmar Press, New York (1985)
6. Hoyle, G. "High Speed Steels", Butterworths, London (1988)
7. King, A.G. and Wheildon, W.M., "Ceramics in Machining Processes", Academic Press, New York (1966)
8. Klamecki, B.E., and Weinmann, K.J., "Fundamental Issues in Machining", PED-Vol. 43, (Proceedings presented at the Winter Annual Meeting of ASME in Dallas Texas), The American Society of Mechanical Engineers, New York NY (1990)

9. Komanduri, R., "Symposium on US Contributions to Machining and Grinding Research in the 20th Century". Reprinted from Applied Mechanics Reviews Volume 46, Number 3, March 1993 and available from the ASME Book Number AMR 126. (1993).
10. Komanduri, R., "Tool Materials" in The Kirk-Othmer Encyclopedia of Chemical Technology" Fourth Edition, Volume 24., John Wiley and Sons, New York (1997)
11. Loladze, T.N., "Toughness and Wear Resistance of Cutting Tools", Machinostroenie, Moscow (1982)
12. Oxley, P.L.B., "The Mechanics of Machining: An Analytical Approach to Assessing Machinability", Halsted Press, New York (1989)
13. Shaw, M.C., "Metal Cutting Principles" Oxford Series on Advanced Manufacturing Oxford Science Publications, Clarendon Press, Oxford (1991)
14. Stephenson, D.A., and Stevenson, R., "Materials Issues in Machining III and the Physics of Machining Processes III" TMS Press (Minerals, Metals and Materials Society) Warrendale PA., 15086 (1996)
15. Zorev, N.N., "Metal Cutting Mechanics", Pergamon Press, Oxford (1966) (English translation)

15.3 GENERAL TEXTS THAT INCLUDE METAL CUTTING

1. DeGarmo, E. P., Black, J.T., and Kohser, R. A., "Materials and Processes in Manufacturing", 8th Edition, Prentice Hall, New York (1997)
2. Groover, M.P., "Fundamentals of Modern Manufacturing", Prentice Hall, Upper Saddle River, New Jersey (1996)
3. Kalpakjian, S., "Manufacturing Processes for Engineering Materials", Third Edition, Addison Wesley Longman, Menlo Park, CA (1997)
4. Koenig, D.T., "Manufacturing Engineering: Principles for Optimization" Hemisphere Publishing Corporation, Washington, New York and London (1987)
5. Merchant, M.E., "The Factory of the Future - Technological Aspects" In Towards the Factory of the Future, PED-Vol. 1, American Society of Mechanical Engineers, NY. pp 71-82 (1980)
6. Pressman, R.S., and Williams, J.E., "Numerical Control and Computer-Aided Manufacturing", Wiley and Sons, New York (1977)
7. Schey, J.A., "Introduction to Manufacturing Processes", McGraw Hill, New York.
8. Wysk, R.A., Chang T.C., and Wang, H.P., "Computer Aided Manufacturing", 2nd Edition, Prentice-Hall Inc, New Jersey (1998)

15.4 SELECTED TEXTS ON RAPID PROTOTYPING

1. Beaman, J. J., "Solid Freeform Fabrication: A New Direction in Manufacturing: with Research and Applications in Thermal Laser Processing", Kluwer Academic Publishers, Austin (1997)
2. Jacobs, P.F., "Rapid Prototyping and Manufacturing: Fundamentals of Stereolithography", Society of Manufacturing Engineers, Dearborn, MI (1992)

3. Jacobs, P.F., "Stereolithography and other Rapid Prototyping and Manufacturing Technologies", Society of Manufacturing Engineers, Dearborn, MI (1996)
4. NSF Solid Freeform Fabrication Workshop on "New Paradigms For Manufacturing", Held in Arlington VA., May 2-4 (1994)
5. NSF Solid Freeform Fabrication Workshop II, on "Design Methodologies for Solid Freeform Fabrication", Held at CMU, Pittsburgh, PA., June 5-6 (1995)

15.5 SELECTED TEXTS ON FORMING

1. Backofen, W.A., "Deformation processing", Addison-Wesley, MA. (1972)
2. Chryssolouris, G., "Laser Machining" Springer-Verlag New York (1991)
3. Johnson, W., and Mellor, P.B., "Engineering Plasticity", Van Nostrand Reinhold, London (1973)
4. Kobayashi, S., Oh, S., and Altan, T., "Metal Forming and the Finite Element Method", Oxford University Press, Oxford (1989)
5. Pittman, J.F.T., Wood, R.D., Alexander, J.M., and Zienkiewicz, O.C., "Numerical Methods in Industrial Forming Operations", Pineridge Press, Swansea, UK (1982)
6. Richmond, O., "Concurrent design of products and their manufacturing processes based upon models of evolving physicoeconomic state.", Simulation of Materials Processing: Theory, Methods and Applications, ed. Shen and Dawson, Balkema, Rotterdam, pp. 153-155 (1995)
7. Rowe, G.W., "Principles of Industrial Metalworking Processes" Arnold, London (1977)
8. Sachs, G., 1955, "Principles and methods of sheet-metal fabrication", Reinhold Publishing Corporation, New York (1955)

15.6 SUGGESTED BACKGROUND READING

1. Ayres, R.U. and Miller, S.M., "Robotics: Applications and Social Implications" Ballinger Press, Cambridge MA (1983)
2. Womak, J., Jones, D., and Roos, D., "The Machine that Changed the World", Rawson Associates, New York (1990)
3. Rosenberg, N., "Perspectives on Technology", Cambridge University Press, Cambridge (1967)
4. Thomas, R.J., "What Machines Can't Do" University of California Press, Berkeley, Los Angeles, London, See in particular Chapter 7, The Politics and Aesthetics of Manufacturing, pp. 246-258 (1994)
5. Ulrich, K.T., and Eppinger, S.D., "Product Design and Development" McGraw Hill Inc., New York (1995)

15.7 RECOMMENDED SUBSCRIPTIONS

1. The Economist magazine which often includes special "pull-out sections" on "high-technology": for example, see the June 20th 1998 copy which contains "Manufacturing".
2. International Journal of Machine Tools and Manufacture

3. Journal of Manufacturing Science and Engineering (ASME)
4. Journal of Tribology (ASME, New York)
5. Journal of Materials (ASME, New York)
6. Journal of Manufacturing Systems (SME, Dearborn)
7. Journal of Manufacturing Processes (SME, Dearborn)
8. Transactions of the North American Manufacturing Research Institute (SME, Dearborn)
9. Machining Science and Technology (Marcel Dekker)

15.8 WEB-SITES of INTEREST

1. "Merchant's shear angle solution" by Sandstrom, D., <http://www.halcyon.com/sandstr/ApTool.html>
2. "Modular fixturing on the WWW" by Goldberg, K., <http://riot.ieor.berkeley.edu>, and then click on "Fixturenet". Also see:
   R. Brost and K. Goldberg, A Complete Algorithm for Designing Modular Fixtures Using Modular Components, IEEE Transactions on Robotics and Automation, **12**, (1), February, (1996)
   R. Wagner and G. Castanotto and K. Goldberg, FixtureNet: Interactive Computer Aided Design via the WWW, International Journal on Human-Computer Studies, **46**, 773 August (1997)
3. "EMSIM for predicting form errors" by DeVor *et al.*, <http://mtamri.me.uiuc.edu/software.testbed.html>. Click on "EMSIM"
4. "Green machining analyzer & environmental impact analysis" by Sheng, P. and colleagues, <http://greenmfg.me.berkeley.edu/gshell/> (to obtain a password send email to psheng@me.berkeley.edu)
5. "A manufacturing analysis service" by Smith, C.S., <http://cybercut.berkeley.edu> If necessary, click on "Research" and then "The Manufacturing Analysis Service"
6. "Comparisons between EMISM and experiments" Stori, J.A., and Wright, P.K., <http://madmax.me.berkeley.edu/webparam/demo2>
7. "CODEF, The Consortium on Deburring and Edge Finishing" by Dornfeld, D. *et al*, <http://dnclab.berkeley.edu/codef/>

15.9 WEB-SITE for "Metal Cutting"

A site for this edition has been created at:

<http://madmax.me.berkeley.edu/metalcutting>
It contains the links to the other sites above. All readers are invited to send related links and topics for inclusion. Thank you in advance!
pwright@robocop.berkeley.edu

CHAPTER 16   INDEX